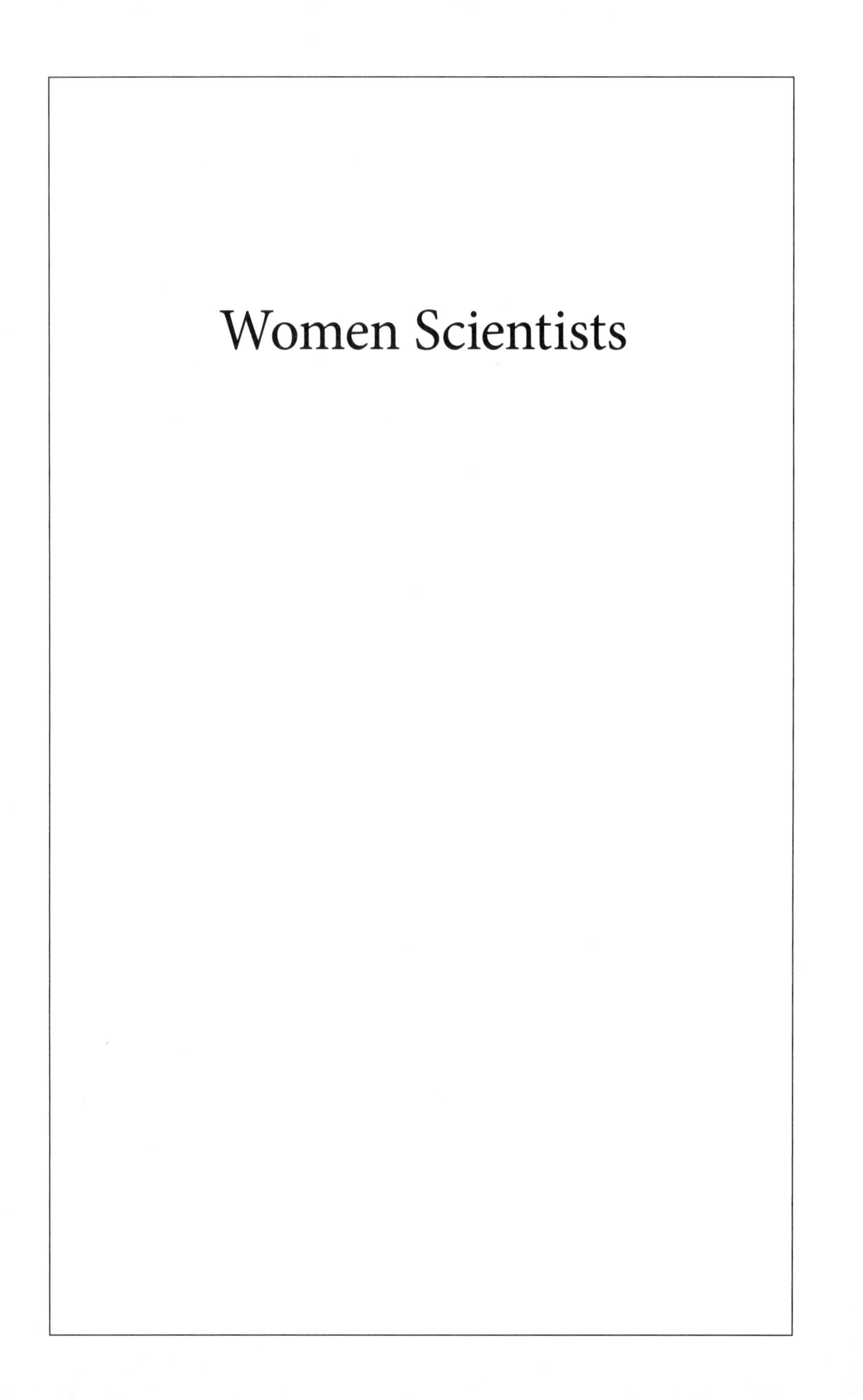

Women Scientists

Also by the Author

Great Minds: Reflections of 111 Top Scientists (Oxford University Press, 2014, with B. Hargittai and I. Hargittai)
Symmetry through the Eyes of a Chemist 3rd edition (Springer, 2009, with I. Hargittai)
Visual Symmetry (World Scientific, 2009, with I. Hargittai)
Candid Science IV, VI: Conversations with Famous Scientists (Imperial College Press, 2004, 2006, with I. Hargittai)
In Our Own Image: Personal Symmetry in Discovery (Plenum/Kluwer, 2000; Springer, 2012; with I. Hargittai)
Symmetry: A Unifying Concept (Shelter Publications, 1994, with I. Hargittai)
Cooking the Hungarian Way, 2nd edition (Lerner, 2002)
The Molecular Geometry of Coordination Compounds in the Vapor Phase (Elsevier, 1977, with I. Hargittai)

Edited Volumes

Candid Science I, II, III: Conversations with Famous Scientists (Imperial College Press, 2000–2003, with I. Hargittai)
Advances in Molecular Structure Research, Vols. 1–6 (JAI Press, 1995–2000, with I. Hargittai)
Stereochemical Applications of Gas-Phase Electron Diffraction, Parts A and B (VCH, 1988, with I. Hargittai)

Women Scientists

Reflections, Challenges, and Breaking Boundaries

Magdolna Hargittai

OXFORD
UNIVERSITY PRESS

Oxford University Press is a department of the University of Oxford. It furthers the University's objective of excellence in research, scholarship, and education by publishing worldwide.

Oxford New York
Auckland Cape Town Dar es Salaam Hong Kong Karachi
Kuala Lumpur Madrid Melbourne Mexico City Nairobi
New Delhi Shanghai Taipei Toronto

With offices in
Argentina Austria Brazil Chile Czech Republic France Greece
Guatemala Hungary Italy Japan Poland Portugal Singapore
South Korea Switzerland Thailand Turkey Ukraine Vietnam

Published in the United States of America by
Oxford University Press
198 Madison Avenue, New York, NY 10016

Library of Congress Cataloging-in-Publication Data
Hargittai, Magdolna, author.
Women scientists : reflections, challenges, and breaking boundaries / Magdolna Hargittai.
pages cm
Includes bibliographical references and index.
ISBN 978-0-19-935998-1
1. Women scientists—Interviews. 2. Scientists—Interviews. I. Title.
Q130.H376 2015
509.2′52—dc23
2014028839

1 3 5 7 9 8 6 4 2
Printed in the United States of America
on acid-free paper

To the memory of my mother, Magdolna Reven Vámhidy;
circumstances prevented her from getting the higher education that she strived for.

CONTENTS

PREFACE

This book grew out of my encounters with famous women—physicists, chemists, biomedical researchers, and other scientists—during the past fifteen years. They are from eighteen countries on four continents. For a number of years, we had a family project of recording and publishing conversations with famous scientists, and we collected most of them in our six-volume *Candid Science* book series.[a] Each volume contained at least thirty-six conversations, and at least half of them were with Nobel laureates. There were too few women in this collection, although we had no bias in selecting our interviewees. This led me to the realization of what others had noticed long before me: the unjustifiable underrepresentation of women at the higher levels of academia. When I was a student, majoring in chemistry, there were about the same number of women and men in our class. Then, moving up the academic ladder, the balance kept changing in favor of men. I myself could have fallen off this ladder at two important stages of my career, but luckily, I did not.

My future husband, Istvan, and I started seeing each other when he was a beginning researcher in a laboratory of the Hungarian Academy of Sciences in Budapest and I was a sophomore in my university studies. We got married just before my senior year. His enthusiasm for his project was contagious, and I joined him for my diploma work (master's degree equivalent) and continued after my graduation. He was an independent researcher who had initiated his own project and needed associates. When he showed success, it generated appreciation as well as jealousy among his colleagues in the laboratory.

One day—this happened in the early 1970s in pro-Soviet Hungary—the so-called quadrangle of the laboratory, the director, the party secretary, the chief of personnel, and the trade union secretary, invited Istvan for a talk. They told him that it was not right for a husband and wife to work together. They were not belligerent and told him candidly that had he been less successful in his project, it would have not generated such interest in his circumstances. Istvan calmly responded that he understood and would start looking for a new job for himself. He later told me that his quick response surprised not only the others, but himself as well, because he knew that he could not have found another place with the level of independence and support for his ideas that his current position offered. Looking for a position outside of Hungary

[a] *Candid Science*, Vols. I–VI, Imperial College Press, London, 2000–2006. A compilation of excerpts from a selection of interviews appeared recently: B. Hargittai, M. Hargittai, and I. Hargittai, *Great Minds: Reflections of 111 Top Scientists*, New York: Oxford University Press, 2014.

was unthinkable in those days; the borders were closed. Of course, the members of the "quadrangle" had taken it for granted that the wife—that is, I—would be the one to go, as indeed would have been logical to expect. After this exchange, they never brought up the issue again, and since there was no official rule excluding husband and wife working together, we kept doing so for quite a few years.

This arrangement gave me a tremendous advantage when our children were born. I stayed at home for about six months each time, but was not left behind in my work, because we never separated our activities at home and the laboratory.

Before a too idealistic picture emerges from my account, I must admit that our family developed in a traditional way in that Istvan's career was our focus. I found it natural that running the family was my duty, even though he helped in every way. However, the expression "helped" already gives the situation away; we were not equals in our duties at home. While our children were small, I put my career on the back burner; I did not want our son and daughter to suffer for having a scientist mother. I was slower in earning my higher scientific degrees than might have been the case had I not focused so much on raising our children, and my career took off around the time when our younger child entered high school. This was when I embarked on developing my independent research direction.

I had not wondered about the difficulties women scientists have to overcome until we were already well into our hobby, the *Candid Science* project. Eventually, though, I started wondering why there were hardly any women professors among my teachers, why I never encountered a woman dean let alone rector or president during my studies, and why there were so few female Nobel laureates. Once I made these observations, I decided to look for patterns and the roots of the problem. I sought out prominent women scientists with whom I could discuss these issues in addition to learning about their science. I started giving talks about women scientists, and the interest these talks generated encouraged me to continue these activities and led to the idea of this book.

I have recorded about one hundred conversations with famous women scientists. All these women have excelled in their scientific fields. Some of them decided at one point in their career that they would take another challenge, and became involved with science administration. This aspect is of additional interest because such leadership roles have also been traditionally male-dominated.

Literature on women scientists is extensive and has fulfilled a useful mission in alerting people to the potentials of women and their achievements, as well as to the challenges they are facing even today. My book puts emphasis on the scientific, geographical, and social diversity of its women heroes. I was privileged to learn a great deal of exciting science from my excellent partners in conversation, and I tried to convey some of this as well.

My mother was a most intelligent woman, but her circumstances prevented her from getting a higher education. She was determined that I should get one. My daughter has found it natural to feel comfortable in academia. I do not know whether my granddaughter will have academic aspirations. If so, I hope she will have no barriers. I thought of them a great deal while writing this book.

ACKNOWLEDGMENTS

Putting together such a volume requires the involvement of many people.

First, I have to express my thanks to all those women scientists who accepted my invitation and took the time to talk with me about their life, their science, and the difficulties they faced during their career. I appreciate their candor in sharing with me their thoughts about the question of women in science.

I received invaluable assistance from many people in the most diverse capacities—in giving me information about their relatives or former colleagues, arranging for meetings, helping with pictures, discussing with me the situation of women scientists in their country, and reading and commenting on certain chapters: Ernest Ambler, Mátyás Baló, Anders Bárány, Annarita Campanelli, Charusita Chakravarty, Janet Denlinger, Natalia Engelhardt, Richard Garwin, Rohini Godbole, Boris Gorobets, Kolbjørn Hagen, Evans Hayward, Drahomir Hnyk, Dale Hoppes, William Jenkins, Jan Kandror, Roger Kornberg, Karl Maramorosh, Shobhana Narasimhan, Oleg Nefedov, Ramakrishna Ramaswamy, Ladislas Robert, Sobhona Sharma, Manfred Stern, Svetlana Sycheva, K. VijayRaghavan, Pal Venetianer, Clara Viñas i Teixidor, Brigitte Van Tiggelen, Olga Valkova, Larissa Zasurskaya. Bob Weintrub, and Irwin Weintrub read the entire manuscript and made valuable suggestions. I am grateful to them all.

The loving interest of our children, Eszter and Balazs, and our spirited discussions concerning this project meant a great deal to me.

My special thanks go to my husband. Ever since Istvan and I met, decades ago, we have been partners in all aspects of life. We have faced hardships together and have shared fantastic adventures, many of which were connected with science. As a student, I learned from him that it is worthwhile to be a scientist only if we adhere to the highest standards in research. I learned from him the pleasure of making a discovery and to look beyond one's narrow field of research to see the larger picture. It was also through him that I became interested in the lives of scientists, which eventually led me to this project.

I appreciate the continuous support of the Hungarian Academy of Sciences and the Budapest University of Technology and Economics. I thank Senior Editor Jeremy Lewis and Assistant Editor Erik Hane of Oxford University Press (New York) for encouragement and assistance.

Women Scientists

INTRODUCTION

For centuries the expression "woman scientist" could be regarded as an oxymoron—two words appearing together that are contradictory. However, already in ancient times there were women who had interest in and talent for natural philosophy. One of the first women recorded for her involvement with mathematics and astronomy was EnHedu'Anna, around 2350 BCE, a priestess in Babylon. According to legend, Xi Lingshi was a Chinese empress around 2700 BCE who figured out how to make silk and thus originated the silk industry in China. Two women from ancient Egypt are well known. Maria (sometimes called Mary or Miriam) the Jewess lived in the first century CE and is considered the first woman alchemist. For many centuries, she was one of the greatest authorities in this profession.[1] She designed and built chemical instruments, for example the water bath—the double boiler is still called "bain-marie" in French. Her name is also preserved in the pigment "Marie's black," a lead-copper sulfide compound synthesized by her. The best-known woman from this age is Hypatia (ca. 370–415) of Alexandria, a mathematician and astronomer, who had a chair at the Neoplatonic Academy. She was also versed in other sciences, such as physics, chemistry, and medicine.

The Middle Ages spanned about a thousand years in human history during which science did not prosper. What was known by then was preserved mostly at the monasteries and convents. There were a number of physicians among nuns who were well respected for their knowledge. The German nun Hildegard of Bingen (1098–1179) was one of them. Jewish women doctors were also quite common throughout Europe for centuries. In the eleventh century, a medical school opened in Salerno, Italy; it was the first university in Europe, and women were also allowed to study there. Trota of Salerno, also known as Trota Platearius (eleventh century), who is thought to be the first woman professor of medicine, received her education there. She specialized in obstetrics, gynecology, dermatology, and epilepsy.[2] She published a book on women's diseases—known as the *Trotula*—that became the standard text for many centuries to come. The thirteenth century witnessed the foundation of universities in Bologna, Paris, Oxford, and other cities. Bologna and Salerno were still open for women students, but universities in other European countries did not allow women to enroll. Gradually, women were excluded from the medical profession.

During the time of the Scientific Revolution, women were still excluded from science and were not allowed to study (a few Italian universities provided exceptions). The new discoveries in science, however, stirred the imagination of some women of the aristocracy, who participated in the discussions in their salons and decided to learn from private tutors. The Englishwoman Margaret Cavendish (1617–1673) studied astronomy and mathematics and wrote books on a variety of scientific topics. In France, Émilie du Châtelet (1706–1749) was famous for her talent for languages, mathematics, and physics; she translated Isaac Newton's new theories into French. In Italy, Laura Bassi (1711–1778) received her doctoral degree in Bologna in 1732 and became the first woman to get a teaching position and professorship in physics at a university in Europe. Another Italian, Maria Agnesi (1718–1799), spoke seven languages and studied mathematics—including calculus—and held a chair in mathematics and natural philosophy at the University of Bologna.

Marie Paulze Lavoisier (1758–1836), Antoine Lavoisier's wife, became passionately interested in her husband's work—he was the leader of the eighteenth-century chemical revolution. She received formal training in chemistry and in drawing.[a] She learned English and translated the important literature into French for her husband. They spent much of their time in the laboratory together. This helped him understand that the then prevailing phlogiston theory of burning was wrong and to learn about the discovery of oxygen by the Englishman Joseph Priestley and the Swede Carl Wilhelm Scheele. Marie recorded all the experiments carried out in their laboratory, complemented with detailed pictures of all instruments and experiments. Her records helped future generations learn about the state-of-the-art chemical techniques and equipment of their time. His famous book, the *Elementary Treatise on Chemistry*, considered the first real textbook of chemistry, contained her thirteen engravings of laboratory instruments. Lavoisier was executed in 1794 during the French Revolution. His property, including all his scientific work, was confiscated. Marie was incarcerated for several months. Eventually, she recovered all his notes and published his *Memoirs*, which she had to complete. She held scientific salons where she explained the new chemistry to her guests.

During the seventeenth and eighteenth centuries, the relatively large number of women astronomers is noteworthy. In Germany, for example, about 14 percent of all the astronomers were women. Astronomy was ideal for women because it could be conducted at home. Most of these astronomer women were wives, sometimes sisters, of astronomers; the women worked as assistants to their husbands or brothers. The German Elisabetha Hevelius (1647–1693) was married to Johannes Hevelius, a well-known astronomer thirty-six years her senior, and she worked with him for twenty-seven years until his death. They discovered many stars. After his death, she completed and published their joint work in a book which catalogued and gave the position for almost 1,900 stars. It was their joint work, though the book carried his name alone.

Another German astronomer, Maria Kirch (1670–1720), was the first woman to discover a comet. She married the famous Gottfried Kirch and became his assistant at the Berlin Academy of Science for two decades until his death. Afterward, she tried to continue her work, but was not allowed to stay at the Academy. Sophia Brahe

[a] Her teacher was Jacques-Louis David, who also created the famous painting "Monsieur Lavoisier and his Wife" (1788), which is at the Metropolitan Museum of Art in New York (see http://www.metmuseum.org/collections/search-the-collections/436106).

(1556–1643) was a legendary Danish astronomer, working together with her brother, Tycho Brahe. Caroline Herschel (1750–1848) was a German-born British astronomer who worked all her life together with her brother, William Herschel; together they discovered many comets.

The nineteenth century brought major progress in the sciences—scientists gradually replaced the earlier natural philosophers. Great changes in women's position in society started taking place about one hundred and fifty years ago, in the second half of the nineteenth century, when the first waves of women's movements swept through Europe and the United States. Women demanded certain rights, among them the right to a higher education. Universities gradually opened their doors to women, a historic step. There were new women's colleges, such as Mount Holyoke (1861) and Smith College (1871) in the United States and Girton College at Cambridge (1869) and Lady Margaret Hall at Oxford (1878), both in England. Soon, some older universities became coeducational. Women were allowed to enroll at universities in several European countries, starting in the mid-1860s. Zurich and the German universities became favorite venues for women from places where they did not yet have this possibility. Within a decade, countries on other continents, such as New Zealand, Chile, and Australia, followed. I mention two women from this period. The American astronomer Maria Mitchel (1818–1889) discovered a comet and became the first astronomy professor at Vassar College in 1865. The Russian mathematician Sofia Kovalevsky was the first woman appointed to a professorial chair at Stockholm University in 1889.

The beginning of the twentieth century was marked with an enormous feat when in 1903 Marie Curie won the recently established Nobel Prize in Physics together with her husband, Pierre Curie. This success received great publicity and brought attention to women in science. It inspired many young women, but women still had enormous difficulties if they wanted a career in science. Although universities had opened for them a few decades before, to get a teaching or research position at a university was another matter. Numerous examples illustrate this. Emmy Noether, a successful mathematician, did not get a position at the University of Göttingen in the 1910s in spite of strong support from her colleagues. Hertha Sponer (1895–1968), by then already an associate professor at the same university, was fired in 1934 due to the Nazis' position against women in academia. The future Nobel laureate Gertrude Elion had great difficulties in finding her first job in the United States. We will see a number of other examples.

The most famous woman scientist of this period, besides Marie Curie, was Lise Meitner (1878–1968). She is well known not only for her groundbreaking achievement as a nuclear physicist but also for the fact that she was left out of the Nobel Prize awarded to her colleague, Otto Hahn. She received her doctoral degree in physics in 1905 from the University of Vienna, the second woman to have done so. Marietta Blau (1894–1970) was another Austrian physicist who made discoveries in nuclear physics, for which Erwin Schrödinger nominated her for the Nobel Prize, which she did not get. The prize went to Cecil Purcell, who further developed the method that

Blau and her coworker Hertha Wambacher had devised. Clara Immerwahr (1870–1915) was a German chemist, the wife of the famous chemist Fritz Haber. She was the first woman to receive a doctorate at the University of Breslau, but as the wife of a professor, she could not get a position—she could only assist her husband. After the poison gases developed by Haber were first used in World War I, Clara committed suicide.

During the first part of the twentieth century, more and more women studied sciences and received degrees, but they still faced difficulties when looking for jobs. The number of women graduates increased considerably in the second part of the century, but the popular expressions "leaky pipeline" and "glass ceiling" did not lose their validity. "Leaky pipeline" refers to the fact that the proportion of women falls off at every step of the way up the academic ladder. The "glass ceiling" is the point at which the advancement of women is blocked in an institution. In 1999, Professor Nancy Hopkins and her colleagues at the Massachusetts Institute of Technology (MIT) published a report, "A Study on the Status of Women Faculty in Science at MIT."[3] It showed how badly women were discriminated against at that institution. The report had a strong effect nationwide and brought about major changes in the working conditions, salaries, and other aspects for women in faculties. A few years later, the president of Harvard, Larry Summers, shocked the public by announcing that women's innate aptitude was the reason why there were so few women at the top. But it was a sign of change that a large outcry made him resign his presidency.

The bulk of this book is about women scientists whom I have met in person. I included a few others as well, because they usefully augmented the rest. In these cases, I tried to add something to their stories already known, to make them more complete. Thus, for example, Edward Teller's extended correspondence with Maria Goeppert Mayer showed the role other scientists played in Goeppert Mayer's life and vice versa. In case of Chien-Shiung Wu, her research area has always fascinated me, and my probing into her story made me see that the widespread lamentation about her not having received the Nobel Prize is misdirected. I did not include some other famous women scientists, like Lise Meitner or Rosalind Franklin, whose stories are no less fascinating, but where I did not have anything substantial to add to the accounts already available.

The women in this book represent a variety of scientific fields, mostly physics, chemistry, and the biomedical sciences, and to a lesser extent mathematics, astronomy, engineering, medicine, and other fields. Of course, there is considerable overlap between various fields, and such overlaps often characterize the activities of the scientists appearing in this book; hence, the designation assigned to them often does not describe the totality of their activities, in which they have invariably excelled.

An important feature of this book is its broadly international character. My heroes come from different backgrounds and different geographical regions, eighteen countries on four continents. A little less than half of them are from Europe, a third from the United States, and the rest from Asia and Australia.

I grouped the chapters into three loosely defined sections. Perhaps because of my own initial experience working jointly with my husband, I am interested in other "scientific couples." As much as this working arrangement is a source of joy, it has certain disadvantages for the woman in the pair. It often happens that the credit for her work goes to him—rarely the other way around. The first three female Nobel laureates in the sciences were all part of such scientific couples. The scientist husband and the joint interest meant an advantage when women had barriers in building a career in science during the first half of the past century. This also applied to the fourth Nobel laureate woman scientist, Maria Goeppert Mayer, whose husband was also a professor, which made it possible for her to work—even if her prize-winning discovery was not part of joint research with him. I start the presentation in the first section "Husband and Wife Teams" with the first three Nobel laureates, followed by the others in alphabetical order.

The next section of the book is titled "At the Top." As the Nobel Prize is the highest recognition a scientist can receive, I started this section with a short summary of the women Nobel Prize winners. During the 113 years of the Nobel Prize, altogether sixteen women have received a science award;[b] I mention briefly the achievements of the seven who will not appear later in the section in more detail. This brief section is followed by the introduction of a large number of scientists from different countries and professions in alphabetical order. At the end of this section, I introduce women scientists of three countries separately, because relatively little has been written about the women representatives of the science of these countries; namely, Russia, India, and Turkey. Curiously, there is very little information about women scientists in Russia outside the country, in spite of the fact that since the 1917 Bolshevik Revolution women have been involved with science in large numbers. Gender issues have hardly surfaced there, not even after the political changes in the early 1990s—is this so because there are no gender issues to discuss? For India and Turkey, I found it intriguing to look into how women could succeed as scientists, considering the traditional role of women in their society. In the last section, "In High Positions," I grouped those women scientists who at one point in their career got involved with science administration and became leaders of universities or big research institutions or occupied other key leading positions.

[b] There were seventeen Nobel Prizes given to women but Marie Curie received two.

HUSBAND AND WIFE TEAMS

In this section, I discuss husband-and-wife scientists who worked closely together for at least part of their careers. The best-known examples are the first Nobel laureate women in the sciences. It is rather telling that the first three women who received science Nobel Prizes were all part of such couples.

Marie Curie's first Nobel Prize was shared with her husband, Pierre Curie. The two received half of the 1903 Physics Nobel Prize for their joint researches on the radiation phenomena discovered by Henri Becquerel. Marie Curie (née Sklodowska) received another Nobel Prize, in 1911, in Chemistry—by then her husband was no longer alive. The next woman to receive a science Nobel Prize was their daughter, Irène Curie, who shared the 1935 Nobel Prize in Chemistry with her husband, Frédéric Joliot. The third couple, Carl and Gerty Cori, shared half of the 1947 medicine prize. It took another fifteen years for another woman, Maria Goeppert Mayer, to receive a science Nobel Prize, in Physics, in 1963. Although her prize-winning research was not done together with her husband, her opportunity to do research was due to his professorial appointments.

The fact that Marie Curie and Gerty Cori received the Nobel Prize with their husbands was no coincidence.[a] During the first half of the past century, even getting a university education for a woman was not at all straightforward. It was even more difficult to get a position at a university. There are many examples to illustrate this. Emmy Noether (1882–1935), was an outstanding German mathematician. In the 1910s, only male candidates could be granted habilitation, the higher prerequisite degree for lecturing at German universities. Two famous mathematicians, Felix Klein and David Hilbert, wanted to invite her to teach at the University of Göttingen, but there was strong opposition. The nonmathematical members of the philosophical faculty argued that the soldiers coming back from the war should not find themselves "being lectured at the feet of a woman." To which Hilbert angrily replied: "I do not see that the sex of the candidate is an argument against her admission as a Privatdozent. After all, we are a university, not a bathing establishment."[1]

Being married to a successful scientist opened up the opportunity to work in science for Marie Curie and Gerty Cori. The situation proved "lucky" for their husbands as well. Their life stories tell us that both partners mutually benefited from the joint work. The blending of their talents enhanced their joint output in such a way that it was much larger than would have been the sum of their independent contributions had they worked separately.

[a] Irène Curie was different; she grew up in the limelight of her parents' success and fame.

THE CURIE "DYNASTY"

Physicists and Chemists

Marie and Pierre Curie in their laboratory. (courtesy of the Oesper Collections in the History of Chemistry, University of Cincinnati)

Marie Curie has long been the number-one role model for young women interested in a career in science. She was a "first" and an "only" woman in many aspects:

- First female lecturer, professor, and head of laboratory at the Sorbonne (1906)
- First woman science Nobel laureate
- First person to receive two Nobel Prizes
- Only woman ever who received two Nobel Prizes
- Only person who received two *science* Nobel Prizes in two *different* categories
- Only person whose daughter also received a Nobel Prize
- First woman who was buried in the Pantheon in Paris for her own merits

Her story has fascinated people, and there have been many accounts of her life. I mention only a few aspects that are especially relevant to our topic of the scientific couple.

We might ask the question, who was the greater scientist of the two, Marie or Pierre? This is impossible to tell, and it also depends on how we define a "great scientist." Pierre was eight years older than Marie and already had made important discoveries before they met. He and his brother, Jacques, collaborated for years and discovered the piezoelectric effect. Pierre Curie alone made a most far-reaching observation concerning symmetry and discovered laws describing the transition from ferromagnetism to paramagnetism (the temperature of transition is called the "Curie point").

Nonetheless, when Pierre met Marie, he was not yet well known, nor was he a member of the French Academy. He did not publish much, did not care about priority, and was not interested in recognition and fame. Much of what we know of his early works is due to Marie Curie's writings.[1] It was Marie who started to work on radioactivity (a word she coined), and Pierre joined her when he saw its importance. Ultimately, it was their joint work on this topic that brought about *his* recognition as well. Due to the success of their joint work on radioactivity, he gave up much of his ongoing studies on magnetism and crystals. Thus, Marie had a crucial role in making Pierre a more recognized scientist than he might have been without her.

The success in their joint work was in that they famously complemented each other. He was a withdrawn person, given much to contemplating, one who took time to formulate his thoughts and conclusions. The great French physicist Paul Langevin noted that Pierre Curie never showed a sudden outburst of enthusiasm over an idea, and the only time he took quick action was when he decided to marry Marie. According to the biographer Helena Pycior, Pierre was a "familial" scientist.[2] In contrast to the idea of a lonely scientist, he thrived on cooperation and liked to work with a family member. At first, this partner was his brother, Jacques, who was also uninterested in disseminating their results. It was soon after Jacques left Paris that Pierre met Marie, and he recognized in her a perfect partner—both in private life and in science. Marie had the ability to bring out the successful scientist in Pierre. She was his opposite, quick in thinking and not inhibited about making bold conclusions and declaring them, and she was very interested in getting recognition for her work. The combination of their two sets of traits proved to be the right recipe for success.

There is a general attitude that blurs the results of a junior partner of a well-known famous scientist and tends to attribute them to the senior person. This attitude was named the "Matthew Effect,"[3] referring to the Gospel: "For whomsoever hath, to him shall be given, and he shall have more abundance; but whomsoever hath not, from him shall be taken away even that he hath" (Matthew 13:12). This effect often occurs when male-female cooperation is involved, and science is no exception. Margaret Rossiter suggested calling this the "Matilda Effect."[4a] With respect to the Curies, Pierre is generally recognized as the "thinker"—and he was a great one indeed. The question is how Marie should be described. If Pierre was the thinker, she must have been the worker, the person who gets the things done. This image rhymes with the internationally renowned physicist Valentine Telegdi's comment: "She was a great scientist. She did not have a great mind but she was a great scientist. Her husband was a great scientist and also had a great mind. But she was a great scientist by what she had done . . ."[5]

The Curies' daughter, Irène, characterized her mother as the "thinker-doer," who "more often directed toward immediate action, even in the scientific domain,"

[a] Named after Matilda Joslyn Gage (1826–1898), American women's right activist, who first observed that women's achievements were often attributed to their male colleagues.

while her father besides being an "excellent experimenter . . . [was] also a thinker."[6] Marie's favorite image of Pierre was one in which he was looking ahead and rested his head on his hand. She could have associated this image with the famous sculpture *The Thinker* by their friend Auguste Rodin.[7] She even posed for a picture in which she also rested her head on her hand, just as in Rodin's sculpture and Pierre's photograph.[8]

Given the historical backdrop of the fin de siècle world of science, we might even be a little surprised that she did receive her due recognition right from the start. In order to facilitate this, Pierre and Marie were—possibly due to her foresight—very careful about their publication policies. They always delineated in their publications who did what, and Marie had publications under her name alone as well. This way, it was impossible to ignore her when the question of the Nobel Prize came up. In his Nobel lecture, Pierre Curie specifically stated which of the results were those of Marie by saying "Mme. Curie showed," "Mme. Curie has studied," or "Mme. Curie then made the assumption, . . ." When it was their joint work, he said "Mme. Curie and I . . ."[9]

The story of their joint Nobel Prize in 1903 is interesting in itself. According to the Nobel rules, documents (but not the deliberations, which are not even recorded) can be studied fifty years after the prize, so theirs have been available for quite some time. It appears that for the 1903 physics prize only Pierre had been nominated. However, a member of the prize-awarding Royal Swedish Academy of Sciences (RSAS), Gösta Mittag Leffler, a mathematician, wrote a letter to Pierre about the situation.[10] Apparently, Mittag Leffler had a strong personality and on occasion intervened in questions about the Nobel Prizes. He was also a strong friend of France. So Mittag Leffler wrote to Pierre in 1903. Pierre wrote back that Marie's contribution to the work was essential. She, in fact, had been nominated in 1902 by a French foreign member of the RSAS. This foreign member had again been asked for a nomination for the 1903 prize, but on this occasion failed to submit a nomination. It was then argued that this foreign member still wanted Marie to receive the prize (without formal nomination, nobody can get the Nobel Prize). Just to suppose what one might have wanted, in general, would not satisfy the rigorous rules of nomination, but from time to time the prize-awarding institutions afford themselves some slight flexibility, so Marie Curie was duly included in the award.

As to the 1911 chemistry prize, Marie received two nominations. One was from France, the other from the Swedish Svante Arrhenius. She was announced to receive the award, but when the love story between Marie and Paul Langevin surfaced, the Academy requested Arrhenius to write to Marie and ask her not to come to Stockholm in December to attend the award ceremony. She responded that she would go because she felt that the prize was given for her scientific work and not for her private life. Indeed, she went to Stockholm for the ceremonies with Irène. The Nordic women's rights associations strongly supported Marie Curie in standing up for herself.[11]

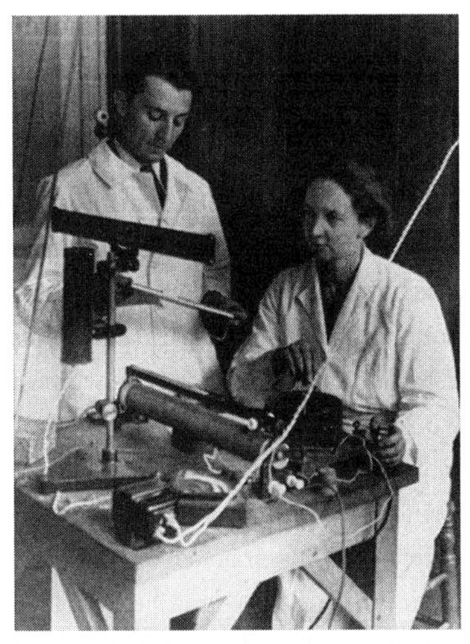

Irène and Frédéric Joliot-Curie in their laboratory. (courtesy of the Oesper Collections in the History of Chemistry, University of Cincinnati)

The Curies' elder daughter, Irène, grew up in a world for which, beside family, science was the most important thing. She was born in 1897, just a year before her parents discovered radium. Eugene Curie, Pierre's father, moved in with them soon after Irène's birth, and it was essentially he who brought her up. When Pierre and Marie Curie received their Nobel Prize, Irène was six years old, and she lived through all the publicity that her parents received. She was still a teenager when World War I broke out, and she accompanied her mother to the front with their X-ray machines to help the doctors diagnosing the injured.

For Irène, choosing a profession was not a question; it was obvious that she would follow in her parents' footsteps. After her father's death, she grew yet closer to her mother. Of course, Irène had an easier path to science and recognition than Marie; it was quite natural for her to continue the family tradition. After World War I, she worked with her mother in the Radium Institute for a proper salary. There Irène met her future husband, Frédéric Joliot, and after they got married in 1926, her mother bought a new apartment for them.

In a few years' time, Irène and Frédéric started to work together. Their joint work lasted merely a few years, but it was during this period that they discovered artificial radioactivity. When they bombarded aluminum with alpha particles, they noticed that by moving the source of these particles farther away, the neutron emission diminished, but the positron emission continued. They understood that bombarding aluminum with alpha particles produced radioactive phosphorus, which eventually decayed to silicon and a positron. In 1935, they received jointly the Nobel Prize in Chemistry.

There were marked similarities as well as differences between their duo and that of her parents. Like Pierre, Frédéric was very good at constructing instruments. On the other hand, of Irène and Frédéric, he was the more public person and she was the more introverted one.

There were several occasions during Irène's and Frédéric's career when they missed making important discoveries that they could have made based on their own experiments. This does not diminish the significance of their outstanding achievement in discovering artificial radioactivity; it only shows the enormous capacity they possessed for creativity. We mention here only one of their nondiscoveries, perhaps the most conspicuous one. By all evidence, Irène and her group in an experiment in 1938 observed nuclear fission—but she did not recognize it. By this time, after the Nobel Prize, Irène and Frédéric did not work together any more. Their daughter, Hélène Langevin-Joliot, herself a noted physicist, heard her parents saying: "maybe if we had worked together, we could have discovered fission!"[12]

Their most productive scientific period was the time when they worked together. In their joint work, initially, a most unusual situation developed, in sharp contrast with the usual scientific-couple relationships. At first, Irène was the leader and the authority, and so she appeared to the outside world as well. Frédéric had to prove that he was on equal footing with her in their science. Irène and Frédéric both appeared, for different periods of time, in public positions as well. Frédéric was a dedicated communist, a great friend of the Soviet Union, and was much involved with politics. He found it prudent to add the Curie name to his own and has become known as Frédéric Joliot-Curie. Irène died at the age of fifty-eight from the effects of radiation, just as did her mother. James Chadwick, the Nobel laureate discoverer of the neutron, paid tribute to Irène and Frédéric in her obituary: "a collaboration of husband and wife in scientific work rivaling in productive genius even that of her parents."[13]

Irène and Frédéric Joliot-Curie had two children, Hélène and Pierre Joliot. Both became scientists, she a nuclear physicist and he a biophysicist. Hélène married Michel Langevin, the grandson of Paul Langevin. It is worth mentioning that up till now three Curies have become members of the French Academy of Sciences: Pierre Curie, Frédéric Joliot-Curie, and Pierre Joliot. None of the women, equally successful and well-known scientists, got in; not Marie Curie, not Irène, and not Hélène Langevin-Joliot. Both Marie and Irène had been proposed, but refused. The first woman who got into the French Science Academy, in 1962, was one of Marie's students, Marguerite Catherine Perey, who in 1939 discovered the element francium.

GERTY AND CARL CORI

Biochemists

Gerty and Carl Cori around 1946 in their laboratory at Washington University in St. Louis, MO. (courtesy of the Oesper Collections in the History of Chemistry, University of Cincinnati)

"May I express my deep gratitude for the signal honor conferred upon me by the award of the Nobel Prize. . . . That the award should have included my wife as well has been a source of deep satisfaction to me. Our collaboration began 30 years ago when we were still medical students at the University of Prague and has continued ever since. Our efforts have been largely complementary, and one without the other would not have gone as far as in combination."[1] These were Carl Cori's words in the traditional brief speech delivered during the banquet following the Nobel Prize award ceremony in 1947. The two Coris shared the distinction for understanding the catalytic conversion of glycogen.[a]

Gerty Cori was born Gerty Theresa Radnitz in 1896, into a Jewish family in Prague, Bohemia, at that time part of the Austro-Hungarian Empire (today the capital of the Czech Republic). She attended medical school, where she met a fellow student, Carl Cori. Both were interested in biochemistry, and they started doing research together right away. They published a joint paper, the very first of their numerous shared scientific publications. They got married in 1920, and had one son, born in 1936.

As she was Jewish, with the rising anti-Semitism in Central Europe, the Coris decided to emigrate. Their former associate and 1959 Nobel laureate Arthur Kornberg commented: "I understood the trauma of anti-Semitism that forced both of them to

[a] The third awardee, Bernardo Alberto Houssay, received his prize for an unrelated discovery.

leave Europe in 1920–21; Carl was concerned that she would never have any academic advancement in those turbulent times in Europe."[2] In 1922, Carl received a position at the Roswell Park Cancer Institute in Buffalo, New York. It took about half a year for her to get an assistant pathologist's job in the institute. Beside doing their work, they were free to do any kind of research they wanted—and they enthusiastically made the most of this opportunity. During the nine years they spent there, they published extensively, mostly with joint authorship, and they alternated the role of first author in their papers. There they started the research that eventually led to their Nobel Prize.

The complementarity in their research style was obvious already then. Mildred Cohn (see elsewhere in this book), a distinguished scientist and one of their many students, observed: "They were remarkable in that way. He would start a sentence and she would finish it. They were completely complementary. In personality, they were very different. He was aloof and she was very vivacious, she was outgoing and he was not, although he was very insightful about people. They let me read their Nobel speech before they presented it, for my comments. (Carl had written half of it and Gerty had written half of it.) I showed it to my husband, and he could tell exactly where one stopped and the other began."[3]

As their fame grew, but years before their Nobel recognition, Carl Cori received attractive job offers, but none of them included the possibility of a job for Gerty, which was in line with the antinepotism rules of the time. The University of Rochester, for example, stipulated that he would get the job only if he stopped collaborating with his wife. When he declined, they told Gerty that it was un-American for a man to work with his wife and that she should understand that by insisting on this, she was ruining his career.[4]

Finally, they received an offer from Washington University, in St. Louis, Missouri, which being a private institution did not follow the antinepotism rule. He became full professor and chair of the pharmacology department, and she received a position of research associate. In 1946, after about fifteen years of being a research associate, just a year before their Nobel Prize, Gerty Cori became full professor when Carl Cori was made chair of the new biochemistry department. During their time at Washington University, they continued working together and producing significant results.

From the beginning, their research was aimed at understanding how energy is produced and transmitted in the human body. In particular, they wanted to find out how the sugar stored in our body gets into the muscle and turns into the energy that we need for exercise; this is called carbohydrate metabolism. After countless experiments, they finally understood what happens. Sugar (glucose) is stored in our body in the form of a polysaccharide, called glycogen, in the cells of the liver and also in the muscle. When we exercise, the glycogen in the muscle breaks down and besides the needed sugar it produces lactate that diffuses from the muscle to the bloodstream. The blood carries it to the liver, where it turns into glucose that is either stored there as glycogen or is transferred back to the muscle. The Coris called this order of events the "cycle of carbohydrates," but now it is known as the Cori cycle.

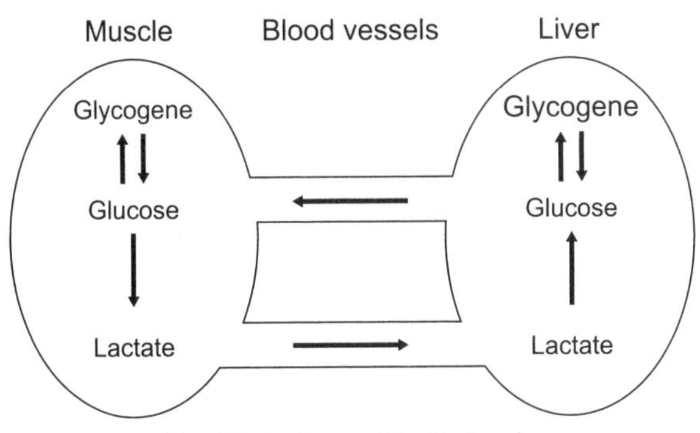

Simplified scheme of the Cori cycle.

In the late 1930s, the Coris turned to enzymology. They recognized and isolated the enzyme called phosphorylase, which induces the breakdown of glycogen into glucose, together with other enzymes participating in these reactions. Hugo Theorell said in his presentation speech at the Nobel ceremonies in 1947: "For a chemist, synthesis is the definite proof of how a substance is built up. Professor and Doctor Cori have accomplished the astounding feat of synthesizing glycogen in a test tube with the help of a number of enzymes which they have prepared in a pure state and whose mode of action they have revealed. This synthesis would be impossible by methods of organic chemistry alone . . . The Cori enzymes made this synthesis possible, because the enzymes favour certain modes of linkage."[5]

The Cori laboratory in St. Louis became one of the most important centers in enzymology. Talented researchers from all over the world came there to work for a while. Eventually, six of their colleagues received Nobel Prizes as well. Arthur Kornberg noted: "There were few laboratories [at that time] in the United States that were hospitable to Jews and refugees: Carl Cori in St. Louis, Hans Clark at Columbia, and perhaps some others. Carl Cori, one of my heroes, made a place for [Severo] Ochoa, Herman Kalckar, Luis Leloir, and many others."[6]

In 1947, Gerty Cori was diagnosed with an incurable disease, a special form of anemia that took her life ten years later. During her life, she had to face plenty of discrimination. Although the two Coris worked and published together from the very first moments of their career, for decades it was only Carl Cori who was recognized. He received many awards and distinctions alone; for example, he was elected to the US National Academy of Sciences (NAS) already in 1940, while Gerty only after their Nobel Prize, in 1948. An interesting episode illustrates how frustrating rendering her a "second-rate researcher" must have been for Gerty. At Washington University, the dean once asked Gerty to write a report about their work that was to be attached to a grant application to the Rockefeller Foundation. In this report, she wrote about "the Coris" work and about what "they" discovered. The dean sent this report in unchanged, but someone at the Rockefeller Foundation altered the style and inserted "Dr. Cori" in place of "the Coris" and "he" in place of "they."[7]

According to Osamu Hayaishi, a distinguished scientist at Osaka University:

> They were not only great scientists but great human beings as well. They were very helpful to me, perhaps because of their own experience having immigrated from Europe when they were young. . . . The Coris had a broad vision and he was a great leader but he could not have succeeded without her as she was not only a good scientist but also an excellent experimentalist. Because of his fame, Carl was always very busy, but Gerty carried on in the laboratory. She was never too busy to talk with me when I had problems. If I had to see Carl, Gerty always arranged the meeting for me. They were excellent partners.[8]

ILONA BANGA AND JÓZSEF BALÓ

Biochemist and biomedical scientist

József Baló, Mátyás Baló, and Ilona Banga in 1954, in Budapest.
(courtesy of M. Baló, Budapest)

In 1952, Ilona Banga received a telegram from the Parliament of Hungary that she had been awarded the Kossuth Prize—the highest honor a scientist could receive in the country. However, when she learned that she was so honored alone, without her husband and partner in research, she declined it. The omission of her husband from the recognition must have been an oversight, but the authorities were not about to admit the error and warned her that her act would be considered hostile, inviting repercussions—this was at the time of Soviet-type dictatorship in the country. Her response was that they worked together, and their achievements could not be separated. After Stalin's death in 1953, there was some relaxation in the political situation, and the previous oversight was corrected: in 1955, Ilona Banga and her husband, József Baló, shared the Kossuth Prize for their discovery of elastin and the enzyme elastase (see below).

József Baló (1895–1979), an MD, had done extensive research in bacteriology, immunology, and parasitology. He discovered what is known as Baló's disease, a rare pathological condition of the brain. In 1940, he was elected a corresponding member of the Hungarian Academy of Sciences, and in 1948, a full member. However, he, along with many other members of the Academy, was downgraded to advisor in 1949. This was part of a general revision of the membership of the Academy as a consequence of the communist takeover of the country. Incidentally, the memberships of most of those downgraded members were reinstated—alas, in most cases posthumously—at the time of the political changes in 1989–1990. As a rare exception, in 1956, Baló was re-elected to the Academy. His wife never became a member of the Hungarian Academy of Sciences.

Ilona Banga was born in 1906 in Hódmezővásárhely, a mostly agricultural town in southeastern Hungary. She wanted to study to become a medical doctor, but her mother thought that it would not be a proper profession for a woman, so Ilona decided to study chemistry instead.[1] She started her university studies in Szeged, not far from her hometown; continued them at the University of Vienna; and completed them at the University of Debrecen, in Hungary. She received her MSc degree in chemistry in 1929, and started as an assistant to Albert Szent-Györgyi, the future Nobel laureate, at the Institute for Medicinal Chemistry of the University of Szeged.

Albert Szent-Györgyi and Ilona Banga in the 1930s–1940s at the University of Szeged, Hungary. (image from the National Institute of Medicine, courtesy of Gábor Tóth, and the Klebelsberg Library of the University of Szeged)

She was Szent-Györgyi's first associate, and one of his most successful. They worked together for about fifteen years and had twenty-five joint publications, many of them in prestigious international journals. She actively participated in Szent-Györgyi's research, leading to his Nobel Prize in Physiology or Medicine in 1937. Their research involved two related topics. One was the investigation of respiration in muscle tissue and establishing the importance of fumaric acid in the biological combustion process. The other was the study of the role of vitamin C and its preparation in large quantities from Hungarian paprika. Szent-Györgyi and Banga were the only two authors on the paper describing the large-scale preparation of vitamin C.[2]

Szent-Györgyi was full of ideas and he found a perfect experimenter in Ilona—Iluska, as he called her. She was committed to her work; she was both meticulous and ingenious about experiments, and ran many parallel ones to ensure the reproducibility of the observations. The work was complicated and often tedious. For example, in order to prepare a sufficient quantity of vitamin C she had to use about a ton of Szeged paprika. Szent-Györgyi's Nobel Prize drew international attention to the Szeged research group, and his colleagues, including Ilona, received invitations from abroad. She spent some time in Liege, Belgium, and Oxford, England. She worked together with another future Nobel laureate, Severo Ochoa, in Oxford, studying the biochemistry of vitamin B1.

Following the Nobel Prize, Szent-Györgyi was at the crossroads: he had to decide whether to become a great statesman of science or charge ahead in research. He chose the latter and started a new direction, muscle research. His new line of interest was catalyzed by Engelhardt and Lyubimova's paper in *Nature* in 1939 (see elsewhere in this book).[3] The Russian authors showed that the material responsible for muscle contraction was the enzyme myosin. Its role was not merely in splitting adenosin-triphosphate (ATP) and hence releasing energy—it was myosin itself that actually triggered this activity. This discovery suggested possibilities for further discoveries in the field. Szent-Györgyi recognized that understanding the interaction between myosin and ATP might lead to the understanding of muscle contraction. At the start, Szent-Györgyi and Banga repeated the Engelhardt-Lyubimova experiment. For a long time, Ilona was Szent-Györgyi's principal coworker in this project.

The experimental facilities of the Szeged laboratory were quite rudimentary, but the dedication of its members more than compensated for it. They prepared myosin threads, hung them in saline solution, and then added ATP. What they saw was that the threads suddenly began to shrink to about one-third of their original size. This was a most exciting moment, observing lifelike motion in a flask. Szent-Györgyi later called it "perhaps the greatest excitement of my life, to see motion in the bottle of more or less known substances for the first time."[4]

Ilona was carrying out these experiments, which always started by putting minced rabbit muscle in saline solution and after a while extracting myosin from it. Once, late in the evening, she did not have time to do the extraction, so she just left the minced muscle in the saline solution overnight. The next morning, she was surprised at what she observed in the bottle. Usually, the myosin was a thin liquid layer, while this time it looked like a thick viscous jellylike substance. They immediately understood that they were onto something important. They analyzed this new substance, and it was also myosin, but different from the previously known thin liquid. Eventually, they determined that if they added ATP to this "new" myosin, which they called myosin B (and the original one, myosin A), it became the thin "old" myosin.

This was a turning point in their research, and Szent-Györgyi decided to "give" this research to another of his associates, F. Bruno Straub (1914–1996). Szent-Györgyi was the professor, and he decided everything about who did what in his laboratory. Straub eventually determined that myosin B contained another protein beside myosin A, which they named actin. They renamed myosin B actomyosin, and the "old" myosin kept its name simply as myosin.

It is noteworthy that all this research was being carried out in Szeged during World War II. Szent-Györgyi understood that even under war conditions they had to publish their results. Although they kept receiving western scientific journals—thanks to the generosity of the Rockefeller Foundation—they could not publish their own manuscripts in them. Instead, they published their results, in English, in the periodical of the University of Szeged.[5] It is puzzling that Szent-Györgyi did not include Banga's name among the authors of the two articles that included her experiments.

Many consider Szent-Györgyi's work on muscle his greatest contribution to science. In 1954, the Lasker Foundation recognized him with its Albert Lasker Basic Medical Research Award, whose citation specifically mentioned the discovery of actomyosin. The Lasker Foundation prides itself that its awardees often receive the Nobel Prize, but it seldom awards scientists that had already received the Nobel Prize. Szent-Györgyi's recognition was one of the few exceptions, and he was recognized for his muscle research, that is, the achievement different from what he had received the Nobel Prize for.

There has been a lot of discussion about the history of the discovery of the proteins responsible for muscle contraction.[6] Why did Szent-Györgyi give the experiment that Banga had been carrying out so successfully suddenly to Straub? According to Ilona's son, the renowned medical doctor dermatologist Mátyás Baló, Straub had just recently returned from Cambridge, where he learned new methods and Szent-Györgyi hoped that would be useful in determining what this new substance was. Szent-Györgyi used to emphasize Straub's role in these discoveries, but by 1973, he seems to have changed his mind when he declared during a visit to Hungary that it was Banga who under his supervision discovered actin and actomyosin.

Straub and Banga had two very different careers in postwar Hungary. Straub was elevated to high positions; he was elected to the Academy of Sciences and eventually served as its vice president. Toward the end of the communist regime, he was even appointed to be the figurehead president of Hungary. Banga continued as a recognized researcher. Straub did not seem to have facilitated her—or, for that matter, his other former colleagues'—higher recognition.

Ilona Banga was not only an outstanding scientist; she was also a brave woman and patriot. Toward the end of World War II, after Germany occupied Hungary, Szent-Györgyi was hiding from the Gestapo, which was out to arrest him for his anti-Nazi activities. Banga cleverly saved the equipment of the Institute for Medicinal Chemistry. She had notes posted on the door of the Institute in Hungarian, German, and Russian that in the institute they carried out research on infectious materials. The announcements indicated the receiving hours when such materials should be submitted. This kept the departing German and the arriving Soviet troops, as well as Hungarian thieves, away.

Ilona Banga married József Baló in 1945, in Szeged. When Szent-Györgyi left Szeged for Budapest, Banga followed him, and continued her research at the Institute of Pathology of the Medical School. She worked there until her retirement in 1970, studying the biochemical processes of aging. In 1947, Szent-Györgyi left Hungary; he had made arrangements for jobs abroad for his coworkers, but Ilona was one of those who did not go.

She became the head of the chemistry laboratory in the Institute of Pathology. She collaborated with her husband in a research area that was new to her, the study of arteriosclerosis. They wanted to understand the causes of fiber degradation in vein walls. They suspected that it was something that the organism itself produced. After many painstaking experiments, she found that the culprit might be an enzyme that the pancreas produces, which they named elastase. The discovery was of major importance, but it sounded so new that many experts in the field doubted its validity. After another long stretch of experiments, she managed to crystallize elastase.[7] This convinced all doubters. This was the third major research area in which Ilona Banga produced exceptional results—this time it was a joint project with her husband.

Ilona Banga died in 1998. Many of her colleagues and followers in her field feel that she should have received more recognition for her many accomplishments, although she did receive some. In 1940, she became the first female Privatdozent (comparable in rank to an associate professor) at the University of Szeged. In 1950, she received her DSc degree, a prerequisite for professorial appointment, but she was never made a professor. In 1986, she received the first medal that the University of Szeged established in memory of Albert Szent-Györgyi.

She wrote two books, one of them in English, *Structure and Function of Elastin and Collagen*.[8] In 1962, she was elected to the Leopoldina Academy (Halle, East Germany). She had a happy demeanor; from an early age, she engaged in what she loved best, challenging laboratory projects, and she found ample rewards in these activities—significant discoveries. She used to say: "Research is my life motif, and it gives me fulfillment."[9]

RITA AND JOHN W. CORNFORTH

Chemists

John and Rita Cornforth in 1997 in their garden in Lewes, East Sussex, UK. (photo by I. Hargittai)

John W. Cornforth (1918–2013) received half of the 1975 Nobel Prize in Chemistry for his research on the stereochemistry of enzyme-catalyzed reactions.[a] He started his research career during his studies in Sydney, Australia, together with Rita Harradence. Later, in Oxford, she wrote, "Our collaboration then just continued forever and finally Kappa was my boss."[1b] All their life, they worked together; about three-quarters of her publications were together with John. Nobody doubted the decision of the Nobel Committee—but it is nice to read what John said in his Nobel lecture: "my wife Rita Cornforth, with patience and great experimental skill, executed much of the chemical synthesis on which the success of the work was founded. To her, in this as in other ways, I owe more than I can well express."[2]

Rita Harradence (1915–2012) was born in Sydney, Australia. Her parents were both of English descent; their parents had migrated to Australia in the early 1880s. Her father was a carpenter and her mother worked as a seamstress in a department store in Sydney until she married.

In school, Rita was always at the top of her class. In high school, she fell in love with mathematics. Chemistry as a separate subject started only toward the end of high school, but then she was inspired by a very good teacher. She had not even heard

[a] The other half of the prize went to Vladimir Prelog.

[b] Kappa (the Greek letter κ) was John Cornforth's nickname starting from their university years back in Sydney, because he liked to etch his initials on his glass apparatus in the Greek form ιωκ, the nearest he could get to JWC.

of a university before she started high school. In the Leaving Certificate examination, she got first-class honors in chemistry and mathematics, coming top of the state of New South Wales in chemistry and top of the girls in mathematics. This gave her automatic entrance to the University of Sydney at age seventeen. It was during her studies there that she encountered organic chemistry—and in the last year, she started to do research.

Rita first met John in 1934 when she was in her third year; she was one year ahead and two years older than he was. By that time John Cornforth was totally deaf due to otosclerosis, a progressive disease that started when he was about ten years old. She had broken the side arm of a precious flask and a friend suggested her to ask John for help because he knew how to blow glass. They got to know each other and had common outside interests, like geology and bushwalking in the Blue Mountains during long weekends. They chose chemistry as a profession independently of each other.

After receiving her MSc degree, there was no possibility for her to study further in Australia at that time. Her parents could not support her, so she needed a scholarship to go abroad for further studies. But there were very few overseas scholarships. There were only two "1851 Exhibition scholarships" for the six Australian universities, and in the year 1938 she was too shy to apply. But next year both she and John applied. They both wanted to go to Oxford and work with Sir Robert Robinson. As John said: "Robert was on the selection committee and he must have been very persuasive because he got both scholarships for us."[3] Since she had seniority, she had been recommended as the first choice and John second. Later, Robert joked, "If I wanted Cornforth I had to have Harradence as well!" By that time he knew about John Cornforth's outstanding ability from their professors at Sydney.

In Rita's narrative:

> We were not attached to one another when we left Sydney on 6 August 1939 on the Orient liner *Orama* (it later became a troopship, and was sunk). War broke out when we were half way between Colombo and Aden and the ship was diverted round Africa. I asked Kappa if he thought we should return from Capetown, as many passengers did. His reply was "You realise if we don't go now we never will; I'm going on." So, of course, I did too. We docked at Southampton after 11 1/2 weeks at sea and late for the beginning of term in Oxford."

They got engaged in 1941, while they were writing their theses, were married in September of that year, and stayed in Oxford till 1946. Because of John's deafness, Robinson arranged for them to go to the National Institute for Medical Research (NIMR), then under the direction of Charles Harington. They had started to do research together already back in Sydney; they had two publications on organic synthesis along with their professors. In Oxford, both of their theses were concerned with the synthesis of steroids.

The most important project of her research career was the part she played in elucidating the chemistry, and especially the stereochemistry—the three-dimensional arrangement of atoms in a molecule—involved in the biosynthesis of cholesterol.

They started this work during their Oxford days, and when they moved to the NIMR they continued it. At that time they had their second child and Rita thought she probably would not go back to work for a while, but Harington "actively encouraged me to do so; scientific wives of staff members were offered part-time work, so I returned to work during 1947." Throughout the next fifteen years she used her skills at the bench to advance their research. They had to use several techniques to understand which part of a molecule reacts during a chemical process. One of the crucial techniques they used was isotope labeling. This means that they had to change one or another atom in the molecule with its isotope (different isotopes of an atom have the same number of protons, but the number of neutrons is different). They tracked the path of the labeled atoms during the reaction in order to understand how it happened. She labeled the atoms in dozens of ways—which also meant that she had to invent new synthetic routes. They wanted to understand how cholesterol is formed in our body, and eventually succeeded in reproducing the fourteen-step procedure by which our enzymes build up a large molecule, called squalene, the precursor of cholesterol.

John was the head of the group and there were many other colleagues involved, but she was the genius of the lab, although Rita told me "I want to emphasize that neither George [George Popják, their most important collaborator in this work] nor I could have done any of this exciting and challenging work without Kappa's ingenuity and three-dimensional thinking."

When we met, Rita appeared to me very modest, even shy. She did not have much recognition for her work in chemistry, just an honorary doctorate from the University of Sussex, and there is a research fellowship for women at the Australian National University named after her. I asked whether she received proper recognition for her achievements and she said she believed that her scientific peers knew what she had done, and that was what was important to her. She never cared for fame or publicity.

The Cornforth family at Trafalgar Square in 1949 in London. (courtesy of the late J. Cornforth)

She must have had a difficult time with bringing up three children, taking care of their home, and working in the laboratory on challenging tasks that needed a great deal of invention and painstaking laboratory work. John never learned lip-reading, so she had to be his ear as well. When their children were young, they had nannies; she took a year off twice, and sometimes she worked half-time. "It was difficult, of course, and by the time I had three children I sometimes thought: 'If I hadn't embarked on this I wouldn't, but I can't give up now.' I was not a superwoman and could not have combined an independent career with bringing up children. It was possible because we worked so closely together. I found it easier to put chemistry out of my mind when I was at home than to put the children out of my mind when I was in the lab."

JANE M. AND DONALD J. CRAM
Chemists

Jane and Donald Cram in 1995 in their home in Desert Palm, California. (photo by I. Hargittai)

Donald J. Cram (1919–2001) shared the 1987 Nobel Prize in Chemistry with Jean-Marie Lehn and Charles J. Pedersen for their discoveries that made it possible to design molecules with "holes" where other molecules or ions could then bind. The achievements of the laureates catalyzed the development of many branches of synthetic chemistry. Cram called this new area of chemistry "host-guest" chemistry; Lehn called it "supramolecular chemistry."

Jane Lewis Maxwell (1924–) received her PhD degree at Emory University in organic chemistry. She was professor of chemistry at Mount Holyoke College in South Hadley, Massachusetts. Her first interaction with Donald Cram happened through a project of hers; she used a particular research topic of his as a subject for a seminar at Emory University. Later she spent a sabbatical at the University of California at Los Angeles (UCLA) with him. Eventually, in 1969, they got married; Jane was his second wife.

After their marriage, Jane never had a formal job. She did not have an appointment at UCLA; as her husband said, "her only appointment is to keep me reasonable."[1] That did not mean at all that she said goodbye to chemistry. She turned to writing on chemistry together with her husband. They wrote review articles and books together, an elementary textbook as well as a monograph on host-guest chemistry, *Container Molecules and Their Guests.*[2]

This is how Donald Cram described how their joint book writing worked: "This was a real challenge! For co-authors to be of use to one another, each has to provide criticism to the other. Yet Jane and I love each other. Reconciling these two roles led to some real battles. I am blessed by having a very forthright, analytically minded wife. Everyone needs a critic and she is mine. She is probably more intelligent than I am, but she is not as creative and daring in her thinking. We complement one another." His conclusion sounds like a recurring observation about couples working together. "Our science is so demanding that one person seldom has all of the qualities needed for success. . . . We have a sort of Gilbert and Sullivan relationship; one supplies the lyrics and the other the music. It is hard to sing solo."[3]

Donald Cram always acknowledged the importance of their interactions, her ideas, and her criticism to his success. He mentioned this in his Nobel lecture and expressed it in our conversation: "Every new idea that I have had since 1970 has enjoyed the benefits of her criticism. . . . I would never have received the Nobel Prize without her. That is very clear."[4]

MILDRED AND GENE DRESSELHAUS

Physicists

Mildred and Gene Dresselhaus in 2002, in Cambridge, MA. (photo by M. Hargittai)

Mildred (known to everyone as Millie) sits down with her friends and plays chamber music most evenings. She does this no matter how busy her day has been with research, teaching, mentoring, and public service. The day we met, February 5, 2002, they played Brahms quintets. Millie Dresselhaus[1]:

> I got into science through music. I grew up in a family that was far away from science. As a child, I had a music scholarship, but somehow I got interested in science and I was mostly self-taught in science all the way. When I was about ten years old, I read Paul de Kruif's book, *Microbe Hunters*, which was a strong influence, yet for quite a while I did not think about science as a career. When I was an undergraduate . . . I had physics, but I had chemistry and math also, and I could have gone in any of those three directions. Math almost happened because I applied for graduate school in physics and math. I was accepted in math and I accepted a graduate fellowship in math to become a graduate student here, at MIT. I never did it because in the meantime I got a Fulbright scholarship, which was in physics. It seemed like a great occasion to go abroad. That's how I became a physicist and not a mathematician.

Millie's parents came to the United States from Poland in the 1920s. She was born as Mildred Spiewak in 1930 in Brooklyn. She started her university education at Hunter College in New York City, where she was greatly influenced by one of her teachers, Rosalyn Yalow (see elsewhere in this book). Yalow had a hard time in finding a research position, but her difficulties did not discourage Millie: "I didn't see it [the difficulties]. I am ten years younger than she is and I didn't see the negatives, and maybe Rosalyn didn't see the negatives so much herself either. Our expectations were not so high. She was happy to have some kind of job and she loved the research."[2]

After graduating from Hunter, Millie went with her Fulbright scholarship to the Cavendish Laboratory in Cambridge, UK. Her master's degree is from Radcliffe College and her PhD from the University of Chicago. Enrico Fermi, the great physicist and teacher, was still there, and she attended his course on quantum mechanics. She believes that Fermi's course taught her to think like a physicist. Unfortunately, not all the professors were like Fermi; her faculty adviser, "who was the only adviser in my area of physics, didn't believe women should go to graduate school. He didn't believe that I should be doing what I'm doing. He was very unhappy every time I got a fellowship or any kind of recognition. He said it was a waste of resources. This was the person who was supposed to take care of me."[3] It is a testament of her determination and her abilities that, in spite of the contrary circumstances, she did an excellent job there, in the field of superconductors, a hot topic in physics that time.

Millie met her future husband during her doctoral studies. Eugene (Gene) Dresselhaus came to the University of Chicago as a postdoc after having received his PhD at the University of California at Berkeley for his work on semiconductors and semimetals. They married in 1958, the year when she got her PhD. Gene was a theoretical physicist, and he got a job at Cornell University. Cornell did not employ her on account of the antinepotism rule but she stayed there, supported by a two-year National Institutes of Health (NIH) grant. There were other signs of discrimination as well. It happened that a professor, who taught electromagnetic theory left at the beginning of the semester and they did not have anyone to teach that course. This is how Millie remembers the event[4]:

> I volunteered to teach the course, with no pay, because I had a fellowship. There was a big uproar. The faculty met every day for a whole week to decide not whether I was qualified to teach this course, but whether the young men would pay attention to me as a young woman. I had a lot of experience in the area of electromagnetic theory. There were no women in the course. It was difficult for the senior faculty to comprehend and deal with having a young woman teach young men. Maybe they decided that it was OK because I was married and already had a baby. I never knew exactly what went on behind the closed doors. However, I did get a chance to teach the course and I did a very good job with that course. I found out more about that many years later when various students who were in the course came to me (they met me in other ways at a later time), and they remembered my teaching, because it was somehow different for them. They told me years later how much that course meant to them.

When the two years were over, she had to find another job. Gene decided that it was more important for her to be able to work than for him to have a faculty position;

so they decided to look for a possibility where they both could work. This turned out to be the Lincoln Laboratory, a defense lab, at MIT, where besides their not too heavy duties they could do whatever research they wanted to do. She chose a field in which there was no such fierce competition, magneto-optics. It is the study of the effect of electromagnetic waves on a material that is put into a magnetic field. For the object of investigation, she selected semimetals and, on her husband's suggestion, graphite, which proved to be an excellent idea. Their research was successful, and they had many interesting results. Even without strong competition, her life was rather complicated with four children; they were born between 1959 and 1964. It did not help that her supervisor kept complaining that she was late each morning. Again, in Millie's words[5]:

> There was always a problem in organizing one's family in the morning. My supervisor at Lincoln Lab complained about me so much that I got tired hearing of all the complaints, because I was doing the best that was humanly possible. So I was looking for a year off from all this unpleasantness in my life. It wasn't that I wasn't productive; nobody ever complained about quality, quantity, anything about my work. They didn't like that I came to work at 8:30 instead of 8 o'clock. My oldest child was less than five years. I had a baby essentially every year and it was very hard to make everything work out for an 8 am arrival. The people who were judging me were all bachelors.

Fortunately for her, after seven years at Lincoln Lab, she got a visiting professorship at MIT that eventually lead to a regular appointment for her, which she still holds. Gene stayed at Lincoln Lab for another ten years, but then he started to feel unhappy because he could not do the research that would truly interest him. This happened when Millie became the director of the Materials Center at MIT; her research program was so huge that she could not imagine how to manage it. Thus, they decided that it would be beneficial for both if Gene moved to Millie's laboratory at MIT. They have been working together ever since. They have published numerous research papers together and have written books widely used by researchers in their field. Currently, she is Institute Professor Emerita at the Department of Physics and Gene is on the faculty of the Francis Bitter Magnet Laboratory, both at MIT.

Millie's most important research projects have been related to carbon, and specifically to intercalation compounds in which different elements or molecules are inserted between layers of graphite. She and her group have been instrumental in the development of nanoscience. She joined this field in the early 1970s. Soon afterwards, they became interested in fullerenes and nanotubes. In 1985, it was big news that a research group in Texas found a new form of carbon; they published the discovery in *Nature*.[6] They called it buckminsterfullerene, in short "buckyball," a beautiful symmetrical molecule consisting of sixty carbon atoms and looking like a soccer ball. By that time, the Dresselhaus group had already been involved with related research for many years[7]:

> About the buckyball research, we did a paper in the early 1980s, before Smalley and all the other people, on some related topics, as did others, I am sure. We figured out that what comes off of a carbon surface when you bombard it with a laser is a big chunk of carbon atoms. It couldn't be just one or two carbon atoms, it had to be a big chunk, a hundred atoms, I used to say. Many people laughed at that. They thought it was impossible. We didn't, however, do the key experiment on that emission. The key experiment was the mass spectroscopy measurement. We just didn't think of doing that experiment.

Nothing shows better the importance of carbon research than the fact that there have been two recent Nobel Prizes for discoveries in this field: one in 1996 to Robert Curl, Harry Kroto, and Richard Smalley for the discovery of buckminsterfullerene, and the other in 2010 to Andre Geim and Konstantin Novoselov for their discoveries concerning graphene. Graphene is a two-dimensional carbon layer that can be looked at as a single graphite sheet or as an opened-up nanotube that the Dresselhaus group has studied for years.

In 2012, it was Millie's turn to be recognized for her role in carbon science. She received a most significant award, the Kavli Prize. The Norwegian Academy of Science and Letters awards the Kavli Prize every second year for outstanding achievements in three fields, astrophysics, nanoscience, and neuroscience; 2012 was the third time it was awarded. The Kavli Prize is not yet well known, but its weight is indicated by the one million US dollars given out in each category; Millie received it alone. After the announcement of the prize, the *US News and World Report* called her "The Queen of Carbon Science."[8]

Just a few of the other prestigious awards that Millie has received: in 2012, the Enrico Fermi Award, presented to her by President Barack Obama; in 2007, the L'Oréal-UNESCO Award for Women in Science; in 2005, the Heinz Award; and in 1990, the National Medal of Science from President George Bush. She is a member of the US National Academy of Sciences (1985), the National Academy of Engineering (1974), and the American Academy of Arts and Sciences (1974).

After many years of being directly involved in scientific research, at some point she thought that she should help young researchers and women scientists building their careers. She accepted high-level positions, serving as president of the American Physical Society, president of the American Association for the Advancement of Science, and treasurer of the NAS. For a while, she held the position of director of the Office of Science at the Department of Energy. Thus, she was a member of the national administration, and she was responsible for reversing the trend of decreasing support for the physical sciences in the federal budget. Serving in high positions, however, never stopped her from doing science.

Another area in Millie's activities is women's issues, to which she is wholly dedicated.[9] This started at the beginning of her affiliation with MIT, where she promoted scholarship in science and engineering for women. She was asked to mentor women students. At that time, the proportion of women among the MIT students was 4 percent. She was asked to help in the evaluation of undergraduate applications. In working on this project, she realized that it was much harder for women to enroll at MIT than for men. Dormitory space was limited, and women did not usually perform academically as well as the men, which she saw as due to social factors—discrimination and harassment. Eventually, in the late 1960s, she prepared a motion to adopt equal requirements in admission, which was accepted. Later, she and other women colleagues started a women's forum that has become active. She was proud of all the changes that she initiated. By now, the number of women at MIT has increased almost tenfold compared to what it was in the 1960s. In the late 1990s, another MIT professor, the biologist Nancy Hopkins, looked into the question of women scientists at MIT and prepared a report, which has become known throughout the United States.[10] At that time, Millie thought that the problems they faced decades ago had been more or less dealt with. But when they sat down to discuss this question and she saw the data, she realized that there were still many inequalities. She believes that since then the situation has improved a great deal.[11]

Mildred and Gene Dresselhaus in 1988. (courtesy of M. Dresselhaus)

The Dresselhaus couple is unusual in that in their duo it is Mildred who is more successful and more famous than her husband. This is how she sees how it happened for them[12]:

> It didn't start out that way. It started out exactly the opposite. When we married, he [Gene] was very well known and I was not known. He had a good position and I was just a graduate student and then a postdoc. When we came to Lincoln Lab, it was not that way either. Even though we were only one year apart, he was a senior person and I was much more of a junior person. I was the person who had the breaks because there was discrimination and he did not have so much discrimination. I worked the discrimination into my benefit; it wasn't something I planned that way. I am sort of a more of a natural teacher and my talent

> for teaching was very helpful; students liked me. I was a more outgoing person. I don't know how it happened, it just happened like that. Maybe being a woman was an advantage.

She finds it unfair that people give more credit to her than to him. Perhaps this is because she is a more outgoing public person and he is rather shy and withdrawn. They have been working together most of their lives. The ideas to solve a problem come from both of them; often they arrive at them together after long discussions. The general observation that the output from a scientific couple is much larger than it would be had they worked separately applies to the Dresselhauses as well. Millie added: "It's also a lot more pleasant that somebody close to you understands all your craziness. Devotion to science, as we do it, is a kind of craziness."[13] They talk about science in the lab and at home. When their children were small, the children used phrases without knowing what they meant—they had simply heard them from their parents when they discussed science.

Managing the household, the research, and her administration duties was possible only because she could rely on her husband. "Firstly, I had a husband who did half of the work. He helped me with everything. That made it possible. Also, I had a babysitter, the same woman for 29 years. That was very helpful. At that time, four children was very common among academics. For men, of course, it did not matter too much to have four children in the family, because mothers generally stayed at home with children."[14] She knows that she never would have been able to do what she has been doing without Gene.

GERTRUDE SCHARFF AND MAURICE GOLDHABER

Nuclear and particle physicists

Trudi and Maurice Goldhaber. (courtesy of the late M. Goldhaber)

"Throughout her life Gertrude Scharff Goldhaber had to struggle against tyranny and discrimination, as a child during World War I, as a Jew in Nazi Germany, as a woman in a scientific discipline when there were few such practitioners, and as the wife of another scientist at a time of strict nepotism rules. That she was so successful is a testament to her talent, drive, and will"—thus begins her biographical memoir published by the US National Academy of Sciences.[1]

Gertrude (Trudi) Scharff (1911–1998) was born in Mannheim, Germany. Already at the age of four she "fell in love with numbers," and as a teenager she decided to study mathematics and physics, because she "wanted to understand what the world is made of."[2] Her father wanted her to become a lawyer, but he supported her in her determination to become a scientist.

In Germany, at that time, the students could go each semester to a different university. Trudi started in Munich, and at one point she went to Berlin, where she met her future husband, Maurice Goldhaber (1911–2011). Eventually, she returned to Munich for her PhD work on the magnetic behavior of materials. She was in Munich when the Nazis came to power in 1933 and the persecution of Jews began. However, her professor, Walter Gerlach, let her complete her doctoral studies.

By the time she received her PhD, Maurice had already fled from Germany and was working in Ernest Rutherford's Cavendish Laboratory in Cambridge. In 1935, she followed Maurice to England, where she worked for a short while with the 1937

Nobel laureate G. P. Thomson at Imperial College in London. In 1938, Maurice left for a job at the University of Illinois at Urbana-Champaign, but the next year he returned to London to marry Trudi and they went back to the United States together.

Because of the antinepotism rule, she could not have a job, not even lab space to do her own research, at the University of Illinois, in spite of her qualifications. Her only option was to join her husband and work with him, unpaid, on his project in nuclear physics. During the war years, he produced important papers that were published in *Physical Review* only after the war because of their classified contents. The Manhattan Project made use of his discoveries, but Maurice Goldhaber could not get clearance for actually joining the project on account of his wife's parents living in Germany. They learned only after the war that already in 1941, the Nazis had murdered Trudi's parents—the Goldhabers found out "the exact day when they did it, because the Germans kept good records!"[3]

She also produced valuable results for nuclear physics. While still in Urbana-Champaign she was the first to show that neutrons are produced from spontaneous fission of uranium. This paper was submitted in 1942, but due to the related war efforts it appeared only four years later; a footnote of the published paper indicates that it was voluntarily withheld from publication until the end of the war.

After the war, the Goldhabers moved to the Brookhaven National Laboratory, where she became a researcher using grant money from the US Navy. Maurice remembered: "The nepotism rule at the Brookhaven National Laboratory was much milder; it was only that husband and wife could not work at the same department. First we came here on a temporary basis. Then, in 1950, we came for good. When they dropped even this weak nepotism rule, we became the first couple at the same department. Now there are many. First we both worked in nuclear physics, but eventually I moved to the physics of elementary particles."

Their collaboration in nuclear physics lasted for about twenty years. Their most important joint discovery concerned the nature of elementary particles. They proved that the so-called beta particles were identical with electrons and were produced in radioactive decays. They directed beta rays onto lead and measured the energy released. Had such a release occurred, it would have indicated that the beta rays were not identical with electrons. On the other hand, if they were identical with electrons then, according to the Pauli exclusion principle, the beta particles could not occupy positions as deeply bound electrons, and there would be no energy released. The experiment did not release any energy, and thus the identity of beta particles and electrons was proven.

During the late 1940s and early 1950s, there was much interest in the structure of the atomic nucleus. Various models had been developed to describe it, and their applicability was not universal. Thus, for example, it seemed that the Nobel Prize–winning shell model of Maria Goeppert Mayer and Hans Jensen (see Maria Goeppert Mayer, elsewhere in this book) describes only the ground state of the nucleus and is not applicable to its excited states, that is, when the nucleus is brought to a higher energy level than its ground state.

Gertrude Goldhaber had been interested in studying the properties of nuclei when they were slightly excited and determined their properties. "Her early studies of such systems were an important component of the background for the collective theory of nuclear motion, which earned a Nobel Prize for Aage Bohr and Ben Mottelson."[4] In Maurice Goldhaber's words:

> She built a model, which serves as a good guide for theory. This is a model of the first excited states of even-even nuclei [nuclei that contain even numbers of protons and neutrons]. The heights of the columns are proportional to energy. This illustrates the need both for the shell model and the collective model. At a closed shell the energy is very high and in between there are some very low energies. My wife discovered the first very low energies at a time when people thought that even-even nuclei could not have low energies. Such discoveries change concepts, but are usually quoted only in tables.

The figure shown here is a photograph of the model that Trudi built.

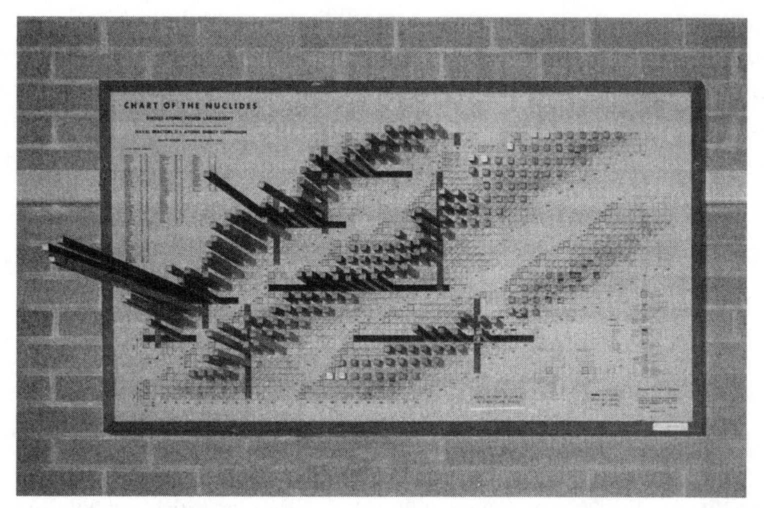

Three-dimensional model of the first excited states of even-even nuclei on the wall at the Brookhaven National Laboratory (photo by M. Hargittai). The horizontal line shows the number of the neutrons in the nuclei and the vertical line the number of the protons in the nuclei. The heights of the wooden bars perpendicular to the chart correspond to the energies of the nuclei. They are largest for the ones with magic numbers, showing the validity of the model.

To her great disappointment, Gertrude Goldhaber had to retire at the age of sixty-six from the Brookhaven National Laboratory—in the 1970s, mandatory retirement age was still in effect. She was full of energy and ideas, and in her retirement she participated in numerous activities, advising and consulting for a variety of governmental and private agencies.

Even though the start was hard for her, eventually Trudi received proper recognition for her work. Already in 1947, while she was still without official employment at the University of Illinois, she was elected a fellow of the American Physical Society. In 1972, she became the third female physicist to be elected to the US National Academy of Sciences after Maria Goeppert Mayer (1956) and Chien-Shiung Wu (1958).

The Goldhabers had two sons, one of whom is also a physicist with whom Trudi did some research, leading to two publications that might be the first mother-son cooperation in physics.

She felt strongly about the position and opportunities of women in science. She was one of the founders of the Brookhaven Women in Science program. There is a prize named after her, the Gertrude S. Goldhaber Prize in Physics, which is given out annually to women scientists who are students at Stony Brook or have done their work at Brookhaven. There are other prizes named after Trudi and Maurice.

Discussing the place of women in science, Maurice commented that there might be a difference in how men and women relate to working in science—or at least this might have been so when Trudi and he were young. At that time, for a woman to become a scientist proved to be so much more difficult than for a man that only the extremely dedicated women tried to do it. This intense dedication and commitment showed in their work as well. "Those women who are deeply involved in science pursue it even more intensely than men, as far as their time permits it. Meitner, Wu, Yalow, and my wife were all very committed." However, Maurice felt that sometimes their joint work was too intense. "This was after many years of being involved day and night. From 1939, we worked together for about 15 to 20 years. Then I deliberately moved into fundamental particles, in which I had become interested." He characterized her as "very meticulous in working on experiments and thinking about the results. I was more jumping around, although I also pursued some things to the end. To some extent we complemented each other."

ISABELLA AND JEROME KARLE

Crystallographers

Jerome and Isabella Karle in their office at the Naval Research Laboratory. (courtesy of I. Karle)

For me, it is especially rewarding to write about the Karles, as they have been our friends for decades. They are among the pioneers of the field in which my husband, Istvan, and I have been doing research, the determination of molecular structure, primarily by gas-phase electron diffraction. Although I had heard about them for years from Istvan, it was only in 1978 that I met them for the first time. It happened at an international meeting in Pécs, Hungary. It struck me at once how friendly they were. Although they were "big names" in our field, they were ready to discuss everything patiently with young people.

The Karles and the Hargittais during the 1978 meeting in Pécs, Hungary. (photographer unknown)

They certainly qualify as a "scientific couple." They met at the University of Michigan. Jerome Karle (1918–2013) was a few years older than Isabella, but they started their PhD studies at the same time with Lawrence Brockway in gas-phase electron diffraction. Jerome described how it happened[1]:

> I met her the first day that I went to school at the University of Michigan. I was one of those people who, when we had to set up an apparatus, for example, would go and do it, rather than wait until the lab period started. The places in the teaching laboratory were assigned according to alphabetical order. Her last name started with an L [Lugowski] and mine started with a K. When Isabella arrived, she saw me and the entire apparatus set up to carry out an experiment. I do not remember what she said, but she was surprised. That is how I met Isabella. The first year I did not see her very often, except in class. We went for an occasional walk in the evening. Once she was not feeling well and I brought material that she had missed in class to her lodging. The next year we started to have lunch together and by the end of that school year we were married.

Isabella Lugowski was born in 1921 in Detroit. Her father was a housepainter and her mother a seamstress; both were immigrants from Poland. Isabella only learned English when she started school. She decided during her high-school years to pursue chemistry. After Isabella and Jerome graduated from the University of Michigan, they both worked in the Manhattan Project at the University of Chicago. Both were involved with finding procedures to make the fission fuel plutonium without impurities. Jerome found a clever method to produce the metal from its oxide, and Isabella succeeded in making pure plutonium chloride. After the war, for two years they both worked at the University of Michigan. There was such a shortage of instructors that the university was happy to have them both in spite of the antinepotism rule. In 1946, they were again fortunate to each get a position at the Naval Research Laboratory, and they were even allowed to work together.

According to Isabella, they "worked together separately." By this she meant exactly what made these famous scientific couples successful—they complemented each other. As she put it, "for much of what Jerome does, he would like to have experimental examples and he uses the results of many of the structures that I have determined. This has not happened by design but it turns out that way."[2]

Originally they both were interested in experiments, and that was what their graduate work with Brockway was about. They made major improvements in the experiments and data analysis in the relatively new gas-phase electron diffraction technique that is used to determine the three-dimensional structure of gas-phase molecules. Gradually, Jerome moved toward theoretical work and Isabella stayed with experiments. Their constant efforts to extract more information from the experimental data helped in formulating better methods for the analysis.

They improved the experiments in a way that made it possible to determine finer details of molecular structure. The structure of a molecule can be imagined by a geometrical model in which each atom of the molecule is represented by a small sphere and these spheres have fixed positions in the three-dimensional space. The electron diffraction technique yielded information of such a structure. However, in reality these structures are not rigid constructions; rather the atoms are in permanent motion about their positions. Jerome and Isabella made the technique of electron diffraction more sensitive and enabled it to provide information about the motion of the structures.

Curiously, one of the molecules in which they faced the consequences of motion was a very simple and quite rigid molecule, the common carbon dioxide. Its molecule consists of three atoms, a carbon and two oxygen atoms, CO_2, and we can depict its structure as O=C=O. The Karles wanted to determine its geometry accurately. There was no reason to suppose that the three atoms would not align along a straight line. Yet when they finished their analysis, it appeared as if the distance between the two oxygen atoms might be a little shorter than the sum of the two C=O distances, which would indicate a slightly bent shape. I am saying "might," because the experiment did not allow this to be stated unambiguously. The Karles might have just left this puzzle alone and ascribed the discrepancy to the uncertainty of the experiment. However, the problem kept bothering them, and finally they came to the recognition that the shorter distance they observed between the two oxygen atoms was only an *apparent* shortening—it was the consequence of the motion of the oxygen atoms away from the supposed straight line of O=C=O. The changes were tiny, and it took the rest of the scientific community another decade to fully appreciate what the Karles had noticed. Looking back, it was an excellent demonstration of how their scientific minds did not rest until they understood what they observed as unexpected. Beyond the observation, they came up with an interpretation that moved the whole field ahead toward a more detailed understanding of how molecules are built and exist.

Later, in connection with their work on X-ray crystallography, the theoretical innovation preceded the experiments. In the early 1950s, Jerome and the mathematician Herbert Hauptman together developed the "direct method" for the analysis of X-ray diffraction data. The application of this method was to overcome the major difficulties of X-ray diffraction analysis caused by the so-called phase problem. For a long time crystallographers had accepted that much of the structural information is lost and no rigorously unique solution is possible for crystal structures from X-ray diffraction. Karle and Hauptman showed, however, that although it might not be possible to solve the problem rigorously, there did exist a solution by means of an involved mathematical apparatus. Once Karle and Hauptman provided this mathematical apparatus, the crystallographers acquired a method to solve crystal structures *directly* from the experimental data. X-ray crystallography received a big push from this advancement toward the determination of yet larger systems and with higher reliability. It was this work that eventually brought the Nobel Prize for Herbert Hauptman and Jerome Karle.

But success and recognition did not come easily. For long years, crystallographers were reluctant to accept, let alone apply, Karle and Hauptman's suggestions. This caused a great deal of frustration, and it was Isabella who gave real push for the acceptance of the method. As Hauptman put it: "the reaction from the crystallographic community was skepticism at best, hostility at worst. . . . Isabella Karle determined the structures of fairly complex molecules of 40–60 atoms, which up until that time would have been unthinkable. That convinced a lot of people that there is something here."[3] Indeed, as the years went by and nobody wanted to use the method, Isabella lost her patience and decided to build up an X-ray diffraction laboratory. She worked out the connection between the mathematical description of the new method and her X-ray diffraction data. Thus, she had a major role in making the direct method into a successful tool in X-ray crystallography.

In 1985, on the day of the announcement, when the Nobel Committee wanted to call Jerome about the news, they learned that he was flying back home from Europe. They called the captain of the plane who went out into the cabin and told Jerome about the exciting news. Jerome told me that his first thought was whether Isabella was or was not included in the prize, but the captain did not know. The remaining hours of the flight were most agonizing for Jerome. When he got home, he learned that she was not included and the sadness he felt over this never left him. Many of their colleagues share Jerome's feeling that she should have been included in the prize. Alan Mackay, a noted British crystallographer, said that "Isabella Karle should have been included because it was her work that made the whole thing believable."[4] James D. Watson of the double helix fame expressed similar sentiments.[5] As to Isabella, she feels that although it would have been wonderful to receive the prize together with Jerome, many other prestigious awards have consoled her.

After proving how useful the new scheme for structure determination was, Isabella's attention turned increasingly toward the structure determination of large, biologically important molecules. She uncovered details about the structures of peptides, steroids, and alkaloids, and her results have advanced chemical and biochemical research all over the world.

Isabella and Jerome have three daughters, and it is an obvious question how she could manage. She considers herself lucky in that crystallography is a scientific field in which you can do your science while bringing up children.[6]

> Crystallography wasn't something that you had to watch all the time. You could take it home with you, you could think about it while minding the babies. Most of the projects in crystallography, for example, start with an idea or a substance, and there is an end when you get the crystal structure. This is an isolated thing and in order to have a research project you want a number of related things that would go together, but it was possible to do it stepwise. In other projects there may be so much more interaction.

Of course, that was not enough. When their children were small, they had babysitters[7]:

> We were fortunate that after World War II, there were a lot of women of the grandmother age whose children didn't want to live on the farm anymore. The Virginia Mountains are about sixty miles from here. Many of the children came

to Washington during the war and didn't want to return after it ended. So the elder ladies came and we could have a housekeeper/babysitter during the week who would then be visiting with her children in the city during the weekends. That worked out very well until our children were well enough grown and we didn't need help anymore.

Having children required organization, but I never felt them to be an obstacle for my career. It never interfered with our professional lives. When they got to be a little older, and by a little older I mean at least seven years old, we took our children with us for our summer travels to various meetings in Europe.

One of Isabella's most prestigious awards was from the Royal Swedish Academy of Sciences, the Aminoff Prize (1988), established specifically for pioneers in crystallography. She has received numerous other awards and distinctions, far too many to list them all. She is a member of the US National Academy of Sciences (1978); she was the first woman to receive the Bower Award and Prize for Achievement in Science from the Franklin Institute in 1993, "for determining three-dimensional structure of molecules with X-ray diffraction"; and she received the National Medal of Science from President Bill Clinton in 1995.

We talked about the opportunities and career advancement of women in science, and this is what Isabella had to say in this regard[8]:

Let us look at some other countries and consider crystallographers in particular since I know them best. My women colleagues in France always envied me because of the freedom of research that I have had here. There were some very good women scientists in crystallography, but they could not achieve the status that men had at the universities. Women just weren't appointed to higher positions. There are quite a few women crystallographers in Italy, and they always come up to me and ask me about how I managed because they are usually the glorified assistant but rarely the leader of any group. In England, the atmosphere has been better. There are quite a few women in crystallography who are professors and heads of groups. In the United States, it's mixed. It goes all the way from the Italian way to the British situation. I don't quite know about the present situation because I don't have many young colleagues. Of my generation, the late Shoemakers were a good example. Both David and Clara were excellent crystallographers. He used to be a professor at MIT and she was always living off his grants. Not until they went out to Oregon where he became the department head, was she appointed to her own position as a full professor. This was during the last few years of both their lives. Ken Hedberg and his wife, Lise, were both electron diffraction people and I don't think she ever had a real job. She also worked on his grants. There were several other instances of the same kind. The universities wouldn't hire both husband and wife for any number of reasons. Although these women published extensively and did very good work, either they never succeeded or succeeded only in their later years to become independent. From that point of view, research positions in the US Government laboratories were generally a lot better.

A crucial question for the woman in scientific couples is whether she receives proper credit for her work. When I asked Isabella whether it ever happens that Jerome gets the credit for something that she did, she answered: "I suppose so." The answer to the question whether it ever happens the other way around was "Not often."

In 2009, after more than sixty years (a combined 127 years) of service, Jerome and Isabella retired from the Naval Research Laboratory (NRL). Ray Mabus, secretary of the navy, presented them the Distinguished Civilian Service Award, the highest award a civilian can receive from the navy. Nothing illustrates better how long ago it was that they joined NRL than realizing that "When Jerome Karle began work at NRL, Franklin D. Roosevelt was president, gas was 21 cents a gallon, minimum wage was 30 cents per hour, and a first class postage stamp was 3 cents."[9] During their six decades at NRL, the laboratory underwent a remarkable growth and the development of science and technology was unprecedented. It must have been a wonderful feeling for the duo that they not only witnessed this development but also actively participated in it. Although Jerome passed away in June 2013, they remain a wonderful present-day example of a scientific couple that greatly complemented each other so that their joint output was much larger than it could have been had they worked separately.

EVA AND GEORGE KLEIN

Tumor biologists

Eva and George Klein in 2001 in the Hargittais' home in Budapest. (photo by I. Hargittai)

Eva Klein is a most successful scientist, who is still active at the age of eighty-nine and appears as enthusiastic about her research as ever. In addition to her research career, she managed to bring up three children, a son and two daughters. But before we believe that Eva's life is a fairy tale about a scientist woman's enthralling life, we should get acquainted with her story. She had her share of difficulties to overcome.

Both Eva and her husband, George, are tumor biologists at the Department of Molecular and Cell Biology of the Karolinska Institute in Stockholm, Sweden. They are long past retirement age, which in Sweden is quite rigorously observed, but by special arrangement at the Karolinska, they have carried on in active work, keeping not only lab space but also their respective groups.

When Eva was eighty years old, the department had a celebration for her, and the departmental newsletter asked George to write something about her. His reaction was: "What a request! We have only been married for 58 years, have worked together for 57, have raised a family of three children and watched the growth of seven grandchildren—so what can I say about Eva? Either nothing at all or a novel of Proustian dimensions (at least 17 volumes, certainly no less)."[1]

They have had a fascinating and unconventional life. Both of them were born in 1925 (she as Eva Fischer) into well-to-do Jewish families in Budapest. They experienced the intensifying anti-Semitism followed by the enactment of increasingly harsh racial laws in the country. In 1944/45, during the last months before liberation, Eva and several members of her family were hiding at the Histology Institute of the University of Budapest, where classes had stopped by the fall of 1944. A medical student, Janos Szirmai, helped them and forged masterly documents for them.

Eva and George did not know each other yet. In 1944, he worked as an assistant for the Jewish Council in Budapest and had access to what have become known as the Auschwitz Protocols—eyewitness accounts of the mass murders in Auschwitz. Few people had such access, and even fewer people believed that it could be true, as its horrors defied imagination. George was one of those who found the report credible, and it saved him. When, in the early summer of 1944, the Hungarian authorities put him on a train destined for Auschwitz, he escaped and went into hiding.

After the war, both Eva and George went to medical school. They met by chance at Lake Balaton and fell in love instantly. However, George soon left for Sweden, where he was offered a job at the Karolinska Institute to do tissue culture studies, which he had become acquainted with while a student in Budapest. He then managed to get back to Hungary and marry Eva, and in 1947 the two got out of the country for good.

In Stockholm, they completed their medical training while working at the same time. It helped that both had the same ambitions and could rely on each other, for example, by one working while the other studied. Eventually, they both received their medical degrees, followed by higher degrees, and finally both were elected to membership of the Royal Swedish Academy of Sciences—the highest recognition of scientists by their peers. They both have earned worldwide reputations in their field, tumor biology and immunology. They also became well known in their adopted country, admired by their colleagues and by people at the highest levels.

Eva Klein handing a welcoming bouquet to Queen Louise of Sweden some time in the late 1950s on the occasion of the queen's visit to the Karolinska Institute. (courtesy of E. Klein)

The Kleins have had many scientific achievements, often based on careful and extended experimentation in tumor immunology. Yet their most important scientific contribution may have been an insight rather than an experimental finding. However, the insight was based on their experimental work. It was the recognition that chromosomal translocations in tumors—exchanges between two different chromosomes—reflect oncogen activation effects. This sounds rather technical, but it greatly advanced the science of tumor immunology. The experimental work itself was conducted in two separate groups at the Karolinska. One was dealing with mouse tumors, the other with human tumors. The two groups did not communicate with each other, but they both communicated with the Kleins. Suddenly it became clear that there was a significant common denominator between the behaviors of the two organisms.

There were other seminal findings. Thus, for example, in the 1960s and early 1970s, they showed that chemically and virally induced tumors could induce rejection responses in genetically identical mice. In the early 1970s, together with a group in Oxford, they showed that fusion of normal cells with malignant cells led to the suppression of malignancy. This was unexpected; prior to their work, the opposite was supposed because of the belief that the malignant phenotype was the dominant one. This discovery initiated the research field of tumor suppressors. Eva and her group have been carrying out experiments with mice, but their results are relevant to humans as well. The tumors mice develop are very similar to the ones humans do; moreover, the same genes are usually responsible for them in mice and in humans.

Even mentioning these research items demonstrates that the Kleins' lives have been exciting. But it was not at all easy, and especially for Eva. She had the main burden, and he had the bad conscience. According to Eva, from the time their children were born George never participated in any housework. When the first child, their son, was born, she invited to Stockholm the nanny who raised her in Budapest to live with them in Stockholm. Later, too, they always had live-in and other help, but "still it was a disaster." What made it even more difficult for her was that George not only did not do anything, but on top of that, he hated that she was involved with "nonscientific" activities. Naturally, this made it even harder for Eva. Looking back, she feels that her children probably suffered from her deep involvement with science when they were growing up. But she sees now that it was exactly because she had her children that she had to prove that she could do science as well. This was important for herself and also for George, because she knew that he did not want her be a "housewife."

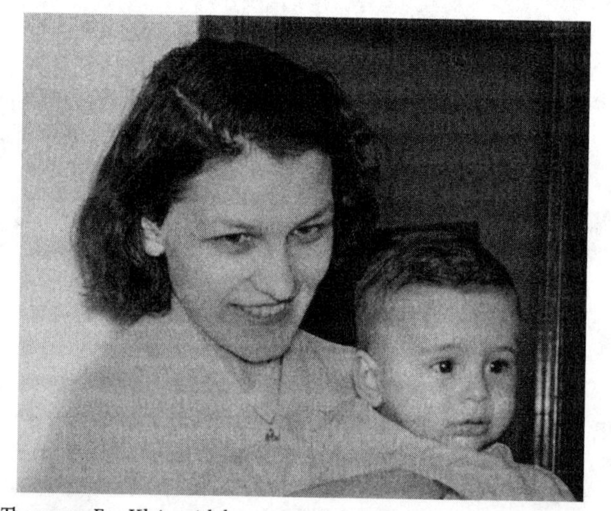

The young Eva Klein with her son, Peter, in 1951. (courtesy of E. Klein)

When George met Eva, she was a vivacious young lady, the center of any company she was in, free of worries even under the trying conditions of the time. She was interested in poetry, in the arts, even in acting, and full of intellectual thirst and dedication. She appeared to him the opposite of his mother. In George's words[2]:

> My mother was a wonderful housewife and was very concerned that I am well fed and well clothed. She was overprotective as Jewish mothers often are. My father died when I was one year old. When I was a young child I had this desperate wish to talk to her. However, every time I started a conversation with her, very soon I would see in her look the worry whether I was properly dressed, whether I was not cold, whether I should perhaps eat something more. This turned me off from housewives forever.... I could therefore never think of marrying somebody who was not an intellectual person. For me it is very hard to understand men who do not want an intelligent wife.

Eva says that of her three children she has always had the best relationship with her son, their firstborn child. Today she thinks that probably she should have explained to her children how important science was for her; they could have understood. But this never occurred to her; she was too busy working, and perhaps the children thought that she did not care about them. She sensed this and always felt inadequate; actually doubly so, both in the laboratory and at home. It felt like always rushing, as if she were trying to catch a train but never reaching it.

Eva and George always worked together; he was her boss. Eva described how different they were in almost every aspect[3]:

> Interestingly, we have very different personalities; he is impatient, I am very patient. We cannot write together. Even in dealing with our coworkers, we are different; he never has the patience to talk things over with them, while I like the "ping-pong" style; always discussing and exchanging ideas. It is a miracle that our joint work actually worked.
>
> We also have different scientific styles; for example, I write very slowly and he does it very fast. He does not like to get involved with details; he is irritated by them; I like to ponder over small details as well because you never know where the solution of a problem comes from. Actually we are so different, we do not think the same way, so our joint work should be ideal. I do remember that my greatest problem was always how to cope. I never felt that I do my best in science, but I felt that I have to try. I have made several small discoveries, but they were all premature and that is not always good; I have never been strong enough to pursue them. I can see it now, when 15–20 years after I noticed something that it is right and accepted now.

In this Eva is probably too modest, as we will see below from George's comments. But first, a few more thoughts from Eva:

> After a while I started to engage with a somewhat different topic from what he had been involved with and that was hard but interesting. However, after a while

> the students complained that the "boss" is not interested in this and that was a problem for them. So the interaction of husband and wife, on the one hand, and the boss and a coworker in the laboratory, on the other, is extremely complex. I can't say that it is good or bad; it is just difficult. During the years, of course, I had my own grants to do research but I never had any kinds of administrative position, I could concentrate on research alone.

Although for a long time Eva and George have had their own separate research groups and research lines, their research is connected through many threads. Her recent project has focused on two related topics. One of them is the so-called Burkitt's lymphoma. This is a disease discovered about fifty years ago in Africa; she has been involved with studying its virological and immunological aspects. Another of her research topics, related to the previous one, is the study of the so-called Epstein-Barr virus, usually the cause of Burkitt's lymphoma. This virus is present in most people without any apparent effect, but it can infect certain cells that are part of our immune system.

George agrees with Eva that their research styles are very different. He always likes to explore new territories, probe into new ideas, whereas she likes to stick to her old topics and look at the same old questions from different perspectives. This is why she is never sure what the next step in her research will be—this depends on the results obtained along the way. George stresses how much she likes to ask questions—all sorts of questions, subtle ones as well as obvious ones—and this approach brings in new ideas. He recalled an episode when decades ago people were trying to find a connection between the lymphocyte tumor interactions and the antitumor responses. To everyone it seemed obvious where they should look for an answer, but not to Eva. She stubbornly wanted to look at what George called the "background," in which no one else was interested. And, as it turned out, she was right and this led to their important discovery of the so-called "natural killer" cells. Today everyone refers to these cells as "NK cells," and this name was also her idea.

With all the difficulties in work and family life, Eva proved not only to George and the world, but also to herself, that she could succeed both in the lab and at home.

SYLVY AND ARTHUR KORNBERG

Biochemists

Sylvy and Arthur Kornberg around 1960 in the laboratory at Stanford University. (courtesy of the late A. Kornberg)

"Sylvy's taste for serious science developed much earlier than mine."[1] This was what Arthur Kornberg (1918–2007), Nobel laureate in Physiology or Medicine for 1959, wrote about his wife in his book *For the Love of Enzymes*. They met when they were students at the University of Rochester.

Sylvy Ruth Levy (1917–1986) was born in Rochester, New York. She became interested in biology during her university studies. Arthur Kornberg mentions that she was one of the few students who commuted from the women's campus to the River campus to attend advanced courses in biology.[2] At that time, Arthur's plan was to become a physician; it seemed more promising for finding a job than doing science, in which he was also interested.[3] She, in the meanwhile, having finished her studies, started to work at the biochemistry laboratory at the University of Rochester.

Eventually, Sylvy moved to Maryland to join the National Cancer Institute in Bethesda. In 1942, Arthur moved to the National Institutes of Health, also in Bethesda, and here they met again. In 1943, they got married. Between 1947 and 1950, they had three sons, and during this period she was a full-time mother and wife. By this time, Arthur was determined to pursue a career in biochemistry. They moved to Washington University, to Carl and Gerty Cori's famous laboratory in St. Louis, Missouri, where Sylvy decided to return to the laboratory. Roger Kornberg, their oldest son, wrote to me: "She worked in the lab with my father during most if not all the time we lived in St. Louis (1953–1959). There she participated in my father's research on DNA synthesis and also worked with him on the discovery of a polyphosphate polymerase, which was important because of great interest in the synthesis of biopolymers at the time."[4]

After Watson and Crick discovered the double-helix structure of DNA in 1953, interest in DNA was on the rise, and Arthur was also drawn to this topic. Sylvy participated in this research, and, in fact, the following years turned out to be her most successful ones scientifically. In 1955, she was using extracts of *Escherichia coli* in their studies of the enzyme DNA polymerase. When she isolated from *E. coli* the enzyme that synthesized polyphosphate polymerase, they discovered the synthesis of polyphosphate, "Poly P." She named it polyphosphate kinase.[5] Then the family moved to California, where she continued working in the laboratory "for a year or two after we moved to Stanford in 1959, and then retired."[6]

Sylvy had been a talented student and was a gifted scientist. According to Arthur: "When we were dating, I mentioned that I'd gotten a perfect grade in chemistry in the New York State Regents' Exam. She said, 'I did too,' and then asked me, 'What did you get in algebra and geometry?' I got 97 but she'd gotten 100 in each of them. But she never had the hubris of her great intellect."[7] Arthur admitted that she had a great disadvantage in science due to being a wife and a mother of three. "She liked to do science and she did it well but unlike Gerty Cori, was not intensely ambitious. Her devotion to me and our young children made it very difficult to maintain the pace in science. She was happy to help, guide and admire our wonderful children."[8]

In bringing up their children she proved to be immensely successful. Their oldest son, Roger, became a biochemist; he is professor of structural biology at Stanford University and received the 2006 Nobel Prize in Chemistry for understanding how cells copy information in our genes in order to produce proteins. Their middle son, Thomas, is professor of biochemistry and biophysics at the University of California in San Francisco; he was the first to characterize DNA polymerases II and III. Their youngest son, Kenneth, is a successful architect, the president of Kornberg Associates, a firm that specializes in designing research and clinical care facilities.

There is some indication that might contradict Arthur's statement about her lacking ambition. After the announcement of Arthur Kornberg's Nobel Prize in 1959, the *Miami News* carried a short article titled: "They Helped Husbands to Nobel Prize." The newspaper interviewed Sylvy: "She is shy, but has a strong mind of her own. 'Who says a wife and mother can't have a career outside of the home?' she demanded. 'Sure, I stayed home when my boys were little, but I kept my hand in at the lab. I carried on a few projects and did some editing on research papers.'"[9]

There is another quote from Sylvy from that time, "I was robbed," and this is used often to indicate that she was disappointed, feeling left out, when Arthur received the Nobel Prize and she did not. Her son, Roger, wrote about this: "She felt bad about the quip 'I was robbed.' It was intended in jest. It meant the opposite. She did not realize at the time how a reporter could take her words literally and embarrass her. Of course, she believed as did so many others that my father was the appropriate recipient of the Prize."[10]

Arthur described this story in his book the following way: "Sylvy was quoted in a newspaper the next day as having quipped: 'I was robbed.' In fact, she had contributed significantly to the science surrounding the discovery of DNA polymerase. I cannot imagine how this work could have gone forward without her unwavering support in the laboratory and at home."[11]

MILITZA N. LYUBIMOVA AND VLADIMIR A. ENGELHARDT

Molecular biologists

Vladimir Engelhardt and Militza Lyubimova, ca. 1950. (photos courtesy of Natalia Engelhardt, Moscow)

Militza Nikolaevna Lyubimova was the wife and longtime collaborator of the famous Russian scientist Vladimir Aleksandrovich Engelhardt (1894–1984), referred to sometimes as "the father of molecular biology in the Soviet Union." They worked together and made an important joint discovery, but only he became famous, and there is hardly any information about her. I tried to collect what is available and then was graciously assisted in completing this sketch by their daughter, Natalia Engelhardt, herself a scientist of professorial rank at the Russian Academy of Sciences in Moscow.

Militza Lyubimova was the daughter of Professor Nikolai Matveevich Lyubimov (1852–1906), head of the Department of Pathological Anatomy of the University of Kazan. He was the first elected rector of Kazan University during the first Russian Revolution in 1905. Militza was born in 1899 and spent her childhood in Kazan. When she was seven years old, her father died. Soon after, in 1910, her mother moved to Moscow with her two daughters. Militza attended school in Moscow, and enrolled at the Medical Faculty of Moscow University. These were hard times: World War I, two revolutions in 1917, years of civil war, and the establishment of the communist dictatorship. Militza was a gifted young woman and had a strong personality; hardship made her yet stronger. She learned very early what hard work was, and that stayed with her throughout her whole life.

In 1926, after graduating from the university, she became one of the first two graduate students of the young researcher Vladimir Engelhardt in the Institute of Biochemistry. This was his first job after having graduated from the Medical University and having spent his time in the army.[1] Vladimir and Militza married in

October 1927. Soon after, in 1929, Engelhardt was invited to be head of the Biochemistry Department of the Kazan State Medical Institute and moved to Kazan, where she joined him the following year. She became his assistant head of department. Beside this administrative position, she continued her research and eventually became a Privatdozent, corresponding to associate professor, without the independence. At this time, Engelhardt and Lyubimova lived very modestly; they occupied two small rooms in a communal apartment.[a] They had many professional friends and had a vibrant intellectual life with regularly held seminars. During vacations, they used to go on long excursions.[2] Engelhardt described their "hobbies" in his autobiographical notes[3]:

> I could mention only a single one, in my younger years, mountaineering. Mostly accompanied by my wife, who bravely shared the difficulties, we visited first the mountain ridges of the central Caucasus, later followed high passes and glaciers of the Pamirs, and then the mountain crests in Tien-Tchang, on the border of China. Fresh in my memory is the day, when . . . we stood on top of a steep ridge, over 4 thousand meters in altitude, which formed the frontier between the USSR and China. We stood with one foot on Soviet soil, the other on Chinese territory, with the hazy sky extending over the great desert of Takla Makan before us.

Their first daughter, Alina, was born in March 1933. Soon after, Engelhardt was invited to Leningrad University and the family moved there. Their second daughter, Natalia, was born in December 1934 in Leningrad (today St. Petersburg).

In the middle of 1935, they moved again. They returned to Moscow, where both Vladimir and Militza continued their professional activities at the A. N. Bach Institute of Biochemistry. In 1937, Militza defended her PhD dissertation and became senior research associate in the Laboratory of Animal Cell Biochemistry.

This was when they made their joint major discovery. In the late 1920s, it was established that the source of energy for our muscles to work is the splitting of a large molecule, adenosine triphosphate (ATP). This molecule is often called "nature's energy store." Scientists had already tried for a long time to understand how exactly this process happens, but without success. When they extracted the muscle tissues with water, they did not find this molecule there. This is why Engelhardt called this molecule "Enzymological Cinderella."[4]

Intrigued by the puzzle, Engelhardt and Lyubimova thought of doing something different; instead of more experiments with the extracts, they took the "leftover" residue, that is, the material that remained after the water soluble enzymes were extracted. In hindsight, this appears logical, but before them, nobody had thought of it. Right away, on their first try, they found an extremely high enzymatic activity in this leftover liquid. Next, they again took an unconventional step.

In trying to isolate the molecule that carried the enzymatic activity, they decided first of all to isolate the protein myosin, which is responsible for the muscle

[a] A communal apartment meant one or two rooms for each family and all families sharing a common kitchen and bathroom.

contractions—this much had already been known (see above, the chapter on Banga and Balo). For this purpose they used highly concentrated salt solutions. To their surprise, Engelhardt and Lyubimova found that the enzymatic activity they were looking for was actually contained in the myosin extract. This showed that, contrary to expectation, the enzymatic activity was actually in myosin itself. They published their results in *Nature* and the article generated great attention. This discovery opened up a new era in biochemistry. In subsequent years, Engelhardt and Lyubimova continued to work on this topic and made further crucial discoveries.

In June 1941, when Nazi Germany attacked the Soviet Union, Engelhardt and Lyubimova were in Moscow. Until October 16, they supervised the evacuation of the equipment and the staff of the Institute of Biochemistry and their families, at first to Kazan, and later on yet further away, to Frunze, in Kyrgyzstan. The Institute of Biochemistry and several other biological institutes remained there until the fall of 1943, when all the institutes returned to Moscow. During this period, besides doing her scientific research, Militza was the head of the trade union committee. She was also engaged in organizing the life of the evacuated scientists and their children.

In 1943, Engelhardt and Lyubimova received the Stalin Prize (later renamed State Prize) for their research in the field of muscles—specifically for their work on the enzymatic properties of myosin and on the mechanochemistry of muscles.

Eventually, other groups also proved that myosin is responsible for muscle contraction and that it is responsible for triggering this act by splitting ATP and thus releasing energy. One of these experiments was carried out in Szeged, Hungary, by Albert Szent-Györgyi and Ilona Banga (see elsewhere in the book). Right after the war, Szent-Györgyi visited Moscow and he met with the Engelhardts at the Institute of Biochemistry. Soon after, Militza translated Szent-Györgyi's book on muscle research[5] into Russian.[6] In 1946, Engelhardt and his wife were nominated for the Nobel Prize for their seminal discovery concerning ATP and myosin, but they were not selected for the award.[7]

Militza surrounded by men in Moscow. On her right is Albert Szent-Györgyi, on her left Vladimir Engelhardt. (courtesy of N. Engelhardt, Moscow)

Militza and Vladimir was a most successful couple—and they were also brave and noble. Here we mention one example. Between 1944 and 1954, Engelhardt was trying to have one of his collaborators, Aleksandr Baev, freed and return him to science. Baev was a talented scientist, one of his coworkers in Kazan in the early 1930s. Baev stayed there after the Engelhardts' departure for Leningrad. After they moved again, this time to Moscow, Engelhardt invited Baev to the Institute of Biochemistry in Moscow. In 1937, during Stalin's great terror, Baev was arrested on false charges and sentenced to ten years in the infamous forced labor camps. Engelhardt carefully preserved the manuscript of Baev's not yet defended PhD dissertation. When Baev was no longer incarcerated but still in exile, confined to Norilsk, in northern Siberia, Engelhardt found him and organized the defense of his dissertation in Leningrad. Afterwards, with great difficulty, Engelhardt helped Baev to get a research position in Syktyvkar in the Komi Republic, west of the Ural Mountains in the north. Sadly, in 1949, a new wave of repression came in the USSR and Baev was arrested again. He was exiled to Siberia together with his family—his wife and two little children. Until Baev's rehabilitation in 1954, Militza and Vladimir actively supported him and the family in spite of the danger such assistance meant to them.[8]

In the late 1940s, Militza began to paint watercolors. These and pencil drawings became her hobby for the following years. This gave her satisfaction and relaxation. She was also fond of gathering amusing natural things, which she used to remake in certain images. Her works decorated the Engelhardts' house. Vladimir was extremely fond of her hobbies and proud of her creativity. She did her best to take care of housekeeping and ensure the best conditions for him to work free of worries.

In the middle of the 1950s, Vladimir and Militza bought a site in a fashionable Moscow suburb, Nikolina Gora. Militza designed their house there based on a Finnish design and planned a charming garden. The family was very fond of this house; Vladimir enjoyed relaxing there, entertaining their friends, and contemplating the many complex problems that so often occurred in his life. Unfortunately, this house burned down soon after Militza's death.

After the war, Militza continued to study the biochemistry of muscle proteins and the mechanism of contraction in green plants. She discovered that creatine phosphokinase activity was not connected with actin. She studied the interaction of ATP with myosin using spectrophotometry. Later she began to study the physical activity of *Mimosa pudica*. Besides doing research in fundamental science, she was also interested, successfully, in applications. She developed a new technology of ATP production and introduced it into Soviet industry, and she obtained several certificates of innovation—these came closest to what patents stood for in the rest of the world. In 1957, she received her higher degree, doctor of biological sciences. She was active all her life and attended several international scientific conferences. In 1969, she was awarded an honorary degree, doctor honoris causa of medicine of Humboldt University of Berlin, which was then in East Germany.

From the time when, in 1943, the Engelhardts returned to Moscow, they continued their work in the A. N. Bach Institute of Biochemistry. Vladimir Engelhardt occupied important positions in Soviet science and was involved with rebuilding experimental biology after the damage caused by the reign of the notorious Trofim Lysenko. In 1959, Engelhardt left the Bach Institute and founded a new one that at first was called the Institute of Radiation and Physico-chemical Biology of the Soviet Academy of Sciences (when the term "molecular biology" was still anathema to the authorities). The name was changed to Institute of Molecular Biology in 1969, and now it is the V. A. Engelhardt Institute of Molecular Biology. Militza stayed at the Bach Institute of Biochemistry, where she headed the Laboratory of Animal Cell Biochemistry. She died in 1975.

The Engelhardts' elder daughter, Alina, became a chemist but after a while moved into the field of scientific information. The younger daughter, Natalia, became an experimental oncologist. Engelhardt writes in his autobiographical essay: "The effect of medical genes seems to continue into the next generation: my younger daughter works as a scientist in the laboratory of the Oncological Center, Moscow, and one of my granddaughters is a postgraduate at the chair of normal and pathological neuropsychology of Moscow State University."[9] Currently, Natalia Engelhardt is a researcher at the Laboratory of Immunochemistry, Institute of Carcinogenesis, Cancer Research Center in Moscow, and she is still active.

IDA AND WALTER NODDACK

Chemists

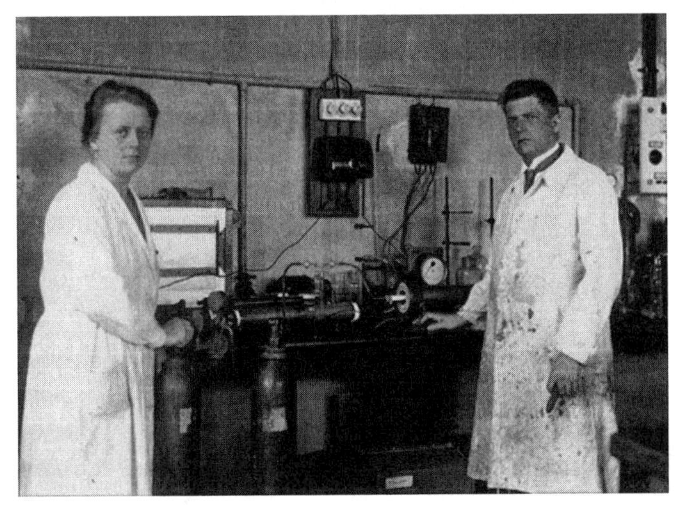

Ida and Walter Noddack in their laboratory. (courtesy of the Oesper Collections in the History of Chemistry, University of Cincinnati)

In 1934, Enrico Fermi and his associates at the University of Rome bombarded uranium (atomic number 92) with neutrons and supposed that the products contained two new elements heavier than uranium. In the same year, the German analytical chemist, Ida Noddack, suggested an alternative explanation of what happened in the Rome experiment. She suggested that the neutron bombardment actually broke the uranium nucleus, yielding atoms of two elements lighter than uranium.[1] Her suggestion was ignored. In December 1938, Enrico Fermi received the Nobel Prize in Physics, in part for the discovery of the new heavy elements. In the same month, in Berlin, Otto Hahn and Fritz Strassmann demonstrated the presence of barium, a known element lighter than uranium, among the products of neutron bombardment of uranium. Soon Lise Meitner and Otto Frisch interpreted the experiment as nuclear fission, thus vindicating Noddack's earlier supposition. The nuclear fission story was not the only one in which Ida Noddack's scientific acumen proved correct.[2]

Ida Noddack was born as Ida Tacke in 1896 in Lackhausen (now Wesel) in Germany. She studied chemistry and metallurgy at the Technical University in Berlin-Charlottenburg as one of its first women students. This was followed by industrial appointments first at AEG and then at Siemens & Halske in Berlin. In 1925, she moved to the Physikalische Technische Reichsanstalt (Imperial Institute for Physical Technology) in Berlin. Here she worked with Walter Noddack, who was looking for elements still missing in the periodic table of the elements. Ida Tacke and Walter Noddack married in 1926.

In 1925, Tacke and Noddack, together with Otto Berg, published the discovery of two new elements and named them masurium (with atomic number 43) and rhenium (atomic number 75). The names referred to two regions of Germany at that time, the Masurian Lakes where Noddack's family came from (today in Poland) and the Rhine region of Tacke's origin, the site of German victories in World War I—later the naming was interpreted as an expression of German nationalism. The discovery of rhenium was accepted by the scientific community, but the story of masurium was something else.

The discovery of masurium was based on the analysis of columbite, an ore containing mostly niobium, oxygen, iron, and manganese. The Noddacks hit a sample of the ore with electron beams and analyzed the emitted X-rays. They concluded that the ore contained a radioactive element with atomic number 43. Their critics pointed out that this element would be too unstable to occur in a rock. Furthermore, it was believed that even if for some reason this supposed element could have appeared in that ore, its amount would be so minuscule that the equipment the Noddacks used could not have detected it. Indeed, while in the following years the Noddacks could extract rhenium from the ores, they never succeeded in extracting masurium from columbite. The masurium story became a major blunder for the Noddacks. It is attributed to Ernest Lawrence that he called the Noddacks' claim "apparently delusional." A few years later, in 1937, Carlo Perrier and Emilio Segrè discovered element 43 in a cyclotron experiment and named it technetium. The name indicated that this was the first artificially produced new element.

Curiously, however, the masurium story was not over yet. In the 1960s, researchers found a minute amount of technetium in naturally occurring ores, thus drawing attention back to the Noddack discovery. In the late 1980s, a process was described by which the presence of element 43 could be explained in the ore that the Noddacks had studied and it was suggested that they be considered the true discoverers of technetium (the element they had called masurium). Others still maintained that the amount the Noddacks claimed to have measured was unrealistically large. Then a "virtual experiment" at the US National Institute of Science and Technology (NIST) simulated the Noddack experiment by using spectral-analyzer software and the best available databases, and found that the 1925 data were consistent with the presence of element 43 in the columbite rock. According to the International Union of Pure and Applied Chemistry (IUPAC), the evidence that what the Noddacks measured in 1925 was indeed element 43 is convincing.

From the time Ida Tacke married Walter Noddack in 1926, she did not have a formal position. In the late 1920s and the 1930s, at the time of the Great Depression, in Germany women were not encouraged to work. Even those who did were often forced out of their jobs if they were married in order to make their places available for men. Ida Noddack continued her research in her husband's laboratory, unpaid. She always stressed, however, that she was never merely his assistant. In 1935, the Noddacks moved to the University of Freiburg, where they worked at the Institute of Physical Chemistry.

In 1941, after Germany occupied the Alsace region in northeastern France, Walter Noddack was appointed professor and director of the Institute of Physical Chemistry at the newly founded Reichsuniversität Strassburg (Strassburg Imperial University; the French university, Université de Strasbourg, had gone into exile). They stayed there until the liberation of France. Here, she was given a paid position, in spite of the general Nazi policy against married women holding jobs. The Noddacks received generous assistance in equipment and other means for their research. Oddly, there were no scientific publications by them from their Strasbourg period, which ended, of course, with the downfall of Nazi Germany. J.-P. Adloff, who started his chemistry classes at the Université de Strasbourg in 1947, years after the Noddacks had left, noted that the display of the periodic table of elements in the lecture room still referred to element 43 as masurium.[3]

It is much discussed in the literature whether or not the Noddacks were Nazis or Nazi sympathizers. At Reichsuniversität Strassburg, which was one of the Nazis' flagship universities, 80 percent of full professors in chemistry were members of the Nazi party, while neither Noddack was.[4] The fact, however, that they were offered positions there indicates that they were trusted members of the Nazi regime, and that they accepted these positions witnessed their willingness to benefit from it.

In November 1944, when the Allied Forces were approaching Strasbourg, the Noddacks packed up their equipment and sent it to Germany. There, eventually, they received permission from the Allies' military government to keep them and continue their research. Possessing the equipment helped them to find a position. At the end of 1946, they moved to the Philosophisch-Theologische Hochschule (Philosophical and Theological College) in Bamberg, which was not ideal for conducting their research in chemistry. He founded a private Institute of Geochemistry with the instruments from Strasbourg. She continued her research in his institute, again unpaid. Eventually, this institute was nationalized in 1956, and Walter Noddack became its director, a position he held until his death in 1960. She continued working there till 1968, when she moved to a retirement community in Bad Neuenahr near Bonn. She died in 1978.

Ida Noddack received several distinctions. In 1925, she was the first woman ever to give a major address at a meeting of the Society of German Chemists. In 1931, both she and her husband received the Justus Liebig Medal from the same society for the discovery of rhenium. The Noddacks were nominated for the Nobel Prize several times, but they never received it.

The most famous among Ida Noddack's publications is the one mentioned earlier, published after Fermi's announcement of the production of transuranium elements. How could it happen that her suggestion was ignored, while research in nuclear physics and chemistry was flourishing? Ernest Hook collected some intriguing reasons, and we briefly discuss them below.[5]

Strictly speaking, Ida Noddack's suggestion was not ignored, it just was not accepted. She did not leave anything to chance and sent copies of her communication to Fermi and Emilio Segrè personally, and probably to others as well. Fermi did do some calculations, based on which he decided that fission was not possible.

Publications of the respected team of Otto Hahn and Lise Meitner at that time supported Fermi's conclusion. In addition to Fermi and Segrè, Hahn and Meitner and Irène and Frédéric Joliot-Curie also missed the early discovery of nuclear fission in spite of the fact that it had been suggested to them. This is a real puzzle.

It may be that since Ida Noddack was an analytical chemist and a geochemist and not a nuclear scientist, her word did not carry enough weight. This was compounded by their previous supposed blunder in connection with the masurium story. Their alleged Nazi sympathies did not strengthen their credibility in the eyes of their peers either. It would be too simplistic to ascribe the dismissive attitude toward Ida Noddack to her being a woman, and not only because of the fame of the Curies, but also because of Lise Meitner's involvement in the field. It is true, though, that Meitner was often underappreciated and viewed merely as Hahn's assistant, which she most definitely was not.

A quite valid reason for dismissing Noddack's suggestion might have been that it could be viewed as premature. When years later Laura Fermi suggested to her husband that they actually must have produced fission in their 1934 experiment without realizing it, Fermi reflected: "This is exactly what happened. We did not have enough imagination to think [of it]."[6] Indeed, fission was a new idea and Noddack merely suggested it without providing hard evidence, and the scientists who read her article considered it as a mere speculation. Whatever it was, in hindsight it was a pioneering idea.

How hard it would have been for Otto Hahn to accept Ida Noddack's suggestion is shown by the fact that in their seminal paper about fission in 1938, Hahn and Strassmann did not even bother to reference her paper from 1934. She felt impelled to publish a short note in the same journal, *Naturwissenschaften*, where Hahn and Strassmann had communicated their discovery, in which she complained about the oversight in spite of the fact that she had personally explained her ideas to Hahn.[7] The journal asked Hahn for a comment, but he refused to comply, and her communication is accompanied by a note from the editor according to which Hahn and Strassmann informed the journal that they had neither the time nor the inclination to respond. Looking back, it is puzzling that the possibility of disintegration of heavy atoms into smaller fragments had been discussed by many authors without drawing any conclusion concerning the experimental findings. They left it to the judgment of others to decide whether or not Ida Noddack's claims and her manner of presentation were justified.

Speculations still abound on why Ida Noddack's idea about the possibility of nuclear fission was ignored. The events that followed the discovery of nuclear fission in 1938–1939 have amply demonstrated its enormous importance. What would have happened had the idea of building an atomic bomb had come four years earlier in Germany? It might have changed world history in a most horrifying way. It may be that we should only be grateful that all the physicists and nuclear chemists were too slow in grasping the implications of Ida Noddack's suggestion.

FURTHER COMMENTS

Concluding the section about scientific couples, I mention two more examples for a particular reason. In 1995, the neurobiologist Lily Jan was elected to the US National Academy of Sciences (NAS), but she was reluctant to accept this great honor. Her motivation was that her husband and partner in work was not offered the same recognition. The bylaws of the NAS stipulate that the nominee has one year to decide whether to accept or decline the honor. Within the year, her husband, Yuh Nung Jan, was also offered membership in NAS, and the two happily accepted their election.

The Jans are professors of biophysics at the University of California at San Francisco. They met during their undergraduate studies at National Taiwan University and moved to the United States, to the California Institute of Technology, for their doctoral work. They first studied theoretical physics, and changed to biology at the urging of the physicist-turned-biologist Max Delbrück. They were already married when they received their PhDs in biophysics and physics.

Although the Jans started to work on separate projects, soon they decided to combine their efforts and have continued working jointly ever since. Their first big discovery was that a particular peptide hormone could act as a neurotransmitter. Another concerned the so-called potassium channels, followed by other significant discoveries in cloning genes and about the development of neurons. Their achievements have been recognized by joint awards, the most recent ones being the 2010 Edward M. Scolnick Prize in Neuroscience, the 2011 Wiley Prize in Biomedical Sciences, and the 2012 Gruber Prize for Neuroscience.

The Jans have raised two children and managed the responsibilities of family just as jointly as their research. They coordinated their schedules; for example, they took different shifts with their experiments so that one of them could be with the children. "Scientific collaboration works very well," Lily has stated: "Two people can really work well together, and what they end up with is not just the sum of when two people work separately." Yuh Nung agreed: "It has worked out very well. We complement each other in our work and in our personality." She is very patient and focused, he explains, while he tends to be a bit more impulsive, with "wild and crazy ideas."[1]

Thirteen years after the Jans' case, in 2008, Nancy Jenkins, a cancer geneticist, was elected to the NAS. However, she argued that it would be impossible to separate her and her husband's contributions to their results, and she could not possibly accept the honor without him.[2] Again, within the critical year, her husband, Neal Copeland, was also elected. The husband and wife team accepted the membership.

Jenkins and Copeland have spent their entire scientific careers together. They met at Harvard Medical School during their postdoctoral training. In 1980, they both went to work for the Jackson Laboratory in Maine, a nonprofit organization conducting mammalian genetic research, focusing on mice. Nancy and Neal got married and they faced a dilemma: "We could either compete with each other for the rest of our careers or collaborate—collaboration won." [3]

They spent three years in Maine, then moved, first to the University of Cincinnati College of Medicine and then to the National Cancer Institute in Frederick, Maryland.

Their main research topic has been modeling human diseases, especially cancer, with experiments on mice. They have identified the genes responsible for cancers and hoped to develop drugs that would block these genes. At the National Cancer Institute (NCI) they built up a colony of mice of about 20,000 animals.

In 2006, they made an unexpected move: they relocated to Singapore. This city-state wanted to develop significant scientific research venues with generous support for molecular biology. A number of top scientists moved there not only because of the favorable financial offers but also because there were "no strings attached" as to the choice of their projects. In Jenkins's words: "We had an opportunity to participate in the building of a new research community, Biopolis, in Singapore, and took it. We had been at the NCI for 22 years and were ready for a change. Also, Singapore was offering a very generous budget and ample mouse space to perform genetic screens for candidate cancer genes. It would have been very difficult to do this work in the US. We have really enjoyed living in Singapore. It is a beautiful city-state with a large expatriate community so one never feels too far from home."

In 2011, Jenkins and Copeland received an exceptional offer from the Methodist Hospital Research Institute in Houston, Texas. As part of a new initiative, which *Science* magazine headlined as "Texas's $3 Billion Fund Lures Scientific Heavyweights,"[4] the couple returned to the United States. He is the director and she the codirector of the Methodist Cancer Research Program; also, he is dean of cancer biology and she is dean of genetics at the Methodist Academy.

Asked how their cooperation works, Nancy wrote: "We share very similar interests and expectations yet have complementing strengths and skills. I am certain this model would not work for most others, we even share an office, but for us it has been terrific!"

The courage of Ilona Banga, Lily Jan, and Nancy Jenkins is appealing and I wondered if this has ever happened the other way around, that is, whether a male scientist working together with his wife and receiving an honor alone would have done the same.[a] Would the male partner in the team feel that he should not get the recognition that the two of them would deserve together?

[a] I tried to find this out at the US National Academy of Sciences but I did not get an answer to my query.

AT THE TOP

The examples of women scientists in scientific couples presented some extraordinary contributors to science. Below we introduce further women scientists of no lesser distinction. They excelled in their respective fields, frequently overcame hurdles, and persevered. Those who are presented here certainly did not disappear into oblivion, but it is reasonable to assume that many others did. Some of our "heroes" have been awarded the most prestigious recognition, the Nobel Prize; they will be blended among the others, as we will follow alphabetical order in our presentation. Nonetheless, due to the high visibility of this award, we will first briefly mention those women Nobel laureate scientists who are not represented by more detailed entries.

There are still very few Nobel laureates among women scientists, although lately their number has been growing more rapidly than during the first century of the Nobel Prize. All Nobel laureate women scientists are emblematic—not because they are necessarily greater scientists than others, but because by virtue of their recognition they provide additional inspiration to would-be scientists. There are also emblematic women scientists whose fame has only gained due to the fact that they were omitted from this special recognition. There are some among our entries, but here we single out two who are not: Lise Meitner and Rosalind Franklin. They have already been mentioned in the introduction, and here I would like to augment that.

By now Franklin's life has been described in great detail, and her achievements have been recognized and appreciated. However, the intensified interest in her was to a great extent due to her unfair portrayal in James D. Watson's otherwise excellent book *The Double Helix*.[1] It was followed by at least two more realistic and thus more favorable presentations.[2] Thus, we may say that Rosalind Franklin could finally take her proper place in the annals of the discoveries she was involved in, and especially that of the structure of DNA.

There were out-of-the-ordinary efforts to assign the proper place for Lise Meitner as well, in the annals of the most important discovery in which she was a participant, the discovery of nuclear fission. There are two monographs devoted to her,[3] but there is also something more. Meitner's missing Nobel Prize had been an issue ever since Otto Hahn received his Nobel award alone in 1945. The Royal Swedish Academy of Sciences was not quite immune to criticism in this connection, and one of its physicist members, Ingmar Bergström, did a critical study of Meitner's omission, and he presented his study to the Academy in 1999. He determined that she could have received the Nobel Prize but did not, and this was one of the unfortunate cases when a deserving scientist was not awarded. The Academy deemed it prudent in the same year to issue a medal honoring Lise Meitner, the first time that a female scientist had been so distinguished. A pruned version of Bergström's lecture was printed in a booklet of the Royal Swedish Academy of Sciences.[4]

From left to right: Plaque on the wall at King's College, London (photo by I. Hargittai); Rosalind Franklin (courtesy of Aaron Klug, Cambridge); and the medal of the Royal Swedish Academy of Sciences honoring Lise Meitner (photo by I. Hargittai)

There is such a small number of Nobel laureates among the many people involved in scientific research that no statistically relevant studies could be conducted, yet there is no question that women are underrepresented among the laureates. Table 1 displays the data for all Nobel Prizes, with more information about the science categories. In addition to physics, chemistry, and the biomedical sciences ("Physiology or Medicine"), there are Nobel Prizes in literature and peace, and there is the "Prize in Economic Sciences in Memory of Alfred Nobel" by the Swedish Bank, usually referred to as the Nobel Prize in Economics.

The distribution among the three science categories is uneven. Physics has the lowest share of women with only 1 percent, while physiology or medicine has the highest with 5.3 percent, with chemistry in between with 2.4 percent. This is in accord with the fact that the largest share of women scientists occurs in the biomedical fields and the smallest in physics. Of all Nobel Prizes, including literature, peace, and economics, the women's share is 5.4 percent—forty-six women in total. Approximately one-third of the Nobel Prizes awarded to women went to the three science categories.

Table 2 lists the names of the Nobel laureate women in science. Nine of the fifteen women Nobel laureates are discussed in this book in detail; the first three Nobel laureates featured in the previous section, and the rest will be introduced in this one. Because of the significance of the Nobel Prize, the remaining six science Nobel laureates are mentioned briefly below.

Table 1 Women's shares in the Nobel Prizes through 2014

Category	Number of Awardees	Number of Women Awardees	Percentage of Women Awardees
Physics	199	2	1.0
Chemistry	169	4	2.4
Physiology or Medicine	207	11	5.3
Literature	111	13	11.7
Peace	103[1]	16	15.5
Economics	75	1	1.3
Science prizes total	575	17[2]	3.0
Total	864[1]	47[2]	5.4

[1]In addition, twenty-five organizations have also been awarded the Nobel Peace Prize.

[2]The number of women laureates is one less because Marie Curie was awarded both a physics and a chemistry prize.

Table 2 Women Nobel Laureates in the Sciences[1]

Physics

1903, *Marie Curie*

1963, *Maria Goeppert Mayer*

Chemistry

1911, *Marie Curie*

1935, *Irène Joliot-Curie*

1964, Dorothy Hodgkin

2009, *Ada Yonath*

Physiology or Medicine

1947, *Gerty Cori*

1977, *Rosalyn Yalow*

1983, Barbara McClintock

1986, *Rita Levi-Montalcini*

1988, *Gertrude Elion*

1995, *Christiane Nüsslein-Volhard*

2004, Linda Buck

2008, Françoise Barré-Sinoussi

2009, Elizabeth H. Blackburn

2009, Carol W. Greider

2014, May-Britt Moser

[1]The names of laureates with entries in this volume are in italics.

From left to right: Three of the Nobel laureate women: Dorothy Hodgkin on a British stamp, Barbara McClintock at Cold Spring Harbor Laboratory (courtesy of Karl Maramorosch, Scarsdale, NY), and Elizabeth Blackburn (photo by M. Hargittai)

Dorothy Hodgkin

Dorothy Crowfoot Hodgkin (1910–1994) was a British crystallographer and is still the only woman among the many British Nobel laureate scientists. Her major contribution to science was the determination of the three-dimensional structure of important biochemical substances, primarily those of penicillin, vitamin B-12, and insulin. For these studies, she had to improve the technique, since it was not yet ready for such complex systems. Insulin is a molecule of close to 800 atoms, and it took her thirty-five years to completely understand its structure—she completed the work five years after her Nobel Prize. She was a most pleasant and popular person, a shy but determined woman, and a wonderful wife and mother of three children. According to one of her close colleagues, the Nobel laureate Max Perutz, "There was a magic about her person. She had no enemies, not even among those whose scientific theories she demolished or whose political views she opposed. . . . Dorothy will be remembered as a great chemist, a saintly, tolerant and gentle lover of people and a devoted protagonist of peace."[5]

Barbara McClintock

Barbara McClintock (1902–1992) received her award for the discovery of transposable elements, also called "jumping genes"—a discovery she had made about forty years before. She studied genetics at Cornell University, followed by brief jobs at other institutions, before settling at Cold Spring Harbor Laboratory in 1940. Her object for genetic studies was maize, also called Indian corn. She discovered that genes could move from one place of the DNA of maize to another, the change being accompanied by mutation or damage of the chromosome. Her discovery was "so far ahead of its time as to border on the heretical."[6] It took a long time for it to be taken seriously. Following her discoveries in maize, about twenty years later similar movable elements were discovered in bacteria.

Linda Buck

Linda B. Buck (1947–) is a biologist at the Fred Hutchinson Cancer Research Center of the University of Washington in Seattle. She investigates how humans and other mammals differentiate the thousands and thousands of odors and the mechanism of how the brain distinguishes them. She shared the 2004 Nobel Prize with Richard Axel (1946–) for their discoveries of odorant receptors and about the olfactory system. They determined which genes control our sense of smell. In our nose, there is a group of smell-receptor proteins "that work in different combinations so that the brain can identify a nearly infinite array of odors—much like the letters of the alphabet are combined to form different words."[7] She has continued these studies since the Nobel Prize, and has extended them to investigate how pheromones induce certain automatic behaviors in animals.

Françoise Barré-Sinoussi

Françoise Barré-Sinoussi (1947–) shared half of the 2008 Nobel Prize with Luc Montagnier (1932–) for the discovery of the human immunodeficiency virus. After her PhD degree and a postdoc year at the US National Institutes of Health, she joined the Pasteur Institute in Paris, where she started to work in Montagnier's group. They were looking for a link between the so-called retroviruses—a type of viruses whose process of transcription is different from other viruses—and cancer. When there was a new epidemic in Africa in the early 1980s, in a few years' time they isolated the virus causing the disease, which was a retrovirus later named the human immunodeficiency virus (HIV). They have been investigating it ever since. As the press release of the Nobel announcement noted: "HIV has generated a novel pandemic. Never before has science and medicine been so quick to discover, identify the origin and provide treatment for a new disease entity. Successful anti-retroviral therapy results in life expectancies for persons with HIV infection now reaching levels similar to those of uninfected people."[8]

Elizabeth H. Blackburn and Carol W. Greider

Three scientists shared the 2009 Nobel Prize in Physiology and Medicine, Elizabeth Blackburn (1948–), Carol Greider (1961–), and Jack Szostak (1952–), for the discovery of how telomeres and the enzyme telomerase protect the chromosomes. The discovery is the result of their joint efforts; Blackburn and Szostak worked separately until the early 1980s, at which time they started to cooperate. Greider, then a graduate student of Blackburn's, soon joined the team. According to the Nobel Committee's press release, they "have solved a major problem in biology: how the chromosomes can be copied in a complete way during cell divisions and how they are protected against degradation."[9] Telomeres are nucleotides that serve as caps at the ends of the chromosomes, protecting them from degradation. Greider and Blackburn also discovered that the enzyme they named "telomerase" catalyzes the production of telomeres. Another of their discoveries was that telomeres are important in relation to aging, in that the shortening of telomeres is one of the factors that lead to aging of the cells and through them of the whole organism. It has been shown that cancer cell division is connected with increased telomerase activity. How this knowledge could be used to fight cancer is now the topic of intense research.

May-Britt Moser

This book was already in production when the 2014 Nobel Prizes were awarded. The husband-and-wife team of May-Britt Moser (1963–) and Edvard Moser (1962–) shared half of the 2014 Nobel Prize for Physiology or Medicine for discoveries concerning the positioning system in the brain. (The other half of the prize went to John O'Keefe [1939–].) May-Britt and Edvard Moser are professors at the Norwegian University of Science and Technology (NTNU) in Trondheim. For their PhDs they had different projects but after that they decided to "join forces" and build a joint laboratory; they have been working together ever since. In a recent interview, she stressed that their complementarity made them perfect partners in work. They accomplished a major step in neuroscience, recognizing how the brain builds its own navigational system (we might call it the brain's GPS). The Mosers discovered a new type of cell that they called a grid cell. These cells form a periodic pattern in the brain that is like a map, helping us to find our way. The Mosers' discoveries have implications for finding cures for Alzheimer's disease and other neurodegenerative disorders.

JOCELYN BELL BURNELL

Astronomer

Jocelyn Bell Burnell, *left*: around 1975 (courtesy of J. Bell Burnell); *right*: in 2002 (photo by M. Hargittai)

"Jocelyn Burnell . . . is a wonderful person and I admire her tremendously. She is always asked if she was not sad for this [not receiving the Nobel Prize] and she always answers, why should I be sad, I made a career of not having the Nobel Prize. With her it is a clear injustice but she is still such a happy person, it does not do her any harm."—This is what the famous physicist and author Freeman Dyson told me about Jocelyn Bell Burnell.[1]

A brief summary of her story: Jocelyn made a discovery when she was a PhD student at Cambridge University in England. Her supervisor, Antony Hewish, a well-established astronomer, planned the experiment and had a major role in explaining the observation. Eventually, he received the Nobel Prize. His award was fully justified, but leaving out Jocelyn was unfair.

Jocelyn Bell was born in 1943 in Belfast, Northern Ireland, as the first of her parents' four children. She was interested in science already in her early school years. After the Soviets launched the Sputnik in 1957, she was caught up in the excitement and she decided to become a scientist. Why astronomy? Her father used to bring home books from the public library and she usually browsed through them. When one day he brought home some astronomy books, she had a different reaction. She took the books to her bedroom and "I was hooked by the scale and grandeur and the excitement that was there in astronomy even in the late fifties, early sixties. I realized that the physics that I was learning in school was a tool that could be applied to help us to understand the cosmos."[2]

She went to school in Belfast, followed by a boarding school in England for her senior years. She attended college in Scotland and received her BSc degree at the University of Glasgow in 1965. Then she went to Cambridge for graduate studies in astronomy. Professor Antony Hewish was just changing his research direction. He decided to study radio galaxies. He designed an extremely sensitive radio telescope and his graduate student, Jocelyn, was actively involved with constructing it. It was a huge structure, built from a large number of poles and kilometers of wires. She was responsible for doing the measurements and recording the thousands of signals coming from the telescope. The instrument became operational in July 1967 and she first noticed some odd signals in November of that year.

Jocelyn Bell in front of the telescope. (courtesy of J. Bell Burnell)

According to Hewish: "The survey was set up to observe several hundred radio galaxies every week. Now, by a very lucky accident the instrument I designed was ideal for detecting a completely unknown phenomenon called the pulsar. It was totally unexpected, unpredicted, and one of those shattering things that science brings up . . .[3] Jocelyn remembers those days vividly[4]:

> This discovery was an accident because such objects had never been dreamt of. They were unimaginable, literally. I was studying quasars, which are very very

> distant objects. The analogy I sometimes use is that you are making a video of a sunset from some vantage point; you have a splendid view of the setting sun. Then along comes a car and parks in the foreground and has its hazard warning lights going and thus spoils the picture that you are making. It was a bit like that with us. We were focusing on some of the most distant things in the Universe and this peculiar signal popped up in the foreground. These turned out to be the pulsars.

The peculiarity of these signals was that they were very sharp and repeated themselves at regular intervals. At first, the discoverers thought that there were problems with their equipment. Then they were afraid that they were picking up some noise, some manmade signals from other sources. She even thought about the possibility that these might be messages from an alien civilization; she called them LGM, "Little Green Men." She knew that if there ever came a message from an outer-space civilization, it would probably be picked up by radio astronomers. Soon they found a second signal that clearly came from another location, and then more and more. They realized that the idea for the source as a faraway civilization should be dropped: receiving messages from two, let alone more, civilizations at about the same time had zero probability. She told me that this realization was a relief. I was astonished to hear that, because establishing contact with aliens would have been a fantastic discovery. But she was practical; she said that exactly this would have been a problem—she could not possibly have finished her thesis during the remaining half a year when her funding would be ending.

Eventually, Hewish interpreted the surprising observations. The signals came from neutron stars. Their extremely high density is comparable to the density of the nucleus of the atom. We can imagine them as if the mass of the Sun were packed into a ball with a radius of ten kilometers. The name "neutron stars" does not mean that they consist entirely of neutrons, only that they are richer in neutrons than anything on Earth. The name "pulsar" is short for "Pulsating Radio Star." The observed pulsars have very strong magnetic fields and their magnetic poles do not coincide with the rotational axes, just as in the case of Earth. As the pulsar spins, the beam that comes out from its magnetic pole sweeps around the sky, like a lighthouse beam sweeping across the ocean. Every time the beam passes over Earth, we can pick up a pulse, and this is how we get a series of regular pulses.

Soon after, other radio astronomers all over the world started to look for pulsars and discovered many. Joseph Taylor at Harvard College Observatory was one of those astronomers, and he was among the firsts to confirm the Cambridge observations. In 1974, Hewish and Martin Ryle (who made a related discovery) were awarded the Nobel Prize in physics. Jocelyn was not among the awardees—and at that time, four decades ago, nobody, including Jocelyn, expected her to be. The recognition of astronomy by this highest award made her happy[5]:

> I was very pleased, mainly for political reasons. I am a strategist, a politician. That was the first time that a Nobel Prize in Physics was awarded to anything

> astronomical. There is, of course, no Nobel Prize specifically to astronomy and physics is the nearest. . . . This was the first time that it was clearly signaled that it [astronomy] was included and that was extremely important. It was the opening of a door to a whole new range of things. So I saw that instantly and I was very-very pleased for that reason. . . . I was content.
>
> At that time there was still around the picture that science was done by great men (and they were men). These great men had under them a group of assistants, who were much more lowly and much less intelligent, and were not expected to think, they just carried out the great man's instructions. Maybe that was the way science was done a hundred years ago or maybe even more recently. What has happened in the last 30 years is that we've come to understand that science is much more a team effort, with lots of people contributing ideas and suggestions. But at the time of the Nobel Prize, there was still around the idea that science was done by great men and the awarding of prizes, any prizes, was consistent with that picture. We did not recognize the team nature of science in those days.

About twenty years after the Nobel Prize for the pulsar discovery, something happened that triggered people's memory and they started to talk about the injustice that happened to Jocelyn. In 1993, the physics Nobel Prize was, again, awarded for astronomical observations, for the discovery of the double pulsars, and it went to Joseph Taylor and Russell Hulse. The discovery of the double pulsars was especially important as it opened up new possibilities to study gravitation.

There were conspicuous similarities and differences between the 1974 and 1993 Nobel Prizes. In the case of the 1993 Nobel Prize, the discoverers were Professor Taylor of the University of Massachusetts at Amherst and his graduate student, Russell Hulse. They constructed the experiment together; Hulse made the observation, and Taylor interpreted it. However, in this case *both* the professor and his former student were awarded the prize. For experts, the dissonance between the two cases was especially noteworthy. According to Taylor, "it certainly is true that . . . the Nobel Prize Committee in 1974 overlooked the significance of the contributions of Jocelyn Bell Burnell. The times have changed over the last several decades in that people are much more aware of the contributions of younger collaborators in large-group efforts."[6] Taylor invited Jocelyn to the Nobel ceremonies: ". . . simply felt that she would enjoy the experience . . . and that it might make up a little bit for something that she came very close to once before but did not quite achieve."[6] In 1974, Hewish did not invite Jocelyn to the Nobel ceremonies, although the laureates often invite their colleagues who participated in their prize-winning work. When I asked Jocelyn about this, she curtly said: "Oh, I was either pregnant, or with a small child, it was at that stage of life."[7]

Of course, the Nobel Prize is decided by the awarding institution and not by the laureates. Therefore, even if we believe that an injustice happened, the actual awardees were not responsible for it. There are examples in the history of the Nobel Prize when the awardees found it possible to acknowledge their collaborators, especially if they might have shared the distinction, but were not chosen. They mention their contributions in their Nobel lecture, invite them to the ceremonies, and sometimes even share the prize money. In 1923, when Frederick G. Banting and John J. R. Macleod received the Nobel Prize in Physiology or Medicine for the discovery of insulin, Banting's student, Charles Best, who was instrumental in the discovery, was not included, and Banting shared his prize money with him.[8] In contrast, when asked about the pulsar discovery, Hewish kept emphasizing "I" did this or that, and had to be pushed for clarification until he was a little more specific in referring to Jocelyn. To the question of whether he had a particular graduate student who made the first observation of the pulsars, Hewish answered: "Oh, yes, I did. She was my student doing observations, which I had designed. She was a good student and worked very hard, helping to build the radio telescope and then carefully analyzing hundreds of feet of chart recordings."[9] For getting her name out, further prodding was necessary.

Hewish was not responsible for the Nobel decision, but he seems to have taken upon himself to defend it. He told the reporter of the *Belfast Telegraph* the following: "You know, in the popular mind, she is the key person in the discovery of pulsars. I'm totally fed up with . . . this stupid business that Jocelyn did all the work and I got all the credit . . . I mean it's just totally wrong. If she's disgruntled about the Nobel, well that's too bad quite honestly. It's a bit like an analogy I make—who discovered America? Was it Columbus or was it the lookout? Her contribution was very useful, but it wasn't creative. And I don't think you do get the Nobel Prize for that."[10]

Not everybody would agree with Hewish's evaluation, and a lovely story illustrates this. It happened during the Nobel ceremonies in 1993. Anders Bárány, physics professor and the longtime secretary of the Physics Nobel Committee told Jocelyn how sorry he was about the events back in 1974. He was sure that such an oversight would not have happened in 1993. Then Bárány added a wonderful gesture. Each year those who participate at the voting of the prize receive a small replica of the Nobel Prize medal. Bárány gave one of his replica medals to Jocelyn as a token compensation.[11]

Getting back to Jocelyn's life story, after receiving her PhD from the University of Cambridge in 1969, she had quite a few positions in succession. She was at the University of Southampton for a few years, then moved to University College London, where she had a professorial appointment. She also worked at the Royal Observatory in Edinburgh. In the early 1990s, she became professor of physics at the Open University. I met her at Princeton, where she was a visiting professor. In 2001, she was appointed dean of science at the University of Bath, and she stayed there for four years. Currently (2013) she is a visiting professor at Oxford University.

Early in her career, when her husband was a personnel officer at the local government in England with various county councils, they had to move often, and this is why she had to change jobs frequently, which hindered her scientific career. When she became pregnant, she went to the chair of the department and asked what kind of arrangements they had for maternity leave. "He said: 'Maternity leave!?! I never heard of it!'" So she just resigned. For eighteen years, until her son went to college, she worked part-time. She feels that for her, being married and being a mother made a very large difference. At the same time, she feels that due to her circumstances, she was able to get a much broader range of skills than she might have had working continuously in research.

While moving around, she managed to keep up with developments in astronomy. She has been involved with various areas of the field, such as gamma ray, X-ray, infrared, and millimeter-wave astronomy. She acquired management training that she found very useful when she accepted the position of dean at the University of Bath. She stated, "If I stayed in academia all my life, I would have had more depth of experience but a much narrower range."[12] She and her husband had been married for about twenty years when they divorced.

Jocelyn is a Quaker. She finds Quakerism a religion that is well suited to a scientist, because it puts less emphasis on holy writings and tradition, and more emphasis on what one can learn about the nature of God and the nature of the world. It is strong on ecology, respect for the Earth and everything living on it[13]:

> One of the testimonies of Quakerism is to live simply not to consume excessively, which . . . is very close to the idea of taking care of the Earth. It is a very exploratory type of religion, . . . You are not required to believe or say a creed or something like that. I think that is why there are so many scientists in Quakerism . . . Interestingly, it formed in Britain in the 1640s, which was just about the same time when science became a recognizable activity and breaking away from theology. That may be a coincidence, I don't know but it is interesting that both got their identities about the same time.

Concerning recognition, apart from the missing Nobel Prize, she has received many prestigious awards and held important administrative positions. She has received prizes from the American Astronomical Society, the Royal Astronomical Society, and the American Philosophical Society. She has received several honorary doctorates; in 1999, she was appointed Commander of the Order of the British Empire (CBE); in 2003, she was elected Fellow of the Royal Society (London); and in 2007, she was elevated to Dame Commander of the Order of the British Empire. Being an FRS (Fellow of the Royal Society) is especially valuable, because it means recognition by her peers in science, but that it came decades after her contribution to the pulsar discovery implies considerable inertia. As to her positions, they have included the presidency of the Royal Astronomical Society and that of the Institute of Physics. It sounds believable when she says that she became more famous by *not* having received the Nobel Prize!

YVONNE BRILL

Aerospace engineer

Yvonne Brill in 2000 in Princeton. (photo by M. Hargittai).

"Each of these extraordinary scientists, engineers, and inventors is guided by a passion for innovation, fearlessness even as they explore the very frontiers of human knowledge, and a desire to make the world a better place. Their ingenuity inspires us all to reach higher and try harder, no matter how difficult the challenges we face." These were President Barack Obama's words when he announced the 2011 recipients of the National Medal of Science and the National Medal of Technology and Innovation, the highest honors bestowed by the US government on scientists, engineers, and inventors.[1] One of the five awardees of the Technology and Innovation Medal and the only woman among them was the eighty-seven-year-old aerospace engineer Yvonne Brill. She was the seventh woman among all the recipients of this award during the twenty-six years of its existence.

Yvonne Madelaine Brill, née Claeys (1924–2012), was born in Winnipeg, Canada. Her parents—neither of them had a college education—were immigrants from Belgium. Yvonne studied science and mathematics, and although engineering always interested her, women were not accepted to study engineering at the time she went to the University of Manitoba in her hometown. She told me that the reason probably was that engineers had to go on a surveying course, camp out in the wilderness for about three weeks, and the university just didn't want to set up special accommodations for women.

Yvonne Brill receiving the National Medal of Technology and Innovation from President Obama, 2010. (courtesy of the National Science and Technology Medal Foundation)

She graduated in 1945 with a degree that was equivalent to a math major in the United States. After that she moved to southern California and worked for Douglas Aircraft Company in Santa Monica. She attended night school working toward a master's degree in chemistry from the University of Southern California. At Douglas Aircraft, she worked in the aerodynamics department. The company won a contract to put up an unmanned earth-orbiting satellite. She worked on trajectories for this satellite, which involved mostly mathematics. The program manager—a real mentor for young professional women—knew that she was working for a degree in chemistry and suggested that she transfer to the chemistry department. That is how she got involved with rocket propellants and then rocket engines and so-called "ramjets," air-breathing jet engines. Nonetheless, she enjoyed engineering, and after having earned her master's degree in chemistry, she decided to pursue a career in engineering rather than going for a PhD in chemistry.

At Douglas Aircraft, the work on the satellite was classified; it was in the framework of Project RAND, which later became the famous Rand Corporation, one of the first think tanks for the US Air Force. During the Cold War, they channeled their efforts toward missiles, and the idea of the unmanned earth-orbiting satellite got pushed into the background. Brill's major task was to calculate thermodynamic properties for very high temperatures, needed to determine the performance of rocket fuels and oxidizers. These data were included in the thermodynamic tables used as the first industry standards.

She worked on this project for a while, but since she preferred experimental work to theoretical, she joined a small company, the Marquardt Corporation, where she ran various tests on their ramjets. There she met her future husband, who was a postdoctoral fellow at UCLA in chemistry. When they both were looking for jobs, there developed some conflicts. She remembered with a smile: "My best opportunities were in the West Coast and his were on the East Coast. So we followed his, of course."[2] They moved to Connecticut, where her husband worked at Olon Industries and she found a job at United Aircraft. But soon her husband decided to change his job and they moved to Princeton, where they have lived ever since. From the mid-1950s, for eight or ten years, she did not have a permanent job; she had their three children, and worked on and off as a consultant in propellants, finding new fuel combinations.

Around 1966–1967, she heard of a job involving spacecraft propulsion at RCA Astro Electronics. Probably her most important achievement is connected with this company. She had never before done anything with spacecraft, so this was a great challenge for her. Nor did the company have experience with propulsion; she was the first and only propulsion engineer on their staff. Her first task was to determine the best means of propulsion for a series of communication satellites. They had tried hydrogen peroxide, but it was very difficult to handle. She heard that at the Jet Propulsion Laboratory at Caltech they had been experimenting with hydrazine but had been running into difficulties. Eventually, she figured out a way to make a reliable and economical thruster, called the hydrazine resistojet or the electrothermal hydrazine thruster:

> I thought there had to be an easier way, so with all of the work I had done early in my career on propellant combinations and the calculations of their performance, I realized that if you look at the equation that is required to calculate the performance you can see that the most important parameter is the square root of the chamber temperature of the rocket over the molecular weight of the products and if you plot performance you get a straight line. I did this as an exercise just for fun once. So in thinking then about hydrazine, the decomposition products are just ammonia, hydrogen and nitrogen. When you get the exothermic reaction you get a considerable amount of heat given off from hydrazine. It seemed to me that if you could heat those exhaust products with just a plain old electrical heater to a higher chamber temperature, the molecular weight of the exhaust is not going to change, but the root of the temperature over the exhaust will give you a higher performance. Actually, when I went through the calculations of the performance, it gives you 30% more, which was very significant. I wrote a disclosure and eventually the company, RCA, applied for a patent and put it on their spacecrafts, there are quite a number of them that uses this thruster. I discovered last summer, checking with a company who builds hydrazine thrusters, that there are 120 spacecraft in orbit using their electrothermal hydrazine thrusters.

The hydrazine thruster is much more efficient than the previously used thrusters; it makes it possible to control the satellites' orbits and to keep the satellites in their proper trajectories longer. Since its first successful use in 1983, it has become a standard in the satellite industry. Brill built other successful propulsion systems as program manager for the RCA/Navy Nova spacecraft project.[a]

Later she worked on the Mars Observer Spacecraft, launched in 1992. She left the RCA Corporation in 1981 and moved to Washington, DC, to NASA headquarters, where for two years she was manager of the Solid Rocket Motor Program in the shuttle project. From 1986, for five years she worked in London at the International Maritime Satellite Organization as propulsion manager for their satellite system. She retired in 1991 but has continued to consult on propulsion systems of orbiting communication satellites, and has served on various panels and boards related to spacecraft technology.

When she started to work in 1945, men were still being drafted into war-related work, so there was a shortage of technical talent, and jobs were open for women. On the other hand, she had had some bad experiences with discrimination in hiring women in chemistry-related workplaces. This experience contributed to her decision not to pursue a PhD in chemistry. Engineering was different: "I decided there were so few women engineers that they were not about to make a rule to discriminate against one person and that proved to be perfectly correct. As long as you showed that you were capable and willing to do the work and not try to get by on some flimsy excuse, you were respected."

The number of women engineers has increased considerably since the time she started to work, but engineering is still a "man's profession." This is why Brill thought there was a great need for role models, and she did her share in inducing young women to get interested in such a career. Just the day before I visited her in 2000, she spent the whole day at a program called "Take your daughters to work."

She was active in the Society of Women Engineers, which was formed in 1950, though she learned about it much later. The basic mission of the society is to make young women aware of the opportunities in engineering, and it has organized a lot of outreach programs. In the late 1970s, she was the executive director in charge of student affairs. She estimated that the programs at the time may have reached about 15,000 high-school girls annually.

Even if the number of women engineers is increasing, there are still only few women in managerial positions. The situation is similar at the National Academy of Engineering (NAE). When I talked with Brill, in 2000, there were only two women members in the Aerospace Engineering Division (AED), she and Sheila Widnall, former secretary of the Air Force and today institute professor at MIT. As of 2012, there were eight women members out of the 211 members of AED at the NAE.

Brill said of her success in aerospace engineering: "I combat the problem at a different angle than, I think, a man would have. I think this is generally true. Or take another example, manufacturing. My daughter who has a degree in mechanical engineering, is in manufacturing and, in general women just know, intuitively, an easier way to put things together and make them work. They may not have been encouraged

[a] The current educational NOVA program has nothing to do with it.

in mechanical things when they were young, but it certainly shouldn't prohibit them from being able to do that work well."

Asked what the greatest challenge in her life was, she first said that she never gave it much thought. But she felt that just managing everything, being able to get to work and provide the inputs that she was required to do during the long hours of work and then getting everything done at home in a twenty-four-hour day, was a challenge in itself.

Her distinctions started with a curious one. In 1980, she received a Diamond Superwoman Award. There was an advertisement in the fashion magazine, *Harper's Bazaar*, looking for nominations for "superwomen." There were suggestions for all sorts of professions—lawyers, doctors, CEOs—and one of Yvonne's friends sent in her name so that an engineer would also be among the nominees. She remembered: "The criteria for being the Diamond Superwoman were to be over 40 and to have stopped your career to have children and then gone back to work and gone right to the top. Of course, I hadn't got right to the top, but I fit the other criteria. It was a really very exciting, fun thing. The award was a 1 karat diamond put out by the DeBeers Corporation and there were 5 Diamond Superwomen selected by a panel of judges and they had a big day for us in New York, with newspaper coverage."

In 1986, Brill received the Society of Women Engineers' Achievement Award. Then she was elected fellow of the American Institution for Aeronautics and Astronautics, which is a peer-group recognition, for her contributions to rocketry and space. Many others followed. She was a member of the US National Academy of Engineering and an inductee into the Women in Technology International Hall of Fame (1999) and the National Inventors Hall of Fame (2010). Among the inductees into the National Inventors Hall of Fame in 2010, the inventors of Post-It Notes generated especially great interest, and the *Washington Post* commented: "there were other very big giants of very obscure industries. . . . You had Yvonne Brill, the only woman to be inducted, creator of the electrothermal hydrazine thruster . . . used to keep satellites in place in space. (Note: It took one woman to invent a rocket thruster, and two men to invent Post-its)."[3]

When Yvonne Brill passed away in March 2012, all major US newspapers carried her obituary. The *New York Times* generated a flood of protest because its obituary started by mentioning her cooking skills and only in the next paragraph did it say that she was "also a brilliant rocket scientist." The paper promptly published a correction.[4]

MILDRED COHN

Biochemist

Mildred Cohn in 1927 (courtesy of the late M. Cohn) and in 2002. (photo by M. Hargittai)

"Here is an example of a woman who had tremendous ability but who was in secondary positions in academia practically until the time when she was elected to the National Academy of Sciences. Mildred continued to make important contributions to enzymology and oxygen-18 measurements that influenced my research over the years."[1] So said Paul Boyer in an interview about Mildred Cohn in 1999. Paul D. Boyer was corecipient of half of the Nobel Prize in Chemistry for 1997 together with John Walker for describing the enzymatic mechanism underlying the synthesis of adenosine triphosphate (ATP). ATP is the energy currency of our organism, so understanding its production was considered of utmost importance. The use of tracers, such as oxygen-18,[a] made it possible to understand the mechanism of the reactions at the molecular level. Boyer was not versed in working with oxygen-18, and it was Mildred Cohn who first introduced him to the oxygen-18 technique. This did not happen through personal interaction; it happened by Boyer reading Cohn's papers in the literature. They met in person much later.

Mildred Cohn (1913–2009) was born in New York City into a Jewish-Russian immigrant family. She had a very good chemistry teacher in high school, and this attracted her to chemistry. When she studied at Hunter College, she found physics even more enticing, but the school did not yet offer it as a major, so Cohn

[a] Oxygen-18 is a rare isotope of oxygen whose common form is oxygen-16.

graduated in chemistry with physics as her minor. She wanted to go to graduate school and applied to twenty, but there was no offer for her, in accordance with the anti-Semitic practices of the time. She used her savings to attend Columbia University, where she could not receive a teaching assistantship because they were available only for men. After the first year, when she ran out of money, she took a job and continued her studies. In 1934, she signed up to be Harold Urey's doctoral student just months before the announcement of his Nobel Prize in Chemistry for the discovery of deuterium. She began working with the isotope oxygen-18 in Urey's lab.

Mildred was Urey's student from 1934 to 1937, and her experience with the variation of the number of women students at the time and subsequently is interesting[2]:

> People are under the impression that there were no women around in those days. The statistics show that there was a larger fraction of female graduate students then than there were in the 1950s and 1960s. There was a real drop in the 1950s and 1960s in the percentage of women PhDs awarded in this country in science. My interpretation is that this drop occurred because the men came back from World War II and the women were displaced. During the war years, women took up a lot of non-traditional jobs. But afterwards to get the women out, they were told that women should get married and have children, and if they did not bring up their children themselves until they were at least five years old, the children would become monsters and so on. That had an effect. It took awhile before women returned to attend graduate school again, not until the women's liberation movement started.

She had superb mentors even after she had graduated from Urey's tutelage. First she worked in Vincent du Vigneaud's group at Cornell University, then in Carl Cori and Gerty Cori's laboratory at Washington University in St. Louis. Urey, du Vigneaud, and the two Coris were all future Nobel laureates. In this Cohn was most fortunate, but it took her two decades before she finally was given her own professorial job at the University of Pennsylvania. She was eventually appointed to the prestigious position of Benjamin Rush Professor of Biochemistry and Biophysics. In 1971, she was elected to membership in the National Academy of Sciences. One of her many distinctions was the National Medal of Science in 1982 from President Reagan.

Mildred's research activities focused on the utilization of isotopes, and she eventually became a great authority in this area of research and in the physical techniques that she used, such as mass spectrometry. This was the continuation of her doctoral studies under Urey. Her research resulted in seminal results both in conceptual and methodological aspects. She contributed to the knowledge of how enzymes work by applying nuclear magnetic resonance (NMR) spectroscopy based on the phosphorus-31 isotope. She studied the enzyme reactions of ATP. She showed that the extent of muscle disease can be followed by measuring the variations in ATP concentration. She also determined the magnesium concentration in the brain.

Mildred's husband, the theoretical physicist Henry Primakoff, was an immigrant from Russia. The two met in 1934 at Columbia University. Her venues of research had been determined by her husband's appointments, but she could have not been employed by the same university because of the antinepotism rules. However, her husband never took a job without making sure that Cohn would find there a research position. He was very supportive; he found it of utmost importance that she should have a career. At the same time, as far as helping at home was concerned—Cohn remembered—he "was very European in this regard. Whenever I asked him to do something, he said, 'Hire somebody.' He never participated in housework, but often played with the children and invented stories for them. . . . I was the practical one in the family. If a child broke a toy, I was the one who repaired it. He never did anything practical, not only traditional women things, he never fixed the car, he was not interested in the garden, and so on. He was very cerebral."[3]

Mildred did not have a faculty position for twenty-one years. Staying in research positions for a long time was disadvantageous financially, but it had the advantage that she could devote herself full-time to her projects. Also, it was easier for her to stay home when their children were sick. However, she was anxious to perform well in the lab, and for thirty years they had an excellent woman helping her to take care of their three children. For the children to have a working mother was quite unusual at the time. Their oldest child, Nina, was the only one in her school whose mother had a job, and she complained about this. In college, Nina majored in psychology,

Family photograph: Mildred Cohn, Henry Primakoff, and their three children, Laura, Paul, and Nina. (courtesy of the late M. Cohn)

and wrote a paper about the effects on children of having a working mother versus a nonworking mother. Her conclusion was that there was no appreciable difference. All three Primakoff children obtained PhD degrees; the two daughters are psychotherapists, and the son is a biochemistry professor.

I was very much taken by Mildred's story about herself as a working mother. When we had our children, I stayed home for about half a year for each of our two children. We came out of this experience very well and our children never complained about it. It is true, however, that we lived in an environment in which most women, especially professional women, tried to return to work even if they had small children lest they fall behind in their work. However, our family pediatrician criticized me severely, and his comments hurt when he predicted that our children would pay back our lack of care in our old age. This is why I asked Mildred to tell me more about her experience:

> The chairman of the chemistry department at Hunter College had told us that it wasn't ladylike for women to be chemists. Why was he teaching chemistry in a women's college? He wanted us to be teachers of chemistry. I got a lot of criticism from relatives. When I had saved up money for my education, one aunt said that it would be better for me to spend the money for straightening my teeth. A great-aunt said that I would educate myself out of the marriage market. After I had my first child, my mother-in-law carried on a campaign to get me to quit working, but she didn't succeed either. One of the reasons my oldest daughter reacted as she did was that when she was in second grade, she was 7 years old, she joined the Brownies and when the woman who ran it found out that I worked she told my child that I was a bad mother. There was a lot of social pressure against the mothers working.[4]

This prompted me to ask the trivial question: "What would be your advice today for a young woman who would like to do science and have a family too?" Mildred responded: "The first thing I would suggest is, marry the right man. That's the most important thing. You have to have a husband who is fully supportive. That he does more than paying lip service to equality. My husband was really a feminist. He liked women and respected them. My second advice is that whatever decision they make they shouldn't feel guilty."[5]

Toward the end of our conversation, Mildred augmented her advice:

> A piece of advice for women scientists. They should stress the scientist more than the woman. I remember in the 1970s being on various committees and I was very unpopular because they wanted women to be on all the committees and have all kinds of administrative responsibilities and so on. I said, let women alone and let women do their science and let them influence others by their example. Even my fellow women scientists did not take it well. One can easily become interested in the power aspect. That's all right for those women who want to have that but they should not be encouraged to take up other activities as a substitute for their

science. When the government decided to have a woman on every advisory committee, I was asked to sit on five study sections, three at NIH and two at NSF. There weren't enough women to go around. I have limited such activities. I have been active in my professional society and at my university I agreed to serve on two committees but no more. I read about a famous physicist who was considering going to Israel, but he hesitated. His wife persuaded him to seek Einstein's advice. So he went to Einstein and Einstein said to him, "I'm a scientist first and a Jew second." The same thing is for women. They should be scientists first if that's what they are interested in, and a *woman scientist* second.[6]

GERTRUDE B. ELION

Chemist, Pharmacologist

Gertrude B. Elion and Magdolna Hargittai in 1996 in Research Triangle Park, North Carolina. (photo by I. Hargittai)

From one of Gertrude Elion's "fan" letters:

> In 1984, our then 5 year old daughter was diagnosed with Acute Lymphocytic Leukemia and was put on a chemotherapy protocol including, among other drugs, 6-mercaptopurine. Our family went through many difficult times, both physically and emotionally. We are thrilled to tell you that our daughter has been in remission for over 5 years, with no relapses, and has been completely off medication for 2 years, 3 months, and 24 days! She will celebrate her 11th birthday in May and is a delight to her parents. When we see the adoration in the media of overpaid sports figures and entertainment figures with inflated egos, we can only think how much more you and your colleagues have contributed to society, with little or no recognition. You are truly a hero.[1]

This was among the many quotes covering the walls in Gertrude Elion's office at Glaxo Wellcome in Research Triangle Park, North Carolina, when we visited her in 1986.

Gertrude Elion (1918–1999) was born in New York City into a scholarly family. Her parents lost everything during the Great Depression. Still, they knew that their children had to get a good education because that was the only way to a better life.

From early childhood, she was a voracious reader, and her favorite book was Paul de Kruif's *Microbe Hunters*. The book has been in print since it was first published in 1926, and has fascinated many young readers about scientists and about careers in science. A number of distinguished scientists of Elion's generation chose a career in science because they had been enthused by de Kruif's stories.

Already as a young student, she wanted to become a chemist so that she could find the cure for cancer. She had a painful personal motivation for her aspirations; her grandfather died of cancer just before she entered college. Later, her determination about finding cures for sick people further strengthened when her fiancé died. He suffered from a disease that could have been cured by penicillin, which was to become available shortly afterwards. Elion succeeded spectacularly in fulfilling her goal, more than she could ever have imagined. This is witnessed by the contributions for which in 1988 she received her share of the Nobel Prize in Physiology or Medicine. It was a joint award together with James W. Black and George H. Hitchings for establishing novel principles for drug treatment.

Her road to success was not easy. She attended Hunter College, a women's college in New York City, where young women who performed well academically were accepted regardless of their economic situation. She graduated with the highest honors. After college, she applied to fifteen graduate schools but none of them offered her support; that she was Jewish and a woman made her unwanted. She could not finance herself, so she accepted teaching positions, and this allowed her to pay for her education. She enrolled at New York University, and she did her research during nights and weekends. She received her master's degree in 1941. She started to look for a job, but her experience was not encouraging: "When I was interviewed . . ., they said: 'Oh, you only have a Master's degree, therefore we should tell you now, this is as high as you can go.' I thought it was a very strange thing to tell me. They had no idea how good I was or if I was any good; I did not ask them how far I could go, it was not even on my mind. That was not what I was working for. I wanted to do something important; I really wanted to cure cancer."[2]

Eventually, she interviewed with George Hitchings at the Burroughs Wellcome Company. He was not interested in whether she had a doctorate or not, only whether she could do a good job.

> When I was first interviewed by him in 1944 and he told me that he was working on . . . nucleic acid derivatives, I really did not know what a purine or pyrimidine was. . . . However, I was very intrigued by what he was telling me. He was interested in the biochemistry of nucleic acids. . . . Hitchings wanted certain compounds to be made. Since I had my Master's degree in chemistry and had all good grades, he figured I could probably do it. I could read German and I could follow the German literature. A lot of it was Emil Fischer's work. I had learned Yiddish at home and then I took German at college. Chemists had to know German. It

> was during the war that women were suddenly able to get a job in chemistry, which they had not been able to do before. I started as Hitchings' assistant.[3]

A beautiful working relationship developed between Elion and Hitchings. He followed the philosophy of the company's founder, Henry Wellcome, who said to his researchers: "If you have an idea, I will give you the freedom to develop it."[4] Hitchings never told Elion what to do or what not to do. "He gave me an opportunity to do as much as I could do, he never said: 'It is not your business to do that.' He never said: 'You are not a pharmacologist, not a virologist, not an immunologist,' so I became all of them."[5] She gradually took on more and more responsibilities in her research.

Hitchings's idea was to block the survival of bacteria, parasites, and tumor cells by replacing certain building blocks in their DNA. He was looking for molecules that were similar to those building blocks. This was a revolutionary approach not only because it was different from the generally followed approach in drug research, but also because it focused on DNA, which not many researchers were yet interested in at that time. He gave Elion the task to work on preparing different purine derivatives. Purine is a simple organic molecule; two of the four bases of DNA are purine derivatives (the other two are pyrimidine derivatives). Purine itself consists of a six-member ring fused together with a five-member ring:

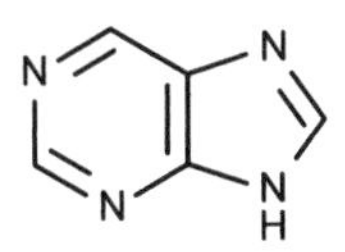

The task was to prepare various derivatives of this molecule and then, by administering it to patients, fool the bacteria, parasites, or tumor cells. Elion was just as enthusiastic about the work as Hitchings, and she not only did countless organic syntheses but also learned related fields so that she herself could test the biological activities of the new substances.

The hard work bore fruit; during the three decades of their joint work, they developed a number of drugs, all of them derivatives of purine. Among the best known ones were:

Thioguanine® against leukemia in children
Purinethol® (mercaptopurine) against leukemia, non-Hodgkin's lymphoma, and other diseases
Zyloprim® (Allopurinol), whose most important role is treating gout
Imuran® (Azathioprine), an immunosuppressive drug that prevents the rejection of new organs during transplantation
Zovirax® (Acyclovir), an antiviral drug, used, for example, to treat herpes virus infections.

Entirely new drugs appear very seldom on the market, and here we have a most impressive list of a whole array. However, Elion and her corecipients of the Nobel

Prize obtained the award not for the discovery of specific drugs but for the principles they established for drug research. The principle that the Nobel Prize citation referred to is called rational drug design. The idea originally came from Hitchings and his early studies of DNA.

After Elion had retired, her younger colleagues continued the work she and Hitchings had initiated. It was during her active days that the first antiviral drugs were produced, and today this is an important area of pharmacology. Even after retirement, Elion remained on hand in case her advice was sought. Antiviral drugs are more difficult to create than antibacterial ones, because the antiviral drugs must be much more selective than the antibacterial ones. In other words, they must be able to distinguish between the DNA of the virus and the DNA of the human organism that the drug is supposed to protect. Under her guidance, her laboratory produced the antiviral drug listed above for treating herpes virus infection. Just after Elion's retirement, her former coworkers created the first drug against the AIDS virus. This was AZT, azidothymidine, and this success pleased Elion no less than when she herself had produced a new drug.

With all her successes, Elion always regretted that she did not have a PhD degree. She was one of only a few Nobel laureates who did not have a doctorate. Actually, at one point during her career, she decided to enroll at the Brooklyn Polytechnic Institute to get her doctorate. After a while, however, she was told that she could only continue her studies if she did them full-time; this would have meant giving up her job, which she loved and needed, and she declined to give it up. Later, when she was becoming successful, she received honorary doctorates; one of them was from Brooklyn Polytechnic.

Gertrude B. Elion and George Hitchings in 1969, when she received her first honorary doctorate from George Washington University. (courtesy of the late G. Elion)

Elion remained active to the end of her life. In 1996, my husband and I visited her in her office in Research Triangle Park. Many photos decorated the walls of her office, all related to her colleagues and to her science. Among them, there was a framed reproduction of the British caricaturist James Gillray's picture *The Gout*. Gout is an unpleasant illness which manifests itself in burning pain in the joints, most often in the big toe. It occurs when there is too much uric acid in the blood and uric acid crystals form and accumulate in the joints. This causes inflammation. Uric acid is a by-product of purine metabolism, and purine chemistry was a focal point in Elion's research. It was almost an accidental discovery that she and her colleagues found a drug for treating gout. This drug then brought a lot of income for the company; it was unexpected, because their efforts were not dictated by profit making.

Elion was seventy years old when she received the Nobel distinction and suddenly found herself in the limelight. At first, she thought the Nobel Prize would not change her life, but it did. She was in great demand to sit on committees and advisory boards and give talks. She found it fortunate that the award came rather late in her life so that it did not interfere with her creative period. To the question what welcome change the Nobel Prize brought to her, she mentioned that it added an activity that she truly enjoyed. Each year she took on a third-year medical student at Duke University to train her or him in research. The goal was not to induce students to choose research for a career, only to broaden their outlook on science.

Elion died in 1999 at the age of eighty-one. She never married; she felt nobody could have replaced the young man she was engaged to but who died of an acute disease, an infection of the heart. She remained active to her last day; she was full of plans, because, in her words, "there were still so many things to do, that research is worthwhile and that here you can really make a difference. You can imagine how I feel when somebody comes to me after I have given a lecture somewhere and says: 'I have a kidney transplant thanks to you and I have had it for 25 years.' I get many 'Thank you' letters and I keep every one of them. They represent the real reward, making sick people well."[6]

MARY GAILLARD

Theoretical physicist

Mary Gaillard in 2004 in Berkeley. (photo by M. Hargittai).

According to the Nobel laureate American physicist Leon Lederman, Mary Gaillard is "one of the tragically few women in physics."[1] Physics in general and theoretical particle physics in particular has been a field where women are hard to find. Mary Gaillard is an exception. Overcoming all the barriers along the way, she has excelled in this field. She is a member of the National Academy of Sciences of the United States, served on the National Science Board under President Clinton, and has received numerous awards and recognitions.

Mary Katharine Gaillard (née Ralph) was born in 1939, in New Brunswick, New Jersey, and grew up close to Cleveland, in Painesville, Ohio. Her father was a history teacher in a small women's college, and her mother taught in high school. There was no one in her immediate family with an interest in the sciences, but Mary liked physics. She went to Hollins College in Virginia, a women's school:

> I majored in physics; there was about one physics major in every two years in that school. During my college years, I spent a year abroad, in Paris, and my physics professor got me into a lab there, so that I would not completely forget physics. This professor was Dorothy Montgomery; originally at Yale, but when

her husband died, they did not keep her there, because she did not have a regular faculty position, and then she moved to this girls' school in Virginia. This turned out to be very lucky for me, because besides getting me into that French lab, she encouraged me to go to Brookhaven as a summer student. I did this twice; after my junior and senior years in college, and that is where I really fell in love with high-energy physics.[2]

Mary went to Columbia University for her graduate studies, where she met a French physics postdoc, Jean-Marc Gaillard; they got married during her first year there, in 1961. Jean-Marc finished his postdoc stint in the same year and they moved to France, to the University of Paris (Sorbonne) in Orsay. She continued her graduate studies there, but had a hard time. She wanted to work on theory but she could not because only a limited number of students were admitted to theoretical physics. She could not join an experimental laboratory either, because—as she was told—only people from special schools could get in. It turned out those special schools at that time took only male students. Thus, her first year was frustrating. Then her husband received a six-year appointment at CERN (European Organization for Nuclear Research) in Geneva and they moved there. Fortunately for Mary, one of her professors from Orsay was also there on leave, and he let her start working on her thesis in theory.

Her stay in Geneva was considered a visiting position; officially, she was first a graduate student at the Orsay campus of the University of Paris, and later employed by the French National Center for Scientific Research (CNRS). The French doctoral system used to have a two-tier arrangement. She received her first doctorate, the doctorat de troisième cycle, in 1964, and the second one, the doctorat d'État, in 1968. During these years, she gave birth to three children; her older son was born in 1962, her daughter was born just as Mary was writing her thesis for the first doctorate, and her younger boy was born just as she was writing her second thesis. It was a blessing that taking care of young children was easier in France and Switzerland than it was in the United States. They lived in a small French village and hired daily help and later even an au pair; all this was affordable.

Mary told me that at CERN she never got a regular paid position: "It was not explicitly spelled out that this was because I am a woman but it was fairly obvious." She found that women physicists were treated rather badly; eventually, she conducted a survey among them and put together a report.[3] She found that among the women physicists only 10 percent were fully paid and 86 percent not paid at all by CERN. Among the reasons given for not paying were: (i) the husband is a CERN employee and (j) priority is given to unemployed men. "Points (i) and (j) are blatantly sexist. Point (j) speaks for itself. Point (i) is not just an administrative matter of nepotism, but carries with it the implicit . . . assumption that the female applicant does not require a salary because she will be supported by her husband."[4] A woman physicist told Mary that a CERN official had told her during the interview that they never gave a salary to physicists who were wives of male physicists; there were just too many of them.

Mary Gaillard in about 1970 at CERN with Murray Gell-Mann. (courtesy of M. Gaillard)

As the years passed, Mary gradually advanced in her position at CNRS based on her work at CERN. She proved that her insistence on becoming a theoretical physicist was justified, achieving significant results that caught the attention of many physicists internationally. Her most important result dates back to her time at CERN, when she and Ben Lee predicted the mass of the so-called charm quark prior to the experimental discovery of this particle.[5] Quarks are elementary particles (that is, they are the smallest building units of matter that are not made up of yet smaller ones). The Greek philosopher Democritus considered atoms indivisible, but later on we learned that atoms consist of protons, neutrons, and electrons. First these were thought to be the smallest building blocks of the universe. Later on, it was determined that although the electron is such a particle, the proton and neutron are not—they consist of quarks. There are six different quarks, one of them being the charm quark. Quarks do not exist in isolation; they can only be observed indirectly, in accelerator experiments. Within a few years' time, Mary and her colleagues made other significant predictions as well. She was gradually promoted, and in 1980 she was appointed to the rank of director of research at the CNRS.

In 1981, the Gaillards divorced and Mary went back to the United States. By then, she was a well-known scientist, and both Fermilab near Chicago and the University of California at Berkeley offered her a job. She opted for Berkeley, and became the first female professor in its Physics Department. She has been there ever since. Her situation now was very different from what she had experienced in France and Switzerland. Of course, about twenty years had passed since her arrival in Europe, and during this time affirmative action changed the situation of women scientists:

> When I left, I was a student at Columbia, that time I had the impression that some people didn't really take me seriously in the sense that they didn't think I would last in the profession, but they didn't outright tell you, "You can't do this." The things some French guys told me when I was trying to get a position in a French lab were just outrageous. But in the meantime, affirmative action in this country has gotten accepted and people don't say things they used to say. . . . Things did change. I was on the National Science Board and if you looked around there, I would say at least 30 percent of the members were women. These were Clinton appointees and he made a big effort in this regard.

Mary's children grew up in France, but they also experienced the American educational system during their parents' sabbatical, and that affected their choices. At one stage or another of their lives, all three of them preferred the greater diversity and flexibility of the American educational system (debates in class, independent projects, and the possibility of picking up where you left off if you quit for a while) as compared with the more rigid French system. All of them got their degrees in a science field or mathematics, but as it happened none of them stayed in science. As to Mary, she married a colleague of hers, the theoretical particle physicist Bruno Zumino.

The Standard Model of what the universe is made of is quite well known. According to this model, everything is made of elementary particles, such as electrons and quarks, and they are governed by certain forces. For a long time, there has been one deficiency of the model, namely that it could not explain why particles have masses. This deficiency was identified as the result of a missing component called the Higgs particle. It could not be found in accelerators because they did not have sufficiently large energy to produce it. The highest-energy accelerator so far, the Large Hadron Collider, was built at CERN with one of its major aims to detect the Higgs; it became operational around 2008.

Back in 1976, Mary, John Ellis, and Dimitri Nanopoulos had predicted the properties of the Higgs. Later, in the 1980s, Mary and Michael Chanowitz pointed out that either the Higgs would be found at enormously large energies or a new physics would emerge different from the physics we know. Their work was taken seriously and influenced subsequent work in the colliders. Eventually, the Higgs particle was detected and confirmed in 2012. The following year, the Nobel Prize in Physics was awarded to the two scientists who first suggested its existence, Peter Higgs and François Englert.

Mary and her second husband, Bruno Zumino, collaborated and published together on topics called supersymmetry, supergravity, and string theory. All these are difficult for a layperson to understand, but all of them help to describe what the Standard Model does. In string theory, all particles and forces of nature are described as vibrations of tiny strings. The Standard Model incorporates the electromagnetic, strong, and weak forces of nature, but it does not consider gravitation. This is what the supersymmetric string theory, called superstring theory, tries to accomplish, and as such, it could be a prime candidate for the "Theory of Everything." Earlier, the Nobel laureate physicist Murray Gell-Mann also entertained similar theories. Presently, Mary is working on predictions of the superstring theory for phenomena that may be detected both in accelerator experiments and cosmological observations. She has contributed to technical developments in supergravity theory and applied them to the study of specific models derived from superstring theory. It does not seem likely that she will find herself without research topics any time soon.

MARIA GOEPPERT MAYER

Nuclear physicist

Joseph Mayer and Maria Goeppert Mayer around 1930. (courtesy of the Oesper Collections in the History of Chemistry, University of Cincinnati)

Three scientists shared the 1963 Nobel Prize in Physics. Half of it went to Eugene P. Wigner and the other half jointly to Maria Goeppert Mayer and Hans Jensen for discovering the shell structure of the nucleus. She was the fourth woman Nobel laureate in the sciences and the second one in physics up to 1963 and so far (as of 2013) the only one in theoretical physics.

Maria Goeppert (1906–1972) was born in Kattowitz (then in Germany; today, Katowice, Poland). When she was four years old, the family moved to the famous German university town Göttingen, where her father, Friedrich Goeppert, was a professor of pediatrics. She was an only child, and she later remembered that her father told her: "'Don't grow up to be a woman', and what he meant was, a housewife . . . without interests. . . . Did I think it strange for him to say such things to me? . . . No, I felt flattered and decided I wasn't going to be just a woman."[1] She became interested early on in mathematics and the sciences, and Göttingen was one of the best places in the world for those fields at the time. The mathematicians David Hilbert, Richard Courant, and Herman Weyl and the physicists Max Born and James Franck taught there, and similarly great scientists or future greats came for visits, such as Paul Dirac, Enrico Fermi, Werner Heisenberg, John von Neumann, Robert Oppenheimer, Wolfgang Pauli, Leo Szilard, Edward Teller, and Victor Weisskopf.

In the 1920s, not many women went to university, and there was not even a public high school in Göttingen where girls could get the education to make them eligible to enroll at a university. Maria went to a private school, which closed down before she could have graduated from it. Nonetheless, she decided to try the university entrance examination and succeeded. She started with mathematics, but Max Born, a family friend, suggested she attend his course on quantum physics—then a revolutionary branch of science. Göttingen played a major role in the development of quantum mechanics, and Born was one of its principal creators. It took only a few lectures for Maria to change her major to physics. She later explained: "Mathematics began to seem too much like puzzle solving. . . . Physics is puzzle solving, too, but of puzzles created by nature, not by the mind of man."[2]

Friedrich Goeppert died in 1927, and this was a great blow for her, as she was very much her father's daughter. Maria decided to continue the family tradition and become a university professor, and she asked Born to be her thesis advisor. The task of supporting the family fell to Maria's mother. She took in boarders, and this is how Maria met Joseph (Joe) Mayer (1904–1983), an American chemist. He had received his PhD from Berkeley and came to Göttingen to work with Born and Franck. Maria and Joe fell in love and married in 1930. She was so much in love and happy with her life that it took some persuasion on Joe's part to convince her to finish her dissertation. Years later, Eugene Wigner characterized it as a "masterpiece of clarity and concreteness."[3]

After she received her PhD, Maria and Joe moved to the United States, to Johns Hopkins University in Baltimore, where Joe had been offered a position. This was a difficult time for women in science, as due to the so called antinepotism rule, Maria could not get a job at the same school where Joe worked. It did not matter that her knowledge of quantum mechanics was superior to all the physics professors' at that time at Johns Hopkins. She was lucky that they let her have some space and participate in the scientific activities of the university. She was enthusiastic and collaborated with several faculty members.

Maria's life in Baltimore was different from the life she enjoyed in Göttingen. There, the Goeppert house had been one of the social centers of the town. While Maria's father was alive, the Goepperts used to give elegant dinners and receptions that the cream of the town attended. Her mother organized everything; they had servants, and Maria never learned to cook and keep a house. This Göttingen lifestyle could not be emulated in America by Joe and Maria, but she was in love, she was full of energy, and she believed she could do anything. Alas, running a home was not her forte; their son, Peter Mayer, writes: "My father's response to my mother's helplessness was, 'As expensive as maids are in the United States, I promise to hire one . . . as long as you remain a scientist!' "[4]

But "remaining a scientist" did not turn out to be easy. In 1933, their daughter Marianne was born, and Maria stayed at home with her for about a year. Then she went back to the university and continued collaborating with faculty members as well as giving courses on physics as a volunteer research associate. Her collaboration with Karl Herzfeld and his student Alfred Sklar was especially successful. They published weighty papers in which she applied her knowledge in quantum mechanics to problems in chemistry. She also collaborated with her husband; together they produced a book on statistical mechanics which became a classic of the field.[5]

With the Nazi takeover in Germany and the exodus of Jewish scientists, several of their good friends came to the United States, among them James Franck and Edward Teller. Franck joined Johns Hopkins, and Teller was appointed professor at George Washington University in nearby Washington, DC. "Maria was naturally drawn to discuss her work with him [Teller]. She found him to be an unusually stimulating physicist and looked to him for guidance in . . . theoretical physics."[6] According to Teller, "In addition to being an extremely able physicist, Maria was also very beautiful. Slender and blond, she had a natural delicacy and grace as well as considerable strength of mind."[7]

In 1938, the Mayers' second child, Peter, was born. In that year, they learned that Johns Hopkins would not give tenure to Joe. Fortunately, Columbia University in New York City offered Joe a tenured position with twice his previous salary. They moved to Leonia, New Jersey, across the Hudson River from Manhattan. Columbia's physics department declined to give Maria any kind of appointment. What one of the faculty members, the soon-to-be Nobel laureate Isidor Rabi, said about women in science reflected the general attitude. According to Rabi, the women's nervous system is ". . . simply different. It makes it impossible for them to stay with the thing. I'm afraid there's no use quarreling about it, that's the way it is. Women may go into science, and they will do well enough, but they will never do great science."[8] Fortunately for Maria, the Nobel laureate Harold Urey, the chairman of the chemistry department, thought otherwise. He was aware of her work and offered her an appointment as a lecturer—without pay.

An important event for Maria's scientific development was the arrival of Enrico Fermi at Columbia University. He was a fresh Nobel laureate, a refugee from Fascist Italy. He suggested to Maria that she work in the emerging field of nuclear physics.

During World War II, most scientists were involved with war-related projects, and Joe worked at the Aberdeen Proving Ground in Maryland. He was away from his family for long periods, so taking care of the children and all household duties fell on Maria. At the same time, the war opened up new opportunities for women scientists. Sarah Lawrence College, a small women's school in New York State, offered Maria a teaching position. On top of this, in 1942 Urey invited her to participate in a secret project to find methods of separating fissionable uranium-235 isotope from the most abundant uranium-238. This eventually became a full-time job, so she had to resign from Sarah Lawrence. For the first time in her life, she was paid for research! "It was the beginning of myself standing on my own two feet as a scientist, not leaning on Joe."[9]

About two years later, Edward Teller involved Maria in another secret job. He was at Los Alamos, working on the Manhattan Project, and he asked Maria to investigate radiation transfer through various materials. She was not informed about the principal goals of the project, but had to swear to secrecy even about the calculations she had to perform. The colonel responsible for Maria's work staying classified warned her that her students involved in the actual calculations should not know that they were dealing with uranium; he thought it would suffice for them to know that the target was element 92. He had to be informed that every student would know that element 92 was uranium. Teller could not tell Maria the purpose of their study, but when he told her the temperature they needed the calculations at, "Maria caught her breath"—obviously she understood the implications.[10]

At Columbia, just as at Johns Hopkins, she taught courses. The Columbia students found her too tough. It was generally noted that she was not a particularly good lecturer; she spoke too rapidly and quietly, and some thought that her mind worked faster than she could deliver her thoughts. She was a heavy smoker and chain-smoked even during her lectures. She often confused her cigarette with the chalk and tried to smoke the chalk or write with the cigarette on the board. The war-related research, her teaching, the sole responsibility of taking care of their children in Joe's absence, and anxiety about her mother and friends in Germany weighed on her heavily, and her love for alcohol gradually increased.

After the war, in 1946, Joe was offered a position at the University of Chicago and the family moved there. Here, for the first time, they promised her a "volunteer professor" position—although they could not pay her. Still, she was grateful because this was "the first place where I was not considered a nuisance, but greeted with open arms."[11] The University of Chicago set up the Institute for Nuclear Studies and most leading physicists of the atomic bomb project went there, among them the Mayers' old friends Teller, Franck, Urey, and Fermi. Their life somewhat started resembling the old golden days in Göttingen—if not in its elegance, in their complete devotion to physics. Soon, the Argonne National Laboratory was established, and its director, Robert Sachs—Maria's first doctoral student back at Johns Hopkins—offered her the position of senior physicist on a half-time basis. Finally, she had a peacetime research job for which she was paid!

Teller was interested in the origin of the elements and invited Maria to work with him. This work required serious mathematical background, and Teller knew that Maria was the right person for it. Their cooperation eventually led Maria to the discovery of the shell structure of the nucleus. She noticed that the cosmic abundance of certain elements, like the number of their isotopes, was much larger than those of others. There did not seem to be any rational explanation for this, but there was something common in the numbers of protons or neutrons in the nuclei of these elements. Several of them had eighty-two neutrons (irrespective of the number of their protons), others had fifty neutrons. Similar stability was found for the elements that had fifty protons. Tin, with fifty protons, has ten different isotopes, while other elements have only a few. There are six elements that have fifty neutrons and seven with eighty-two neutrons, but it is rather rare that the number of neutrons is the same in different elements. Other scientists did not attribute too much importance to this observation. Wigner with his customary biting politeness referred to this observation as amusing nonsense, and called these numbers "magic numbers." Eventually, the label stuck—but no longer in an ironic sense. Later more magic numbers were found; they appeared to be 2, 8, 20, 28, 50, 82, and 126. Maria became obsessed with these numbers; she could not stop thinking about them until she understood their significance.

It is to Maria's credit that she did not let be influenced by the initial skepticism of her colleagues. As Teller was away from Chicago most of the time, she talked about the structure of the nucleus mostly with Joe and with Fermi. She assumed—as had others before her—that the protons and neutrons in the nucleus could be found in shells of different energies. She liked to compare this to the layers of the onion, and Wolfgang Pauli called her "the Madonna of the Onion."[12]

This shell structure resembled the way electrons are positioned in electron shells around the nucleus in an atom, even though the acting forces were of different kinds—long-range for the electrons and short-range for the nuclei. However, no matter how she tried she could not come up with a plausible explanation and a mathematical model that would fit all experimental data. Then one day, when she was having a discussion with Fermi, he was called away to the phone, but before he stepped out of the office, he asked casually, "Is there any indication of spin-orbit coupling?"[13] For an outsider this does not help much—but for Maria, who was so familiar not only with this topic but also with its mathematical formulation, this made everything suddenly clear. She was sure that this was the missing piece in her jigsaw puzzle, and by the time Fermi came back, in about ten minutes, she already had the answer. There are different versions of how their meeting ended. According to one of them, "Maria, when excited, had a rapid-fire oral delivery, whereas Enrico [Fermi] always wanted a slow, detailed, and methodical explanation. Enrico smiled and left: 'Tomorrow, when you are less excited, you can explain it to me.'"[14]

Just about the same time she described the model, a research group in Germany, led by Hans Jensen, figured out the importance of spin-orbit coupling and explained the model the same way as Goeppert Mayer. The two reports appeared about the same time. Instead of jealousy and rivalry, a friendly interaction developed between Goeppert Mayer and Jensen.

In 1960, the Mayers received an offer from the University of California in San Diego for positions for both of them (though with only half-pay for her). By this time she was so famous that, after all these years, the University of Chicago had finally offered her a full-time position with full salary. The Mayers, nonetheless, moved to California because the climate attracted her. By the time her Nobel Prize was announced in 1963, she was not well. A few years before she had suffered a stroke that left her partially paralyzed. She was happy with the award, but she told reporters: "Winning the prize wasn't half as exciting as doing the work itself. That was the fun—seeing it work out!"[15] Even after the stroke, she continued to work till her death in 1972.

It appears to be a characteristic pattern in Goeppert Mayer's career that she liked to work with others. Her colleagues noted that although very bright, she did not seem to try independent research. One of them commented that "her mind is really brilliant and penetrating, although perhaps not very original."[16] She was lucky to have so many top scientists around her as close friends and mentors. Already back in

Göttingen, she was not only a vivacious pretty girl but also the wittiest. Most students in Göttingen were in love with her, and she liked to flirt with men. Her ability to attract them did not diminish in later years. Her son writes: "Women have reported that men would gravitate towards my mother whenever she entered the room. I never noticed, perhaps due to my own gravitation."[17]

The first decisive influence on her—after her father—was Max Born. Victor Weisskopf commented that theirs was not an ordinary professor-student relationship.[18] They used the familiar form of German, "du"—which was quite unprecedented between a professor and a student, especially of opposite sexes. Born had problems with his marriage, and Maria became his confidante. After her departure for the United States they kept up correspondence, in which he wrote "Dearest Maria" and ended his letters with "in old love."[18] He also tried to pave the way for her as much as he could with recommendations in American scientific circles.

Her friendship with Edward Teller was another determining relationship for her scientific development. They corresponded for years, especially between the mid-1940s and the mid-1950s. Although Teller asked her to destroy his letters, Maria did not, and his archival letters helped Teller write his *Memoirs*. His letters are interesting not only because they are testimony to how her science developed but also because little-known traits of Teller's personality shine through them. He considered Maria not only a colleague and a friend but also a soul mate, to whom he could pour out his troubled thoughts during the most difficult period of his life: the big debate about whether the United States should develop the hydrogen bomb, the development of the hydrogen bomb, his testimony at the Oppenheimer hearing, and the initiation of the Livermore Laboratory.[19] For Maria Goeppert, Teller's friendship meant learning from a most imaginative colleague and the possibility of discussing her research into the nucleus structure and her magic numbers with an intelligent and experienced scientist.

There was then her friendship and long correspondence with Hans Jensen, but it was an interaction of a different nature from the previous ones. She looked up to Born and Teller and tried to learn from them as much as possible. By the time she got to know Jensen, she was a recognized authority in the field and thus an equal to her pen pal. Their interactions produced a joint book.[20]

Her most important "coworker" was Joe Mayer. Throughout her life, he was her most dedicated supporter in every way. They talked about science all the time to their mutual benefit. Maria was a theoretical physicist with a strong mathematical background and Joe an experimental chemist from the great Berkeley school. Joe always maintained that everything he knew about quantum mechanics he learned from her, and eventually he became an expert in statistical mechanics, which required a proper mathematical background. During the first decade of their life in the United States, she worked only either with Joe or with his colleagues, who were mostly chemists. Most of her publications were in chemistry. Her familiarity with chemistry proved to be an important factor in her finding the nuclear shell model.

When Maria was young, even though she was very talented and she liked science, it was not obvious for her that she wanted to become a scientist. When she fell in love with Joe, she was not eager to finish her PhD. Even in Baltimore, when Marianne was born, she enjoyed being a mother; she said: "It was such an experience to have a child, such a tremendous experience!"[21] But as time went by and she got involved more and more with research work during their time at Columbia, her priorities changed. Even her old friend Max Born expressed his astonishment when he received from her a long letter with details about everything, but not one word about her children. One of her biographers mentions that from the early 1940s she started losing touch with her children. During the war years, the children were taken care of by nannies, whom they resented. Peter especially suffered from the situation, as he had a reading disability that was only later diagnosed as dyslexia. Marianne said of Maria: "At times she tried very hard to be a mother, but often she wasn't there." Marianne, when she had a daughter, made it a point to stay at home with her. Concerning Maria's relationship with her children in later years, it is telling that she invited her son to the Nobel ceremonies, but he did not go, while her daughter would have loved to go, but Maria did not invite her. During their Columbia years, the Mayers lived in Leonia, a small town in New Jersey. With her growing interest in science, she became impatient at parties when she had to be with the scientists' wives. Talking about science was fun, plus the men clearly enjoyed her company. Laura Fermi wrote: "Faculty wives like me no longer ha[d] easy access to Maria Mayer, because she was always talking to the men and had a too technical conversation."[22]

We may wonder whether she might have become a yet more successful scientist if there had not been the antinepotism rule. Her son, Peter, does not think so. Joe Mayer not only appreciated her as a scientist but encouraged her to keep doing research. The interaction with and appreciation from her colleagues helped. The university atmosphere, the seminars, the discussions, all gave her the background and the intellectual stimuli that she needed. But the most important was probably Joe's presence and continuous support. He was a permanent "sounding board" for her ideas. Their children remember that there was always science talk in their home, even at meals and social gatherings. This was not something the children appreciated, so eventually Marianne asked her parents not to discuss science at family mealtimes.[23]

It is only too human that Maria felt bitter about having to work for free most of her scientific life. In a rare candid moment, she opened up about herself to a young woman just starting her scientific career and warned her how hard it would be to find a job "in the same geographic location as her husband; how her own marriage had given many in academia an excuse not to treat her seriously."[24] However, being a member of a scientific family, Maria succeeded in turning at least some of the disadvantages into advantages.

DARLEANE C. HOFFMAN

Nuclear chemist

Darleane C. Hoffman in 2004 at Berkeley. (photo by M. Hargittai)

There are ninety-two naturally occurring elements, the heaviest being uranium, element No. 92. These elements build up all substances that we encounter in nature. But when we look at any recent periodic table of the elements, it lists well over one hundred. Those heavier than uranium are all human-made; they have been discovered by painstaking investigations. The quest to discover new elements has been fierce, so much so that sometimes frauds occur even in respectable laboratories. Thus, scientists have to pursue the new elements and, simultaneously, watch out for fraudulent claims. Darleane Hoffman has been one of the most respected contributors to this field, along with her mentor, Glenn T. Seaborg, the chemist who has discovered more new elements than anybody else in the history of science. Darleane proudly wears a pin on her lapel displaying the chemical symbol Sg (seaborgium) of element No. 106, named after her mentor. This is a rare distinction that only very few even among the greatest have been accorded. In 1994, Darleane's group published confirmation of the 1974 discovery of element 106 by Ghiorso et al., who then proposed the name "seaborgium" for it. IUPAC delayed confirmation until 1997 because Seaborg was still alive, but no rule prohibiting naming an element after a living person could be found.

When I visited Darleane, she was already formally retired—although right at the time of her retirement she became the charter director of the newly established G. T. Seaborg Institute at the Lawrence Livermore National Laboratory (LLNL). She also retained her professorship at the graduate school in the Department of Chemistry, University of California at Berkeley, and her position as faculty senior scientist in the Nuclear Science Division of Lawrence Berkeley National Laboratory (LBNL).

Darleane Christian was born in 1926 in a small town, Terril, in northwestern Iowa, where her father was a school superintendent. She received her degrees from what is known today as Iowa State University in Ames, concluding with her PhD in chemistry in 1951, and she specialized in nuclear chemistry. She met her husband-to-be, Marvin Hoffman, in graduate school, and they married after she received her doctorate. While her husband was finishing his degree in nuclear physics, she spent one year at Oak Ridge National Laboratory, then followed him to the Los Alamos National Laboratory (LANL). This was the first time she encountered a problem because she was a woman. She was supposed to get a job at the radiochemistry laboratory, but when she arrived[1]:

> I called personnel about it and they said, I am sorry but we don't hire women in that division! . . . I had never run into any overt discrimination because of my sex! I was totally shocked. Then in early January 1953 we went to a party that the director of the lab, Dr. Norris Bradbury, gave for the new hires and I talked to various people. I finally met a man whom I found out, after we introduced ourselves was Dr. Rod Spence, leader of the Radiochemistry Group. He said: "Where have you been? I have been looking for you!" I said, "Oh, I have been here trying to collect my promised job, but I couldn't find it." . . . but I have never trusted personnel departments ever since.

They stayed in Los Alamos for three decades, and their two children were born there. She never stopped working; they had a woman who came every day to take care of the children until her mother moved there to live nearby in 1964, and from then on she helped. During 1978–1979, she worked in Glenn Seaborg's group at the University of California at Berkeley with a Guggenheim fellowship, and finally, in 1984, they actually moved to Berkeley.

One of Darleane's most remarkable feats was the discovery of primordial plutonium in nature. This happened in 1971, but to appreciate its significance, we have to go back to 1940 when two associates at Berkeley, Edwin M. McMillan and Philip M. Abelson, were investigating the consequences of neutron bombardment of uranium. This was right after Otto Hahn and Fritz Strassmann performed their nuclear fission experiment of uranium, which was interpreted by Lise Meitner and Otto Frisch. The Berkeley experiments led to the first artificially produced transuranic element, neptunium. Soon after, in 1941, Seaborg and his coworkers produced the next transuranic element, plutonium, which then played a conspicuous role in the development of atomic bombs. The plutonium bomb was dropped over Nagasaki on August 9, 1945. The discovery of plutonium was only published after the war.

Darleane C. Hoffman with Glenn T. Seaborg in the 1980s in her laboratory. (courtesy of Lawrence Berkeley National Laboratory and D. C. Hoffman)

In 1971, Darleane and her coworkers found remnants of plutonium in nature. It was one of the element's isotopes, plutonium-244.[a] This is the longest living among plutonium isotopes, with a half-life of 80 million years. So uranium may not have been the heaviest element in nature, since the atomic number of plutonium is 94 versus the uranium atomic number of 92 (neptunium is 93). It is not known whether this plutonium isotope was originally formed on earth or came from extraterrestrial sources. Darleane and her team thought that it was primordial, that is, that it has existed since the beginning of the universe.

The discovery was like an exciting detective story; the scientists narrowed the possible location of the isotope, and finally identified it in minute quantities in a mine of the Molybdenum Corporation of America in Mountain Pass, California. This was a significant result for the understanding of the fate of elements on earth. It was also considered important enough to include its mention in the citation of Darleane's National Medal of Science, which she received from President Clinton in 1997. Besides this distinction, she has many others, such as membership in the American Academy of Arts and Sciences (1998) and foreign membership in the Norwegian Academy of Science and Letters (1990). She was awarded the highest distinction of the American Chemical Society (ACS), the Priestley Medal (2000).

[a] The atomic number of an element identifies the element and represents the number of protons in the nucleus of that element. In addition to protons, there are neutrons in the nucleus, and for most elements, their nuclei may contain different numbers of neutrons. Atoms of the same element (same atomic number) but differing in the number of neutrons are called isotopes. The name means they appear in the same place (topos) in the periodic table, even though they differ in their masses.

The plutonium-244 discovery was Darleane Hoffman's most publicized scientific achievement, but she has made scores of other discoveries over the years; two examples are the discovery of symmetric mass division in spontaneous fission and the discovery of heavy, short-lived fermium isotopes. She also got a taste of administration, first when she became the first woman leader of a scientific division at Los Alamos (1979–1984), and then after she had formally retired[2]:

> I retired from active teaching in 1991 and Seaborg was very upset with me. What happened was that in 1991 they had an incentive package for people who were in the California Public Employees Retirement System which meant that I could retire with more than the salary I was currently earning! But I took a different option so that if something happened to me my husband would get the same amount I was receiving, so I took a little less. But it was one of those one-time financial incentives that you could not turn down. But I never saw Seaborg so upset as when I told him about this. I told him that I would continue working the same as I did before but he said, it wouldn't be the same without the power and authority. However, when I retired I took the position as first director in 1991 of the Seaborg Institute at Lawrence Livermore National Laboratory on a half-time basis. Together with Drs. Chris Gatrousis, Tom Sugihara, and Patricia Baisden, I had helped to found this Institute and write its Charter. I retired as Director in 1996. Now I continue to serve as an advisor.

Darleane Hoffman has been first and foremost a scientist, but she has also been interested in women's issues. One of her sabbaticals took her to Norway in 1964–1965. It was a choice motivated by the excellence of nuclear chemistry in Norway and the Norwegian ancestry of Darleane's father. It was a great experience both scientifically and socially, and it had an impact on her: "I found Norwegian women were treated much more equally than women in the U.S. at that time. It was all right for women to go out to dinner alone but men also didn't open doors for them, etc. However, if you asked for help it was willingly given. I learned a great deal and developed a much more independent attitude socially and scientifically after our year there!"[3]

Whether or not it was primarily the impact of her Norwegian experience, in any case, looking back on her career, Darleane Hoffman had a long record of caring about women's issues. She was aware of the low number of women scientists and has noted their increasing number lately. However, she also noticed the so-called scissor effect, manifested in that the large number of women graduates diminishes to a disproportionally small number of women in leading positions, and this appears to be a persistent phenomenon. This is how Darleane described her involvement in women's issues[4]:

> I have been very involved over the years. I was very proud of the fact that while I was division leader at Los Alamos I could do something positive about this. One of the most pleasurable management tasks that I had occurred during the last year I was there. The laboratory had some law suits concerning women not

receiving equal pay to men in equivalent positions and so each division was given an amount of money to bring the average of women's salaries up to the average of men's salaries in the same position. We distributed this money according to merit among the women and I felt that this was one of those win-win situations. Over the years I have been involved with many women's conferences and women in science. I think that overall I had about 30% women among my graduate students at Berkeley. When I first went to Berkeley in 1984, only about 18% of the graduate students in Chemistry were women, and I was only the second tenured woman professor in chemistry out of a faculty of about 40. At that time I felt that if we increased the number of women Ph.D.s in Chemistry, the faculties of the major research universities would also increase in a commensurate manner. However, according to statistics from the American Chemical Society, this has certainly not been the case. Now some 50% of the B.S. degrees in chemistry go to women and more than a third of the Ph.D.s in Chemistry are earned by women, but now [2005] only about 8% of the tenured full professorships at our major universities are held by women. We need to investigate the university climate and the concepts of tenure and why so many women choose not to even apply for positions at the major research universities in the U.S.

VILMA HUGONNAI

Medical doctor

Vilma Hugonnai around 1890. (http://commons.wikimedia.org/wiki/File%3AHugonnai_Vilma_c_1890.jpg, downloaded August 3, 2013. Public domain)

When, after graduating from the Medical Faculty of the University of Zurich in 1879 as a medical doctor and working in a hospital for a year in Switzerland, Dr. Vilma Hugonnai decided to return to her native Hungary, she had no idea what frustration and disappointment she would face there.[1]

She was born in 1847 in Nagytétény, then a suburb of Budapest and now part of its District XXII. At birth, she was countess Hugonnay, but later in her life she preferred the commoner version of her name as Hugonnai. First she studied at home with governesses, and then she completed her basic education in an upscale boarding school for young women in Hungary. She spent her summers with a friend of hers from school and there she met a young man, György Szilassy, whom she married at the age of eighteen. Their marriage was not a happy one; her husband spent his time playing cards, riding horses, and being with his former girlfriend. Vilma was alone, but at least she could spend her time with her son. However, she wanted to do something beyond that with her life.

One day, in 1864, she read in a newspaper that women could attend universities in Switzerland—in the 1860s, this was not yet possible in Hungary. Her husband did nothing to prevent her from enrolling in a Swiss medical school, but refused to finance her studies. She was adamant in her determination, sold all her jewelry and other possessions, and went to Zurich. At that time, there was quite an international gathering of women at the Zurich medical school for similar reasons, though she was the only one from Hungary. She lived very modestly and became a vegetarian because vegetables were cheaper than meat. She was known among her peers as the "pauper countess." Vilma was successful in her studies and in addition she participated in research with one of her professors. In 1879, the University of Zurich conferred the doctoral degree on her, after which she stayed for another year in Zurich to practice medicine in a hospital. Her professors liked her work and offered her an assistant professorship, but she decided that it was time to go home.

In 1882, she tried to get her medical diploma registered in Hungary. However, women still could not attend higher education in Hungary, and the otherwise progressive minister of religion and public education, Ágoston Trefort, refused to accept her foreign diploma. Moreover, he did not permit her to study for examinations at the Budapest medical school, thereby preventing her from practicing medicine. Instead, he suggested that she earn the qualifications for midwife! She was so determined to function in the health system that she was ready even for this. Ironically, to study midwifery first she had to pass high school matriculation, because although she had completed her high school studies before medical school, women at that time had not yet been allowed to pass matriculation. At least, this obstacle was now removed and eventually she became a midwife. Fortunately, the director of the university clinic who was responsible for teaching midwifery, seeing her medical diploma from Zurich, gave her the certificate for midwifery without requiring her to attend the prescribed courses.

With her new certificate, she opened a private midwifery clinic. Later, she taught at a women's boarding school in Budapest. Eventually, on November 18, 1895, Franz Joseph I, Emperor of Austria and King of Hungary, signed the decree that allowed women to receive higher education in the humanities, medicine, and pharmacology. For Vilma, this still did not mean that her medical diploma from Zurich would be accepted—it was not. But she did not mind, and at the age of forty-eight, she passed all the necessary exams in Budapest. Finally, in 1897, she received her second medical degree, this time in Hungary, almost twenty years after having received the first one in Zurich.

In the meanwhile, she divorced her husband and married Vince Wartha, a renowned chemistry professor; they had one daughter. Wartha was a member of the Hungarian Academy of Sciences and for almost thirty years served as the rector of the Budapest Technical University. One of his major achievements was the understanding of the composition of an old glaze, called "eosin," that made the Zsolnay ceramic factory in Pécs world-famous.

Vilma Hugonnai worked as a medical doctor for the rest of her life, mostly treating women and poor people. She was also engaged in many other activities, including women's issues. She taught hygiene in girls' schools, wrote popular books about the care of sick people and childcare, about the women's movement, and about the employment of women. Throughout her life she advocated the importance of women's participation in the health professions. She helped set up women's high schools and took a stand for equal rights of men and women. In 1914, as World War I broke out, when she was already sixty-seven years old, she completed a course for military physicians, and organized services involving eighty-four women physicians and numerous nurses. She died in 1922 at the age of seventy-five, and has become a symbol of independent intellectual women in her country.

FRANCES OLDHAM KELSEY

Pharmacologist

Frances O. Kelsey in 2000, at the FDA in Rockville, Maryland. (photo by M. Hargittai)

"'Heroine' of FDA [Food and Drug Administration] Keeps Bad Drug Off of Market" starts a first-page story of the Sunday, July 15, 1962, *Washington Post*,[1] just as did many other major newspapers with similar headlines in the United States. The horror stories about thalidomide—in Europe called Contergan—filled newspapers all over the world in the early 1960s. The drug was developed in the 1950s by Chemie Grünenthal Company in Germany as a sedative and sleeping pill, and for relieving the morning sickness of expecting women in the first trimester of their pregnancy. Children whose mothers took this drug during pregnancy were born with terrible birth defects, mostly deformed or missing limbs. It is estimated that about 10,000 children were affected by this drug in Europe, Canada, and other parts of the world.

In contrast, very few cases happened in the United States, which is credited to a conscientious and dedicated—or, as some viewed her, stubborn and obstinate—scientist at the Food and Drug Administration. She was Frances Oldham Kelsey.

Frances Oldham was born in 1914 at Cobble Hill, a village on Vancouver Island in British Columbia, Canada. She loved nature and from an early age she knew she would become "some kind of scientist."[2] She finished her high school at the age of fifteen and went to McGill University in Montreal, where she received her BSc degree at the age of nineteen. This was at the time of the Great Depression, and finding jobs was almost impossible. She had a choice of continuing her studies or joining the breadline. She opted for staying at McGill for a master's degree, which she received in 1935, in pharmacology. Then, on the suggestion of one of her professors, she applied to the graduate program of the University of Chicago to continue her studies for a PhD. She wrote to Professor Eugene M. K. Geiling, who had just moved to the newly founded Pharmacology Department at Chicago.

Frances was obviously fond of telling this story. She was happy when an offer came from Professor Geiling, but there was a slight problem. She noticed that the letter started with "Dear Mr. Oldham," and she wondered whether she should write him to explain that Frances with an "e" is a she. Finally, she just accepted the position. She never learned what the professor thought when she appeared in Chicago—nor did she learn whether she would have received the offer had she had an obviously female name.

Dr. Geiling worked closely with the United States Food and Drug Administration (FDA), which in 1937 asked the Geiling group to help them in determining what the problem was with a new drug, called "Elixir Sulfanilamide." Sulfanilamide as a pill had been used for quite some time for treating bacterial infections with excellent results—it was considered something of a wonder drug. But the manufacturer wanted to prepare it in liquid form as well, so that children would take it more easily. At the company, they dissolved the drug in a chemical and without further testing it, put it on the market. This new form of the drug then killed a number of patients. The FDA's request to Geiling was to determine what caused these tragedies. Frances was one of Geiling's associates who carried out the investigation, and they soon found that the solvent was the culprit, diethylene glycol, a toxic material. This study led to the legislation of the Federal Food, Drug, and Cosmetic Act in 1938. This new law required that before a drug is placed on the market, its manufacturers have to prove its safety based on animal experiments, chemical experiments, and clinical studies. Frances received her PhD in the same year.

One of her colleagues at the pharmacology department at Chicago was F. Ellis Kelsey, and Frances and Ellis got married in 1943. After receiving her PhD, she went to the University of Chicago Medical School and got her medical degree in 1950; their two daughters were born during this period. In 1952, Ellis received a job offer from the Sanford School of Medicine at the University of South Dakota, and the family moved to Vermillion. After receiving her medical degree, Frances started to work as an editorial associate for the American Medical Association. She was involved with a large number of papers that were submitted to the journal, an experience that proved to be useful later on. She was also teaching pharmacology at the University of South Dakota.

Science was a constant topic at home for the Kelseys; in fact, Frances and Ellis worked together on several projects and published together extensively. She was also concerned about women getting higher education. During the 1960s, their daughters went to the National Cathedral School in Washington, DC, a girls' school. Knowing that I was interested in women in science, Frances gave me a copy of a talk she delivered at the National Cathedral School on the attractiveness of the medical profession for women.[3]

In 1960, the Kelsey family moved again, this time to Maryland, where Ellis went to work for the National Institutes of Health. Frances again had to look for a job. She remembered that a few years earlier she had received an inquiry from the FDA about whether she would care to join them. She could not at that time, but now it seemed an ideal opportunity, as the agency was close by. In August 1960, she entered the FDA as its new medical reviewing officer. She has been at the agency for about a month when "since I was new, it was decided that I should be given a simple preparation to start on and that's how the thalidomide application was assigned to me."[4]

By this time, Contergan had been used widely in European countries, mostly Germany, often as a nonprescription drug. Because of the European acceptance of the drug, it might have been expected to sail it through the FDA without any problem, but, due to Frances, it did not. Frances and her chemist and pharmacologist colleagues bumped into numerous problems as they were examining the documentation submitted under the brand name Kevadon by the Richardson-Merrell pharmaceutical company. The animal tests were not properly reported, neither were the clinical studies. It caught Frances's eye that the names of the doctors who testified for the drug were familiar; several of them were among those—she remembered—whose articles were regularly rejected by the American Medical Association.

The FDA chemist noticed that it was not indicated which chiral form of the molecule was tested; perhaps it was their racemic mixture? Chirality is a special characteristic of the molecules of many substances. It means "handedness" when there may be two versions of the molecules of the same substance. The two are mirror images of each other but are not superposable, just as our left and right hands. The two forms are called enantiomers. Racemic means that the two forms are present in the drug in equal amounts. It is possible that while one of the enantiomers is a cure for a disease, the other is a poison.

As for thalidomide, it was known that it exists in two enantiomeric forms. It had been supposed that while one of the two provided the effects for which it could be used as a medication, the other form was supposed to be harmless. Subsequently, it turned out that one of the two forms was teratogenic. In fact, the teratogenic effect could not be assigned to only one form, as the forms rapidly interconvert in the organism. But this was learned only much later.

Coming back to Kelsey's investigation, she requested more tests and additional documentation from the company, and there were numerous rounds of communication between them. The company was becoming impatient and there was increasing pressure from them to speed up the process of approval. Then the news came from Europe that there was a chance that cases of peripheral neuritis could be linked to the use of this drug. The company suggested to the FDA that they would indicate this risk on the package as a precaution. But Kelsey was not satisfied, because she had started to worry about another danger. Around this time, there was much discussion about the possibility that the fetus might be affected by drugs that the mother takes during pregnancy. This was an important question, since in Europe this drug had been used for years to treat morning sickness of pregnant women. She was especially worried about long-term use, because the problems with peripheral neuropathy usually only appeared after long-term usage. The company did not have enough case studies from pregnant women, but they argued that had this been a real danger, it would have been reported by European doctors; again they suggested issuing a warning to this effect on the packaging.

As it happened, around this time doctors in Germany had already started noticing the increased number of births with badly deformed legs and arms, but this news was very slow in reaching America. Nonetheless, due to the relatively frequent occurrence of peripheral neuropathy, Kelsey decided not to approve the drug. Eventually, by the end of 1962, the news about the serious birth defects abroad reached the United States, and the Richardson-Merrell Company withdrew the application. By that time, unfortunately, they had distributed "investigative" drug samples to a large number of physicians, but did not require them to keep exact records—they were very sure that the drug was safe. The FDA went out of its way to let the population know that the drug was NOT safe and it should not be used. Consequently, although there were incidents of congenitally malformed babies born, their number was very small, compared to the thousands of poor babies born in other countries due to the teratogenic effect of this drug.

The thalidomide tragedies had an effect on US drug safety legislation, just as did the Elixir Sulfanilamide case twenty years before:

> There are moments even in American politics when heaven and earth seem to align, when the felt pressure for legislation produces quick, consensual action that subsumes and elides persistent disagreements. Thalidomide created one of those moments. . . . thalidomide helped to produce a regulatory regime—the Kefauver-Harris Amendments of 1962 and the Investigational New Drug Regulations of 1963—with stronger regulatory properties than any of the bills previously under discussion in Congress.[5]

Kelsey was put in charge to ascertain that the suggested changes had been implemented. She became head of FDA's Division of New Drugs and later director of the Division of Scientific Investigations. She retired from the FDA in 2005; by then, she was ninety years old.

President John F. Kennedy congratulating Francis O. Kelsey in 1962 at the White House on the occasion of her receiving the Distinguished Federal Civilian Service Award. Senator Hubert H. Humphrey is looking on in the background. (courtesy of John F. Kennedy Presidential Library, Boston)

In 1962, due to the large media coverage, Frances Kelsey suddenly became famous in the United States. Based on a national survey, a Gallup poll put her among the ten most admired women in the world, like such celebrities as Jacqueline Kennedy and Queen Elizabeth II.[6] In that year, she received the President's Award for Distinguished Federal Civilian Service from John F. Kennedy. In 2000, she was inducted into the National Women's Hall of Fame. Exactly fifty years after she started to work on the thalidomide case, the FDA created the Kelsey Award (Dr. Frances O. Kelsey Award for Excellence and Courage in Protecting Public Health) to be given out annually to an FDA employee. The first recipient of the award was the ninety-year-old Frances O. Kelsey.

Over fifty years after the tragedies, the thalidomide story is not over yet. Further studies of the compound have shown that for a variety of ailments the drug is very effective. Already in the mid-1960s, it was found that it treats leprosy well, and later, during the 1990s, it was found that exactly the property of the drug that caused the stunted limbs, namely that it restricts the growth of blood vessels, could be used to treat cancers, such as multiple myeloma. Health officials do their best to make sure that expecting women do not take it—women have to take pregnancy tests before the drug is prescribed to them. But this does not always seem to suffice. Recently, several newspapers reported that in Brazil, where leprosy is a common disease, and thalidomide is widely used, during the past eight years at least a hundred children were born with deformed limbs. This happened in spite of the stern warning printed on every package in which thalidomide is dispensed.[7]

Fifty years ago, when Frances Kelsey and thalidomide crossed paths by chance, she suddenly became famous. This fame has inevitably faded over the years. It was therefore heartwarming to see the commemorative articles at the fiftieth anniversary.[7,8] She very much deserves this for "her dual role in saving thousands of newborns from the perils of the drug thalidomide and in serving as midwife to modern pharmaceutical regulation."[8]

My husband and I visited Frances in 2000 in her FDA office in Rockville, Maryland. She appeared humble, and so modest were her office circumstances that we were moved to ask the other people who worked there whether they knew about her story. They said they did.

OLGA KENNARD

Crystallographer

Olga Kennard in 2000 in Cambridge, UK. (photo by M. Hargittai)

Olga Kennard told me[1]:

> I was 15 years old when we came to England from Hungary; we came in August, 1939, and I did not speak any English. In September, I was taken to the Hove County School for Girls and they wanted to test my English, of which I had very little of. What these ladies did was to give me a story to read and then they wanted to ask me how much I comprehended. I looked at the story and it turned out to be a translation of a Latin fable. Now the ladies never thought that I had read that tale in Latin in Hungary, so it was not very difficult for me to figure out what the contents were and what words to use to explain it. When they found that I understood the story, they decided to put me straight in the matriculation class, for my age group. Somehow I got through the matriculation; although I could answer questions about Shakespeare but I would not have been able to buy anything in a shop because I did not know the right names. After that year we moved and I went to a mixed school in Evesham. I was the only girl in my class. Somehow I took into my head that I was going to Cambridge, and I persuaded the headmaster to let me sit for the right examination—although no one before me had gone to Cambridge from that school. I took the entrance exam and I got in.

Olga Weisz was born in 1924 in Budapest, Hungary. Her father was a banker who had a private bank together with his brother. Her mother did not have a job outside the home; her family had been antique dealers for generations, and she too was knowledgeable in antiques. Olga grew up in a warm and intellectual environment. She went to good schools and had a wonderful life with all kinds of activities. Her father, however, read the signs of the time, the growing anti-Semitism in Hungary and the increasingly brutal anti-Jewish measures, and he was one of the very few people who believed what is so obvious in hindsight, that evil was coming, and thus organized their departure from the country while it was still possible. They left everything behind and went to England at almost the last minute. Several of her relatives perished in the Holocaust.

I wondered what made her interested in chemistry. The reason was her new environment in England. She found that the cultural differences on the art side were enormous, and things looked totally different from the Hungarian and British sides. "Take history, for example, I left Hungary with the 19th century revolution, with Kossuth, and here it was industrial revolution and a completely different set of players. But physics, chemistry, and mathematics are constant and they did not change with countries, so this very permanency of these subjects attracted me." She started in the natural sciences at Newnham College, a girls' college, at Cambridge. At that time, women actually did not get a degree, only a certificate that said that had they been men they would have been eligible for that degree. These women got their degrees about fifty years later, in a belated ceremony. Olga, however, did not stop; in due course, she received an MA and then a DSc, which was a proper degree. She did chemistry, physics, crystallography, and mathematics.

She joined Max Perutz at the Cavendish Laboratory, where they worked on determining the structure of hemoglobin. She stayed there for two years and then moved to another group at the Cavendish. In 1948, she got married and moved to London. She joined the Vision Research Unit of the Medical Research Council, and her job there started with a strange experience. The head of the Unit was an eccentric professor, Hamilton Hartridge. She applied for the job and the professor asked her what she had been doing.

> I said X-ray crystallography. He said, that is fine, you could work on visual purple [rhodopsin]. I asked what the molecular weight was and he said about 4000. I said, I could try. I went home and looked up the molecule and I knew that it would be absolutely impossible to solve. I picked up the phone and called him saying that I can't take the job because I cannot possible determine the structure of this system. He said: "Don't be silly, my girl, you will find something to do, your letter of appointment is in the post." So I went to work there. I tried to solve the structure of Vitamin A, which I could not solve either, but it was the closest.

In 1961, Olga moved to the Chemistry Department of Cambridge University, where she stayed until retirement. She has determined a large number of crystal structures. The studies she is most proud of were on ATP (adenosine triphosphate);

Olga Kennard in 1978. (courtesy of O. Kennard)

several antibiotics, for example Vancomycin; and many of her studies of DNA. These studies included the first determination of a DNA structure with mismatched base pairs.

Olga's first husband was a medical scientist; they divorced in 1962. They had two daughters, and she brought up the girls by herself. Both of them became professionals and are married; Olga has five grandchildren. She got remarried in 1994 to Arnold Burgen, professor of pharmacology, who was director of the National Institute of Medical Research and master of Darwin College in Cambridge. He founded the Academia Europaea (London). They live in Cambridge. Olga was elected a Fellow of the Royal Society in 1987 and was made Officer of the British Empire in 1988.

Olga Kennard's name is forever associated with the establishment of the Cambridge Crystallographic Data Center (CCDC), a nonprofit organization that compiles the Cambridge Structural Database (CSD).[2] As of 2013, CSD has the complete three-dimensional structural data of over half a million organic and hetero-organic molecules originating from X-ray diffraction and neutron diffraction studies. All this started back in the early 1960s.

> I was a very close friend with J. D. Bernal when I was working at the National Institute for Medical Research in London. We started a small project, collecting data on the then known structures because Sage [Bernal's nickname] had an idea that if we could put all this information together, we could get some new insight into how molecules pack together, and what the attractive forces are between them. This was not altogether a new idea because [A. I.] Kitaigorodskii had already published a book on this topic but it had not been done systematically. Bernal and I got a small grant and we had just one person collecting all

> the information on the few hundred structures which were then known. We entered these data into edge punch cards; actually knitting needle cards, where you punch each property and if you want related properties you just pull it all through. Anyway, when we were working on this, the governments of the major countries suddenly realized that information is important and that scientific information was practically the monopoly of the United States and Russia. They were very concerned that this information would not then be available to scientists world-wide.
>
> So they decided to try and set up an international committee and allocate to various countries different areas of science. They appealed to governments and the British government tried to decide what subject they could support. They inquired and then they found our little project; there was this little man sitting in the attics at Birkbeck. They decided that this is fine and this is what they are going to support. Thus, they came to us and Bernal, who was not terribly well by then, asked me to draw up a plan for doing this in a much larger scale. Although I did not really believe that we would get the kind of support we wanted, I drew up the plan and even included in it a building which would house most of the people involved with collecting information and also actual active research workers. The government decided that this is what they wanted to support so we presented this plan in Washington. I presented it rather unexpectedly. And this is how it all started. This was in 1964–65.
>
> Then it gradually grew with the subject; having started with 300 structures now [in 2000] there are two hundred thousand structures. As the computers took off, we latched onto it; crystallographers have been always very keen on using computers of any kind, so we were one of the earliest people to use computers extensively for data collection and data analyses. We developed some of the first programs that are very commonplace now for searching; searching text, searching chemical structures, searching properties. We developed an integrated system, where we provided the data but we also provided tools to use them.

This databank has become well known among people who need structural information. It is a most valuable tool. One example where it can be used is drug design. A particular drug can be compared with compounds that have similar structures, receptors, or other properties. The databank is used by about a thousand universities and other research institutions worldwide. Bernal's idea of putting this databank together and creating new knowledge became a reality. Eventually, Olga realized that they should try to make the databank easily available for researchers. Therefore, they persuaded some governments that information could be treated as a national resource, the same way a big library is. She was able to get money first from the United States and a few other countries, which bought the right to distribute the database locally in return for an annual payment to Cambridge. "Once this started, it was possible to persuade more countries; so finally we had more than 20 countries to support us with different amounts. Thus, part of the cost was underwritten by these national subventions. The positive effect was that in each country all the information was free to use for scientists. This was especially important for young scientists who could not have afforded to subscribe. This was for universities; commercial companies, such as pharmaceutical companies had to lease it for a certain amount."

Regarding women's position in science, Olga remarked that crystallography is probably a lucky field for women because it seems to be neutral in this respect. Indeed, crystallography has a great tradition of women scientists in leading positions—just think of Kathleen Lonsdale, Dorothy Hodgkin, Caroline MacGillavry, and Isabella Karle. Olga has never felt discriminated against. "If you go to a meeting, you always find a woman chair, one or more invited women lecturers and many women in the audience as well. Probably it is a little more difficult to get into the Royal Society but that is just one of those things."

REIKO KURODA

Chemist

Reiko Kuroda in 2000 in Stockholm. (photo by M. Hargittai)

When Reiko Kuroda received her PhD degree in chemistry from the University of Tokyo in 1975, the job market in Japan was not an open system. Usually, the professors found jobs for their graduates—who were almost always men. Her professor told her that the best thing for a woman was to get married and even suggested a possibility.[1] Reiko declined; instead, she took a postdoctoral position in England. Many years later, she became the first woman full professor in natural sciences at the prestigious University of Tokyo.

Reiko Kuroda was born in 1947 in Akita Prefecture and brought up in Sendai, the capital of Miyagi Prefecture in the northern part of Honshu, the main island of Japan. Her father was a professor of Japanese literature and her mother a homemaker. In Reiko's words:[2] "This was very normal in those days. She is very gifted, she writes well, so I think if she were born at least twenty years later, she could have been a very successful career woman, but she stayed at home and cooked and looked after us." Already from a very young age, Reiko was interested in asking the question, why?

Why we can see our face upside down on the concave side of the spoon, but it is the other way around on the convex side, and similar questions. Her father could not answer them, but he encouraged her to be curious. Their house was full of books, but not science books; they were all on Japanese classic literature, many written in quite different Japanese from the modern language. She struggled through a few of them during her high-school days.

In 1975, she started working in London with Stephen Mason at the Department of Chemistry of King's College. Mason was known for the book on the history of science which he produced at the age of twenty-seven (!); the book was translated into many languages. Reiko had known of him because the topic of her PhD thesis was the determination of the absolute configuration of metal complexes by X-ray crystallography and crystalline-state circular dichroism (CD) spectroscopy, in which Mason was a world-renowned expert. He needed a coworker who could determine the absolute configuration of his metal complexes and was happy to have Reiko. She did a very good job and solved the problem he gave her within the prescribed one year. She got an extension and eventually stayed at King's College for six years.

In 1981, she was promoted to research fellow and honorary lecturer. When she arrived in England, her English was not very good and she decided to learn it well. She watched television, listened to her colleagues, and even attended an elocution course. She learned it so well that when the head of the department, the Nobel laureate Maurice Wilkins, took a long leave, he asked her to teach his course on macromolecular assemblies.

Reiko's interest turned gradually toward the molecular basis of biology. She learned new techniques of molecular biology. She became especially fascinated by DNA and chirality (for explanation of chirality, see p. 112). Reiko commented: "Left-right handedness was not yet considered to be crucial to our life when I was studying the absolute configuration of metal complexes in Japan, but now I find this extremely fascinating." This research topic has stayed with her ever since.

Chirality is characteristic of certain molecules building up living organisms. For example, all naturally occurring amino acids are chiral except the smallest one, glycine. All amino acids that constitute proteins in all living organisms are always left-handed. On the other hand, the nucleic acids in all living organisms use right-handed sugars. This chiral selectivity of biologically important molecules is a central puzzle for researchers involved in understanding the origin and nature of life.

During the first years of her stay in London, Reiko thought that soon she would go back to Japan. Somebody had warned her that if she stayed away too long, she would become too westernized and then she would not be accepted by Japanese society. There were a few job possibilities back home, but nothing that would involve research; it was mostly teaching science and English. She loved research and she did not explore these suggestions. She liked her life in London; she was pursuing interesting research, she had good friends, and she did not think there would be any chance to get a professorial position in Japan corresponding to her qualifications. Therefore, she applied for a permanent position in England, and in 1985 she became senior staff scientist at the Institute of Cancer Research in Sutton, Greater London.

However, soon after, she learned of a position opening at the University of Tokyo, and someone suggested she apply.

> I knew that there were internal candidates (assistant professors) and before, they always preferred promoting an internal candidate. This time there were five such candidates waiting for promotion for three available positions. So someone told me, "why don't you apply; if nothing else, just let them know that there are some people like you working abroad, doing well." So I sent in my application. Then, I heard that I was shortlisted. I did not know what to do. Later, someone told me on the phone: "In Japan people never turn down a job offer, if you are offered you really have to be very grateful to get the position. This would be the first time the department offers a job to a woman. If you turn it down it will be very bad on women." So I said, I will think about it. Five minutes later I received an official telephone call from Japan—and I said: Yes, I am happy to accept it!

Thus, she returned to Japan and became the first woman associate professor in natural sciences at the Faculty of Arts and Sciences of the University of Tokyo. Reiko thinks that this probably happened because of a lucky coincidence of several factors. First of all, she fitted the job description, and her experience abroad may have been a considerable plus. She also felt that she had to work hard to prove that offering the position to a woman was not a mistake. After she accepted the job and moved back to Japan, a friend of hers predicted that she would never be promoted to full professor. The friend was mistaken. In 1992, she became the first female full professor in natural sciences at the University of Tokyo. In this connection, Reiko made an interesting comment: "Now they may feel that they don't want to have many women. You know why? It is because their share of administrative work increases as women are busy outside universities. As a woman, I am on many government committees. There is a rule that there must be at least 10 percent women on all committees. As there are few women science professors, I am often approached. It means that I have to go to these committees often and that takes a lot of time."

Earlier, some people thought that there might be a problem with fitting in after spending so much time in the West, but there was no problem at all. Her colleagues sometimes invited her when they went out for a drink in the evening. However, as she was so busy with setting up her lab, doing experiments, keeping up with the scientific literature, she was not always available to join them.

In 1999, she received a large five-year grant from the Science and Technology Agency of Japan as the first woman project director. She had to establish a new laboratory, hire new people, and plan the whole project. She called her project Chiromorphology—she coined it from the words "chirality" and "morphology" to express chiral structures. They are ubiquitous throughout nature, whether macroscopic or microscopic, in both the animate and inanimate worlds. She tried to understand the link between the microscopic and macroscopic domains through chirality, from molecules to crystals and from genes to living organisms.

One of her research topics is related to the chirality aspects of developmental biology. For example, snail coiling is determined by the position of a single gene. She has demonstrated that the gene dictates the relative molecular orientation at the eight-cell embryonic stage. She created healthy mirror-image animals by simply changing the chirality of the embryos through physical manipulation, thereby inverting the gene expression site of a particular set of genes.

L. stagnalis snails that appear in both left- and right-handed versions. (courtesy of R. Kuroda)

I have seen Reiko at international conferences dealing with difficult situations, for example, when the previous famous speaker had encroached on her time. She reacted with humor and dignity that earned respect from the audience. She believes that she used to be very shy.

> Instead of saying something I just kept quiet and complained in the background. But eventually I decided to change because I realized, it is better to speak up with a smile in commenting on something that I felt was unfair. And, in fact, they respected me for that. Also, you have to learn about different cultures. In England, people may say something sharp; but eventually I realized they don't mean it, it is just their culture. When I told them with a smile that they were rude, they said that they were sorry and that they did not mean it. You have to be careful that you do not hurt others.

When she reached the retirement age at her school, she moved to the Graduate School of Arts and Sciences at the University of Tokyo, where she continues her studies of chirality. During her career she has received many awards and recognitions at home and abroad. She received the Saruhashi Prize, the Nissan Science Prize, the Yamazaki-Teiichi Prize, and the Commendation for Science and Technology by the Japanese Minister of Education, Culture, Sports, Science and Technology, and she is a member of the Science Council of Japan (2008). Among her international recognitions, she was elected a foreign member of the Royal Swedish Academy of Sciences (2009) and received the 2013 L'Oréal-UNESCO Award for Women in Science. She has been a governor of the Cambridge Crystallographic Data Centre since 2006 and vice president of ICSU (International Council for Science) since 2008. In 2013, she was appointed to the Scientific Advisory Board of the UN Secretary General as one of its twenty-six members. She has served on various national and university committees and appears on television and in the newspapers. She has become a public personality in Japan. She used to act as master of ceremonies of the prestigious Japan Prize ceremony for many years. She has poise, and with her excellent English she is a perfect choice for that role.

The situation of women in science has changed considerably in Japan over the past two decades. When Reiko was young, she had to make a choice, either science or family. At that time, the young men who wanted to marry her either said that she had to stay at home or that "'All right, you can work if you want to but I am not going to do any house-chores.' That was their attitude. And I thought maybe it would be nice to go abroad anyway." Today, with young couples, it is already happening that they both work and the men help in the household. They are learning from western societies. Even if the woman has a child, if her parents are ready to help, she may continue with her job. This is a big change. Therefore, if a young woman asked her for advice, she would tell her:

> Oh, you can have both now; you can have a career and a family. It is easier now. Parents are full of energy in their fifties nowadays. You can ask them to help with the children from time to time. Preparing food is much easier with half-processed food available in supermarkets. Positions also open up for women. Women who have a baby are entitled to have a couple of months leave. I think that today you can keep your job, but you have to have a very understanding husband. One of my coworkers had a baby and she continued working. The baby is already three years old. Perhaps finding a job after you finished your education is still harder for a woman than for a man. But being a scientist is a very special profession and in this respect it is much better.

Over ten years have passed since our conversation. Recently, I asked Reiko how the situation is now. There have been further changes concerning women in science and in the society in general.

> The Japanese government is keen to achieve a gender-equal society by adopting an affirmative action, designated 2020/30. The campaign means that by 2020, 30% of women will participate in decision-making processes. We now have 4.6% and 9.4% full and associate professors in natural sciences, respectively, in all

> universities in Japan [2011]. The numbers are much higher for the humanities, 20.8% and 35.7%, respectively. Only 13.8% of scientists are women [in 2011], a slow increase from 7.9% in 1992. One target achieved is the number of women in national advisory council and committees, which is now more than 30%. In the age of information technology, women do not need to break off their research to have a family, and, for example, they can make use of Skype and Internet access to university library resources. Becoming leaders is still not easy but the situation is similar in other parts of globe.[3]

When in an earlier conversation we asked Reiko if she could have one wish what it would be, she hesitated for a long time before answering. First, she wanted to be sure that the wish could be anything, however unrealistic. Finally, she said that it would be nice to have a family *and* continue her work as well.[4] On my follow-up questions, years later, she said: "I try not to regret but I think it is very nice to have a family; someone to care and being cared for." With all her busy life, Reiko likes house chores; she likes cooking and cleaning the house. She cooks every day; she loves gardening, so she thinks that if she had had a family, she would have been very good at it. "Children would also be a treasure. But I decided not to regret it because by regretting it you do not get anything. Rather, I am grateful for what I am now."

NICOLE M. LE DOUARIN

Developmental biologist

Nicole M. Le Douarin in 2000 in Paris. (photo by M. Hargittai)

The citation of the most prestigious Kyoto Prize awarded to Nicole Marthe Le Douarin in 1986 stated[1]:

> The process of animal development, which involves dramatic changes starting from the simple form of an egg and growing to an elaborate organization, has attracted the attention of biologists for many years. The total understanding of this process, however, has been insurmountably difficult and constitutes one of the most important themes for biology today. One of the difficulties is to find out the very basic facts about how each cell, the basic unit of an organism, behaves in the process of development.
>
> Prof. Le Douarin invented a new technology to produce quail-chick chimeras from chicken and quail embryos, and devised the quail-chick cell marker system. . . . Prof. Le Douarin has established a new approach to the study of development at the cellular level, and she has made a significant contribution to the

> establishment of a new research technology, called embryo manipulation. In so doing, she has opened a new era in developmental research.

The road of the little girl from growing up in rural France in the 1930s to her success in science, membership in the French Academy and international learned societies, and appointment to the National Order of Merit (Legion of Honour) is worth following.

Nicole Marthe Le Douarin was born in 1930 in a small village in Brittany, in westernmost France. Her father was a businessman and her mother was a teacher in a local school, which Nicole attended. There were few pupils in the classroom, and the children ages six to fourteen were together, since most of those children whose parents could afford the tuition went to religious schools. When she was eleven years old, her parents sent her to a boarding school in Nantes. "The adjustment was very hard because up till then the school was something like the prolongation of my home; I was so protected there, there was no competition; with its liberty it was ideal. The boarding school was very different but I got used to it."[2] This was all during the war. After the war her parents moved to another town, Lorient, and she finished her high-school studies there, graduating in 1949.

Then she went to Paris to study at the famous Sorbonne. At that time she was not yet sure what she would like to do and started in literature. After the first year, she changed her mind and moved to biology. She graduated in the natural sciences in 1954. She then continued to earn a certificate for teaching, and she taught for a few years, just long enough for her to realize that she did not want to teach all her life. She was already twenty-nine years old when she decided to go into research. However, she continued teaching so that she could offer her services in a research lab without compensation. She started to work in the laboratory of Étienne Wolff (1904–1996), a famous embryologist, professor of the Collège de France. For the first two years, she worked there on her thesis. It was no easy life, and she already had two children. Finally, she left the teaching job and became a full time researcher at the CNRS laboratory of Professor Wolff. In 1964, she defended her PhD thesis.

Her first position as a lecturer was at Clermont-Ferrand University in Clermont-Ferrand, about 250 miles south of Paris. She spent two days a week there, commuting from Paris—not an easy time of her life. It seemed that she and her husband got a break when they both landed a job in Nantes, but things did not start well. The dean of the university did not like husbands and wives working at the same place, both as professors.

> They would have taken me as assistant professor but not as a professor; but I did not want to accept that. My earlier professor, Étienne Wolff came to my rescue: he came to Nantes and talked to the head of the school that I have to get the professorial position because I am qualified. He had big authority and they gave me the position—but they did not want to make my life easy; I got a very large teaching load and I did not get any lab space; so I had to use benches in my husband's lab.

> After about five years, Professor Wolff retired and they were looking for a man to replace him but they did not find anybody worthy. Then they asked a well-known American professor for advice and he told them: "there is a woman in Nantes who is very well qualified and she already worked with Professor Wolff earlier." So I got the job—but only as the head of the Institute of Embryology of the CNRS and not the professorship at the College de France. Professor Wolff told me: "Nicole, you are one of my best students but I cannot propose you to be my successor as professor of the College de France because you are a woman and there has never been a woman in the College de France, and I am not going to be the first to propose a woman." So he proposed a man, who was not very good. I finally became professor of the College de France in 1989.

From very early on she liked embryology. Her thesis work was involved with understanding how the liver develops. She carried out her experiments in chick embryos, which she found easy, because all the embryonic development happens in the egg, so if she opened eggs at different times of their development, she could follow the growth of the embryos with her eyes and she could easily manipulate them.

There was a geneticist in France who was very much interested in developing hybrids of the same species. He experimented with quails. Quails produce one egg each day, so he had many quail eggs and suggested she work with quails. The quails are much smaller at birth than chicks, but their embryos are about the same size, even though quail eggs are smaller than chicken eggs. During her experiments, she noticed that the structure of the cell nucleus in the quail egg is different from how it looks in the chick cells. There is a mass of tightly packed DNA in the center of the quail cell, and she found this to be characteristic of quail cells at all stages of their development. This gave her the idea that she would be able to distinguish quail cells from chick cells if she used a special dye to stain densely packed regions of the DNA. She stained the quail cells, removed the corresponding part of the chick embryo, and replaced it with the stained quail cells. Thus, she constructed chick-quail chimeras,[a] in which the cells coming from the quail were always recognizable. With this method she uncovered details of the development of the liver that she could not have learned if she had limited her studies to one species.

Soon Nicole had another idea. She realized that with this combination technique she could study cell migration. In the development of the nervous system, cell migrations are extremely important, and Nicole became a world authority in this research. The particular topic is called the "neural crest," and Nicole explains her relevant research on YouTube.[3] She wrote a book on the topic as well.[4] These studies led to information about what happens during embryogenesis, and the results also have evolutionary interest.

Turning to Nicole's private life: she met her first husband when she was an undergraduate student at the Sorbonne, and they married early. They had two daughters,

[a] Chimeras are organisms created by mixing genetically different embryos; the name comes from Greek mythology.

born when Nicole was in her early twenties. "When they were small, I was teaching in high school and could spend enough time with them; later, when I became involved with research, I had less time for the family. . . . The children never complained but now that they are grown-ups, I have the feeling that they missed me. Recently, we talked about this, but they are very kind and very discreet. One of them is a psychiatrist; she has no children. The other is a gynecologist surgeon and has four children; she has help with the children."

Nicole and her first husband eventually divorced. She married again, and her present husband is a geneticist; his field is population genetics. Nicole is a member of the French Academy, but neither of her husbands is. I asked Nicole if this ever presented a problem.

> How a husband handles the successes of his wife is an enormous problem. Of course, men are obliged to accept certain things, but there always are men who do not accept this. You can be successful academically irrespective of whether you are a member of the academy or not. To have a self esteem does not depend on that. If you are confident in yourself, you are going to tolerate that your wife is a member of the academy, or a minister or something else, or perhaps you are even proud of her. With my first husband, we met in school and we grew together but we had different personalities and that was not so obvious when we were young but became more obvious when we became adults. I met my second husband after I was already well established.
>
> When I started my life, I did not have the goal of making a career. But I was very ambitious, I wanted to learn, I wanted to do research—but to become a professor did not occur to me. When I got married I thought that he would become the professor and I would be working at the CNRS. It turned out differently. When I started to do research, I realized that that was what I wanted to do. The success came to me naturally; I did not do anything for that. When they did not want to give me the position of professor at the University of Nantes because my husband was a professor there, I was not surprised. They really had a negative attitude toward women without any consideration of their performance, their work. It was just a matter of principle and I hated that.

Nicole was recently elected permanent secretary of the French Academy of Sciences. I was wondering if this happened because she is a woman. She agreed that this may have been a factor, because nowadays they want to have women in positions and there are not that many women from among they can choose.

> Yes, this is possible, so that is an advantage now. I don't mind that; I am a realist, I accept facts. This is a sign of the evolution of the society. And they could not have chosen me if I were not good; there were others that I had to compete with. This is an honorific position; many people would love to have that position; for me it just came. My responsibilities are to organize the Academy, to control the budget, other administrative duties, and the organization of the Academy.

I had to retire from my job this year because I am just 70 years old [in 2000]. It is very difficult to accept in my mind that there will not be that active research that I have had for decades; so for that reason it might be good to have this new job that will keep me active. I still have a small group where I will keep up with research. And I am also enthusiastic about the academy job. I also want to write books—I love doing that!

RITA LEVI-MONTALCINI

Developmental biologist

Rita Levi-Montalcini in 2000 in her home in Rome. (photo by M. Hargittai).

When I asked Rita Levi-Montalcini, the corecipient of the 1986 Nobel Prize in Physiology or Medicine, what made her interested in science, this is what she answered: "I don't believe that I have ever been a scientist. My twin sister is an artist, a painter, one of the best in Italy. My brother was an architect, and an excellent sculptor. I believe that my approach to science was from the point of view of the beauty of the nervous system and not just plainly because I was interested. Still now, I don't believe I am a scientist; I approach science more from an artistic point of view than from a scientific one."[1]

Rita Levi-Montalcini (1909–2012) was born in Turin, Italy. Her father was an electrical engineer and her mother a talented painter. She had one brother and two sisters. As she writes in her autobiography, their parents "instilled in us their high appreciation of intellectual pursuit."[2] It was an old-fashioned household, in which the head of the family made all the decisions. Her father strongly opposed professional careers for women. He had two married sisters who were professionals, and their marriages were not happy. "He decided that since it is so difficult to find the balance between family, children, spouse, and a profession, we [Rita and her two sisters] should not get a higher education. I was furious, so I decided that I would never marry, I did not want to have children, but I wanted to study."[3] Interestingly, only Rita's elder sister followed their father's advice. Rita's twin sister, Paola, had artistic talent and became a successful painter. Rita waited a few years, and when she was twenty-one years old, she told her father: "I don't care about being a wife or a mother, I want to study."[3] He still did not agree with her, but let her enroll at the Medical University. Their father died in 1932, just one year after she started her university studies. "Our relationship is best described in the dedication I wrote in my book, *In Praise of Imperfection.*[4] It reads: 'To Paola and the memory of our father whom she adored while he lived and whom I loved and worshiped after his death.'"[3]

At the Medical University, Rita was one of seven girls among 300 students. Two of her colleagues were Salvador Luria and Renato Dulbecco, who also became Nobel laureates. All three of them were students of Professor Giuseppe Levi (no relation), a well-known histologist. In his laboratory, the development of the nervous system fascinated her, and it became her lifelong interest. She learned a very useful technique to facilitate the examination of the nerves under the microscope, the so called silver-staining technique.

Rita graduated in 1936 and started the three-year specialization course in neurology and psychiatry—but then history interfered. In 1938, Mussolini's racial laws forced the Jews to leave the university. In early 1939, she went for a few months to Brussels to the neurological institute, but in December, just before the Germans invaded Belgium, she went back to Italy. She set up a small home laboratory in her bedroom and started experiments in neuroembryology with chick embryos. In this, she was influenced by an article she had read a few years before by Viktor Hamburger (1900–2001) at Washington University in St. Louis, Missouri. Giuseppe Levi soon joined her in this home experimentation. In 1941, due to the heavy bombing of Turin, Rita's family moved to the country, to Piedmont, where she continued her experiments, for which only rudimentary instruments were needed. She went to the farmers to buy fertilized chicken eggs, telling them that they were for her babies because these eggs are more nutritious.[5]

In fall 1943, after the German invasion of Italy, Rita and her family moved to Florence and went into hiding. From late 1944, she worked at the Anglo-American headquarters as a medical doctor. Finally, in May 1945, the family returned home and she resumed her position at the university as an assistant to Professor Levi, who became chair of the Anatomy Department. At the end of the war, Rita and Professor Levi published their joint studies in a Belgian periodical. As it happens, Viktor Hamburger noticed their paper and became interested in it because it contradicted his hypotheses. He invited Levi-Montalcini to join him in St. Louis and work on neuroembryology. She arrived in 1947 for a few months, and stayed for more than two decades. First, she proved that her findings were correct. Soon after, Hamburger showed her an article about a large outgrowth of fibers due to a malignant mouse tumor implanted into chick embryos. Rita repeated this experiment using her silver-staining technique and found that, indeed, nerve fibers appeared everywhere in the embryo's organs. She hypothesized that the tumor released some kind of an agent that induced this fiber growth.

Rita Levi-Montalcini with Viktor Hamburger. (courtesy of the late R. Levi-Montalcini)

She presented her finding at the New York Academy of Sciences in 1951, but she did not stir any serious interest in her results. She wanted to understand what happened, and she decided to try the tissue culture technique that she had learned from Giuseppe Levi. She knew that an old friend of hers, Hertha Meyer, a former assistant of the famous German organic chemist Emil Fischer, had set up a tissue culture laboratory at the Institute of Biophysics in Rio de Janeiro, so she went there. Thanks to this technique, she could prove that her hypothesis was right: the tumors released a growth factor in the culture medium. This is what later became known as the nerve growth factor (NGF). At the time of her return to Hamburger's laboratory in St. Louis, Stanley Cohen joined them, and a period of exceptionally fruitful collaboration followed.

NGF is a protein molecule that induces rapid nerve fiber growth. Rita showed by beautiful photomicrographs that fibers budded out from embryo nerve cells radially, like flower petals or the rays around the sun, if tumor cells were present in the culture. Cohen discovered that the male mouse salivary gland and snake venom are rich in NGF and determined the structure of this protein. Later on, he discovered the epidermal growth factor (EGF). "Though it took decades for their concepts to be fully accepted, Levi-Montalcini's and Cohen's discoveries revealed how cells talk and listen to each other, and paved the way for many other growth factors to be discovered."[6]

Rita Levi-Montalcini and Stanley Cohen shared the 1986 Nobel Prize in Physiology or Medicine for discovering the growth factors. According to the presenter at the Nobel Prize award ceremony, "Rita Levi-Montalcini showed, in a series of brilliantly performed studies, that NGF is not only necessary for the survival of certain nerves but also regulates the directional growth of the nerve fibers. The nerve cells die when NGF is blocked by antibodies. . . . Injections of NGF into the brain cause the outgrowth of specific nerve fibers. This neurotropic effect on NGF offers an explanation of how nerve fibers find their way through the tangle of nerves in the brain."[7]

Rita returned to Italy during the 1960s, first only half-time and then permanently. She became director of the recently founded Institute of Cell Biology of the Italian National Research Council in Rome. Even after her official retirement, she kept going to the Institute and participating in its research program. At the time of my visit, in 2000, the research topic was still NGF. She mentioned that its identification about half a century earlier was just the beginning. This time, her associates were studying the role of this molecule in homeostatic processes, during which the body's internal environment does not change—it is in internal equilibrium. They were investigating the peripheral and central nervous systems as well as the immune and endocrine systems. The new results showed promise that this molecule might be used to treat psychiatric disorders, as, for example, Alzheimer's disease, dementia, schizophrenia, depression, and autism. Similarly, it has been shown to speed up wound healing, and it might be used to treat skin ulcers. Recently, scientists at the University of Pavia found that young people who have recently fallen in love show a much higher level of NGF than those who are not in love or those who have been for a long time. The high level does not remain longer than about a year.[8]

In her book *The Saga of the Nerve Growth Factor,* [9] Rita describes the story of the discovery. According to her "the history of Nerve Growth Factor is more like a detective story than a scientific enterprise." We can only agree with Professor Ottoson from the Karolinska Institute, who writes,[10] "This is certainly true but what is equally true is that the history of Nerve Growth Factor conveys the inspiring testimony of a compassionate devotion to science."

Levi-Montalcini's artistic talent is well expressed in all her books, of which so far only two have appeared in English.[4,9] As a motto for *In Praise of Imperfections*, she quoted two lines from Yeats' poem, "The Choice":

> The intellect of man is forced to choose,
> Perfection of life or of the work.

I asked her whether she felt that for women, it was still an either/or situation? She thought that today this is different; her many women coworkers had all been married and had children. But she added that most of them were divorced. In her opinion: "even now, it is easier to make one choice and not both. I am very very happy about the choice I made, I have never regretted it; I believe I made the best choice I could have made."[11]

She felt that her best book was *Cantico di una vita* (*The Hymn of Life*). In this she edited about 200 of the 1,500 letters that she wrote to her mother during the time of the discovery of NGF.

Rita Levi-Montalcini passed away on December 30, 2012, at the age of 103. When she was 100, the British daily *The Independent* dedicated an article to her. It mentioned that she had apparently been taking NGF for years in the form of eye drops. Whether this helped or not, we can't tell—but according to the article, NGF helps neurons in the brain to survive. At the time of her centennial, she stated that her mental capacity was better than decades before.

Rita was beautiful and was always elegant, and this did not change with age. She was petite, thin, but strong and purposeful. She was a senator for life in Italy, and she took her assignments seriously. In 2006, when she was ninety-seven years old, she voted against decreasing funding for science and thus saved support for science. A *Nature* article wrote, "It was Levi-Montalcini versus Prodi [Romano Prodi, prime minister of Italy at the time]—and Levi-Montalcini won."[12]

Her poetic nature comes to life in her book when she talks about NGF and the Nobel ceremonies in Stockholm in 1986[13]:

> It was in the anticipatory, pre-Carnival atmosphere of Rio de Janeiro that in 1952 NGF lifted its mask to reveal its miraculous ability to cause the growth, in the space of a few hours, of dense auras of nervous fibers. Thus began its saga.
>
> On Christmas Eve 1986, NGF appeared in public under large floodlights, amid the splendor of a vast hall adorned for celebration, in the presence of royals of Sweden, of princes, of ladies in rich and gala dresses, and gentlemen in tuxedos. Wrapped in a black mantle, he bowed before the king and, for a moment, lowered the veil covering his face. We recognized each other in a matter of seconds when I saw him looking for me among the applauding crowd. He then replaced his veil and disappeared as suddenly as he had appeared. . . . Will we see each other again? Or was that instant the fulfillment of my desire of many years to meet him, and I have henceforth lost trace of him forever?

JENNIFER L. MCKIMM-BRESCHKIN

Virologist

Jennifer McKimm-Breschkin in 2000 in her laboratory in Melbourne. (photo by M. Hargittai)

Influenza is a killer all around the world when an epidemic breaks out and millions lose the ability to work or lead a normal life for days. The influenza virus continually mutates, and this is why it is so difficult to find a cure. Although there are vaccines against influenza, if the virus changes just a little it is enough for a vaccine not to work on a new strain. This is what makes research into finding effective drugs so demanding in this field.

Jennifer McKimm-Breschkin is an Australian researcher who has spent most of her career on trying to figure out how to fight these viruses. She was born Jennifer McKimm in 1953 in Melbourne. She was interested in science from an early age, but first she wanted to become a physician. A strange encounter, however, made her change her mind. After she applied to the medical school, a faculty member told her, "Medicine is a bitch of a career for a woman."[1] She thought that if that was the attitude of the lecturers, they were not going to be very sympathetic to female students. Thus, she decided to study science with the goal of becoming a teacher, because at that time if a girl majored in science, it meant a teaching career. However, after her bachelor's degree she decided to continue for an honors degree that included a one-year research project. Here she realized how much she loved working in the laboratory and opted to become a researcher. This also meant that she had to pay back the funding she had received for her university studies from the Victoria government education department.

She graduated in 1974 from Monash University in Melbourne, followed by a Fulbright scholarship that she spent as a graduate student at the Hershey Medical Center of Pennsylvania State University in the United States, leading to her PhD in virology in 1978. At Hersey, she met her future husband, Alan Breschkin, an American postdoc with a PhD from Vanderbilt University. They spent seven days a week in the laboratory together until they completed their stints at Penn State. They married and left together for Australia.

She spent the next few years at Melbourne University, at the Walter and Eliza Hall Institute and the health department of the Commonwealth government, working on virology and immunology. She and her husband had two children during this time. For a while she worked only part-time, but the excellent Australian child-care facilities proved to be a great help. Her husband has always been very supportive, but even so, bringing up the children in a family with two scientists was not easy. Her children often told her that she always came last when she was picking them up after music rehearsals. Fortunately, they understood that science was important for her, and they did not mind that their parents talked a great deal about science at home. In her words: "Twenty years ago, we both worked on the measles virus in Hershey, and we have recently started working on it again. We enjoy working together; this is how I met him, this is how I knew him, and it's natural that we talk about science at home, too. You can't help it because both our lives revolve around it."[2]

In 1987, Jenny joined the Commonwealth Scientific and Industrial Research Organization (CSIRO), Australia's national science agency, with over 6,000 employees and over fifty-five working sites all over the country. It is one of the largest research institutions in the world. Currently, she is a chief research scientist and project leader in virology at the Materials Science and Engineering Division of CSIRO.

She was part of the team that discovered and developed the drug called zanamivir (trade name Relenza), one of the two drugs currently available to fight influenza (the other is oseltamivir—Tamiflu). My husband and I met Jenny in 1999 during our visit in Australia when we visited her laboratory at the Biomolecular Research Institute, a spinoff institute from CSIRO. We learned from Jenny firsthand about an exciting research project that had been going on for almost twenty years by then.

The influenza virus has two types of spikes on its surface; one of them is an enzyme called neuraminidase (NA). The other is a protein, called hemagglutinin (HA), that binds the virus to receptors on the surface of the cell and lets the virus into the cell. Then the virus is reproduced in the cell, and when the new viruses want to get out of the cell, it is the NA spike that cuts off the receptor on the cell and lets the new viruses spread. Viruses very easily mutate, and even if we have acquired immunity to a particular strain of a virus, that immunity would not be there anymore for the new mutant. It appears that these "spikes" on the surface of the virus change over time.

The initiator of the research project at CSIRO was Peter Colman. In the late 1970s, he and his colleagues determined the molecular structure of NA by X-ray crystallography. Their goal was to understand how an antibody binds to the surface of NA. Eventually, they found that although there were many changes on the surface of the virus due to mutations, there was one particular place, a little pocket, that never changed. It was exactly this pocket where the virus bound to the cell to cut off the receptors. Obviously, it had to be specific to recognize the receptor; this is why it could not afford to change. This was the active site of the NA enzyme. They determined the structure of this active site with its receptor-like molecule (called substrate) bound, and based on that information, they set about devising a drug that was similar to the substrate and could interact with the conserved active site of the pocket. This drug would work as a plug, occupying the active site and thus not letting the virus bind to the cell receptors. Jenny inoculated thousands of embryonated chicken eggs to grow the influenza virus in tremendous numbers and then do the complicated biochemistry to purify the neuraminidase from them.

They succeeded in developing a very powerful drug, which is not toxic at all because it is based on the structure of the receptor molecules to which the virus binds, and it binds poorly to any neuraminidases in humans. It works through oral inhalation; it is not a pill. It coats the surface of the respiratory tract where the virus multiplies and acts on the virus outside the cell. Importantly, unlike a vaccine, this drug works against every strain of influenza, since the active site is highly conserved. Her virology group had been working for years on trying to determine whether the influenza virus can develop resistance to this drug. They have found it very difficult to generate resistance to Relenza.

This is how far the project had developed by the time of our visit in 1999. Since then, Jenny has remained involved with the study of viruses and has continued working on the influenza virus drug. Their competitor, Tamiflu, has been commercially more successful, probably because it is a pill (capsule) that has to be swallowed rather than something to be inhaled. It seems, however, that Relenza has excelled and inhaling has its advantage. Inhaling delivers high doses directly to the upper respiratory tract, where the virus is multiplying, and thus it is very effective. Furthermore, it has proved quite successful in fighting the viruses even if they are mutated. Jenny and her colleagues are also studying the resistance of H5N1 influenza viruses; this is the infamous bird flu virus, and there are indications some virus strains remain quite sensitive to Relenza.

Jenny's major goal is the study of resistance in influenza viruses and collaborating with others in designing new drugs to eliminate such resistance. In collaboration with Steve Withers at the University of British Columbia, she has recently described new mechanism-based inhibitors of the influenza viruses. It is a long path to get a new drug to market, but the need for alleviating human suffering and reducing the number of fatalities due to influenza is a major incentive.

ANNE MCLAREN

Developmental biologist

Anne McLaren in 2004 in Cambridge. (photo by M. Hargittai)

"While she made major contributions to studies of mouse genetics and development, her immense strength was in distilling scientific information and communicating it to others, and she worked tirelessly to ensure that sound scientific reasoning informed public policy making." Thus Paul Burgoyne wrote of Anne McLaren in her obituary in *Nature Genetics*.[1] She was the most conscientious ambassador of science wherever and whenever there was a need for this, whether in public debates or official hearings on the question of human reproductive technologies.

According to a story from another obituary, by H. M. Blau, some time in 2003 McLaren was invited to a meeting of the Pontifical Academy in the Vatican on stem cell research and its merits. While all the other invitees talked about their science, about the potential medical applications of stem cells, she talked about ethics and politics, and she was so convincing that the laws for in vitro fertilization have since become stricter in Italy. "The image of this tiny intrepid woman, who spoke her mind despite facing an impossible task, will remain with me always."[2]

Anne Laura Dorinthea McLaren (1927–2007) was born as the fourth child of Christabel McNaughten and Henry Duncan McLaren, 2nd Baron Aberconway, a liberal Member of Parliament and successful businessman. Until World War II, they lived in London, near Hyde Park, but at the start of the war the family moved to their large estate in North Wales. At the end of the war, she finished her basic education at a private school in Cambridge. Although many successful scientists develop their interest in science rather early, it was different for McLaren[3]:

> If I hadn't become a scientist I could equally well become a lawyer or a journalist or maybe many other things because I wasn't particularly motivated to become a scientist when I was young. It was just a convenient thing to do. Because I was good at writing as a child it was suggested that I should read English literature at Oxford, but when I looked at the exam papers it was clear to me that I could never get interviewed at the University on English-literature because I hadn't read all the books one would have had to read. On the other hand, although I had done very little science, the biology papers looked quite easy, so I thought I would do that. For the first two years at Oxford I did zoology, physics, and math and at the end of that I decided that zoology was the area that interested me most. So I went on and became a scientist.

During her studies at Oxford, she worked under such illustrious scientists as J. B. S. Haldane and Peter Medawar. She was especially impressed by the work of the geneticist E. B. Ford, and she decided to work in genetics. Another student there, Donald Michie, was also fascinated by this field, and they started doing research together. She received her PhD in zoology in 1952. She and Michie received a grant to work at University College London; they moved there, and married the same year.

Their joint work turned out to be very successful:

> The project that we were funded to do was investigating maternal effects between two inbred strains of mice. . . . We wanted to find out whether the maternal affect was due to something in the egg, perhaps the cytoplasm, or what would now be called imprinted in the genes. Or whether it was due to an affect exerted by the uterus during gestation. And the obvious way to investigate that was to do embryo transfers between the two strains. . . . We had to work out ourselves the technique of embryo transfer because nobody in Europe was doing it. . . . eventually we got the embryo transfers working. And, we discovered that it was a uterine effect because the embryos that were transferred resembled their uterine foster mothers and not their genetic egg mothers.

During these years, their mouse colony had become so large that there was not enough room for them, so they moved to the Royal Veterinary College in 1955. Her most famous work was in collaboration with John D. Biggers. In her description, it sounded deceptively simple: "John Biggers was there doing culture work in chick skeletal primordia. He was in the lab next door; we got to know him very well. So the time came when he cultured the embryos and I transferred them into the uterus and we got mice born. That was the first time that embryos kept outside the body for 24 hours had been successfully reared into adulthood."

This work made the headlines. But the social and ethical implications of this discovery started to sink in only gradually. In 1978, as a logical consequence of this research, the first "test-tube baby" was born. It is not surprising that she was invited to participate in committees and debates where various implications of in vitro fertilization were discussed. The most important of these was the so-called Warnock Committee, which was appointed by the UK government soon after the first test-tube baby was born. It was charged with studying the social, ethical, and legal implications of human-assisted reproduction. The chair of this committee, Dame Mary Warnock, was a well-known philosopher and writer. Anne McLaren was the only member with scientific expertise in the topic. "The Warnock Committee was set up in 1982 and it was a mixed committee with theologians and doctors and lawyers, different sorts of people. I was the only biologist in that particular area. Mary Warnock was a very good chair for the committee and we produced recommendations, advice, to the government on possible legislation, which included recommendation that the government should set up an authority to regulate IVF (in vitro fertilization), both clinically and also human embryo research."

McLaren's role was essential in steering the committee in the right direction in formulating the guidelines that eventually led to the Human Fertilisation and Embryology Act. Soon the Human Fertilisation and Embryology Authority was funded, in which she served as an active member for ten years. In later years, she was equally active in debates about stem cells. She was also cofounder of the Frozen Ark Project to collect and store the DNA of animals that are close to extinction. Mary Warnock wrote about her: "She taught me what a true scientist should be: a combination of vision and caution, of enthusiasm and a strict demand for evidence. Above all, she had patience, not only with the slow progress of scientific proof, but with the ignorance of her pupils."[4]

Besides her increased involvement in advising on scientific issues, the intensity of her research never diminished. As early as 1959, long before her involvement with science information and advising, she moved to Edinburgh to the Institute of Animal Genetics, where she remained for fifteen years. There she did pioneering work on mouse chimeras (see the chapter on Le Douarin). In 1974, she was asked to be director of a new Mammalian Development Unit of the Medical Research Council at University College London. The establishment of this unit manifested the importance of mammalian development research. She retired from that position in 1992 and moved to Cambridge to the Wellcome Trust/Cancer Research UK Gurdon Institute. There, she continued her research to the day she died. Her latest interest was in mammalian primordial germ cells. In 2004, she told me: "I have always enjoyed my work and I have worked on a number of different things because one thing has always led to another. Now, I am working on the imprinting of genes, which happens in germ cells. My primordial germ cell work led on to this imprinting of genes and also to stem cells which can be made from primordial germ cells. But I wouldn't single out any particular bit of my work as more important."

In spite of her aristocratic roots, she was a socialist all her life, and so was Michie. They joined the Communist Party during the Cold War and supported Soviet and Eastern European scientists.[5,6,7] She was a council member of the Pugwash Conferences. The leaders of the movement remembered her: "Anne McLaren embodied the spirit of the Pugwash community and its fundamental credo . . . of the social responsibility of scientists. Anne's tenure on the Pugwash Council was marked by her liveliness of spirit, independence of thought, and warm collegiality."[8] She was a Fellow of the Royal Society (1975) and became the foreign secretary of the Royal Society in 1991, the first time a woman held this position during the 330-year history of the Society. In 1993, she was president of the British Association for the Advancement of Science. She was made Dame Commander of the Order of the British Empire in 1993. In 2001, she was one of the recepients of The L'Oréal-UNESCO Award for Women in Science. In 2002 she received the Japan Prize, jointly with Andrzej Tarkowski of Poland, for their pioneering work on mammalian embryonic development.

Her talent in science communication was not limited to the debates in committees and hearings. She excelled also in communicating science to the general public. Quoting an example (from another of her obituaries), she said: "When the embryo is outside the woman's body, genetics tells us that father and mother have equal rights. When the embryo is inside the body, physiology tells us that the woman's right is paramount."[9]

McLaren was involved with women's rights and all women's causes. Although she felt herself fortunate in that she never experienced discrimination, she knew very well that this problem existed. She was a charter member and president of AWiSE (Association for Women in Science and Engineering in Britain). She did a lot of advising, at home and abroad, on women's rights. She made it a point that in her laboratory there should be equal numbers of men and women working.

McLaren had three children; they were born two years apart. It never occurred to her to stop working. "As a research scientist, the hours are flexible. When the babies were very young, I used to take them into the lab. And one was able to find au pair girls, it was mainly young women from Norway . . . the children liked them." Her children are now grown; all three of them are professionals. When I asked her what was the most difficult in being a mother *and* a scientist, she answered: "Time. Time. Organization of time."

Concerning family, although she and Donald Michie divorced in the late 1950s, they remained good friends. After their very successful work on the environmental effect on embryonic development, Michie left genetics. He worked on cryptography during World War II at Bletchley Park and in spite of his young age, he was one of the leaders of the project. He was a good friend of Alan Turing, the famous cryptographer and computer scientist, creator of the "Turing machine." After the war Michie had become interested in genetics, and that was how they met. But it was codebreaking and artificial intelligence that were his genuine love.

Early in July 2007, Ann McLaren and Donald Michie were traveling from Cambridge to London, on their way to Edinburgh, where Michie was to get an award. In a tragic automobile accident they both died.

In our conversation in 2004, I asked her what she would like to see happen in her field. It was one particular topic that she was undoubtedly most interested in:

> I would like to know what are the factors in the egg cytoplasm that reprogram the somatic nucleus in cloning and equally, what are the factors in either the germs cells themselves or the tissue environment that bring about the epigenetic changes in germ cells that are involved in imprinting. And I would like to know much more about epigenetic changes in general. That is changes which don't involve changes in the base sequence of the DNA but do involve changes in gene expression because these can often be brought about by the environment and the more we learn about epigenetic inheritance, the more we realize that we have to think carefully, perhaps even we have to reconsider our views on the inheritance of acquired characters because, particularly in plants, there is quite a lot of evidence that the environment can influence the epigenetic state of the organism and that this can be inherited from one generation to another . . . and that's a fascinating area which I'm sure will expand in the future.

Finally, her ability to grasp the broader picture manifested itself in her advice to budding women scientists:

> I think if people are interested in science, then it's worth persevering in their career. On the other hand, if they find, as they may well do that research really is not for them, then there are other possibilities. They can go into teaching. I know several young scientists who've decided that school teaching is really what they want to do, and that's fine. Or, they can leave science and go into business or some quite different career, journalism. And it would be excellent to think in the future that many more people in different walks of life in different professions had a scientific background. It would be excellent if much more of our members of Parliament had a scientific background because at present, very few really understand scientific issues. Just because one starts off working in science, there is no reason necessarily to continue all one's life.

CHRISTIANE NÜSSLEIN-VOLHARD

Biologist

Christiane Nüsslein-Volhard in 2001 in her office at the University of Tübingen. (photo by M. Hargittai)

In late 1979, Christiane was working in the laboratory with her colleague Eric F. Wieschaus (1947–), looking at the results of their latest experiments with mutations of fruit fly embryos, when she exclaimed: "Toll!!"—this German word means strange, fascinating, crazy, all at the same time. She found the new mutant both fascinating and strange because it did not have a "front," only "backs." The name stuck, and today this word is used worldwide for the gene that causes such mutations. Nüsslein-Volhard and Wieschaus published their results in *Nature*, and this was the first step in the experiments for which in 1995 they received the Nobel Prize.[1]

Christiane Volhard was born in 1942 in Magdeburg, Germany. Her parents did not have an academic background, but they supported all their four daughters and one son in pursuing their academic interests. In Christiane's case, this meant science. She has pleasant memories from her high-school years. The biology teacher was especially good; interesting subjects came up, such as genetics, evolution, and animal behavior. Christiane decided to continue her studies in biology. She went to Frankfurt University, but there her experience was disappointing because they were teaching what she already knew from high school. She moved to Tübingen University, where they had just started to teach a new biochemistry class. Eventually, she attended classes also in genetics and microbiology, and she enjoyed them.

After receiving her diploma in biochemistry in 1968, she stayed in Tübingen at the Max Planck Institut für Virusforschung (virus research), where she received her PhD in 1973. She still was not sure what she would like to do in the future. People told her that working on bacteria or in molecular biology would mean staying on a beaten track; she should rather find a new field. Developmental biology seemed to be a good possibility, and there were already researchers at Tübingen working in it. Developmental biology describes how from the smallest units, molecules, particular structures are formed based on a pattern, how cells grow and differentiate. She remembers: "With my experience in molecular biology and in particular with DNA replication using genetics as a tool to dissect a complex process, I tried to find a system where I could combine the question on morphogenesis with genetics and use genetics as a means to get at the molecules as the products of genes that you could mutate and identify."[2] She decided that drosophila (common fruit flies) would be a good subject for her studies. After spending a short period of time in Basel and Freiburg, she accepted a job at the European Molecular Biology Laboratory in Heidelberg. The American Eric Wieschaus, whom she had met in Basel, also came to Heidelberg. They investigated the drosophila mutants for three years and were very successful. Drosophila was an ideal subject, because their larvae had fourteen seemingly equal segments, and yet these segments developed into different parts of the body. How do these segments know what they are supposed to develop into? By excruciatingly hard work, they determined which genes were responsible for what mutations—and identified which ones were responsible for the growth of the different parts of the fly. This is the work for which they received the Nobel Prize together with Edward Lewis, who found how the genes that direct the development of the fragments are arranged in the fly's DNA.

Christiane Nüsslein-Volhard and Eric Wieschaus around 1978 at a meeting in Spain. (photo by Judith Kimble, courtesy of C. Nüsslein-Volhard)

In 1985, she became director of the Max Planck Institute of Developmental Biology, the position she still holds. She and her associates continued working on drosophila, but they also started to use zebra fish for their studies. They isolate their genes, identify them to learn which of them have particular functions, and study their properties. They have already identified which genes determine, for example, the skull, the fins, and the scales of the fish.

She had been married for seven years, but had gotten divorced before she moved to Tübingen. "Nüsslein" in her name refers to her former husband's name, which she assumed after they married (dropping her maiden name); that was the name under which she published. After they divorced, she added her maiden name to it. She said, "As a scientist I did not want to get rid of my previous life."[3]

When I asked Christiane whether she has ever experienced discrimination because she is a woman, she answered emphatically[4]:

> Oh, yes, plenty! . . . First we can go back to the time when I grew up. The general problem then was that women simply were not considered to be professionals; professionals enough to have big important jobs. Therefore, they were just often overlooked and no one entrusted them with important things. I have to say that my science was never discriminated against. So I had no problem whatsoever in getting my science recognized. But in the practical aspects, to get jobs, to get money, to get lab space, women have not been treated equally, I think.
>
> At the time when I was a young scientist, men often had family and children and they got the better positions automatically. The professors always said, "but this is a man, he has to support and nourish a family, so this is why he is going to get the job and you will not." This happened repeatedly to me. The same thing with promotions. They often said, you are a woman and there is a man and he deserves it more. Actually, my worst experience happened with my PhD; I collaborated with a man, I did most of the work, I wrote the paper and then he got first authorship because my boss said; he has a family and for him this is important for his career. You are a woman and married, so it does not matter. This was particularly unjust because he gave up science right after his thesis and went to teach where he did not need the publication at all. Whereas, I did suffer in trying to find jobs later because I did not have this publication with my name as a first author. So discrimination started right away at the very beginning.
>
> When I did my diploma thesis in this institute, it was customary that the diploma students got some stipends, but the big boss said, "She does not need it, she is married." This kind of thing is over by now. But the old professors still think that it is totally legitimate to pay men more if they have families to support. This still happens. If there is a single woman who is good and does a perfect job and there is a man who is mediocre but has a wife and family, he will get the promotion.
>
> In the Max Planck Society they created some jobs specifically for women and they raised the percentage of women, also in the independent group-leader level, dramatically in all their institutes, and filled these jobs in a short period of time. It turned out that all these women are very good in their jobs. This tells that these

> women were simply overlooked previously. It seems that you can do a little push without the danger that these women will not do well. Generally they had been so underrated that it is good to give them better status.

I asked her whether in her opinion a woman has to work harder to get the same recognition as a man[5].

> At least she can't afford to make mistakes, at least not as much as men. When a young men does that, they say, oh, he is ambitious and it can happen to anyone. But when a young woman does that, it is usually overrated. In particular when women are aggressive, people don't like that at all. When men are aggressive, it is perfectly normal; it belongs to their normal habit. When a woman is aggressive and behaves in the same situation as a man would, they always charge that against her. This is bad because aggression in a positive sense is absolutely necessary to succeed in a job. You have to push your point and not that of your neighbor. Unfortunately it is part of the profession.

About a decade has passed since our conversation. It appears that her opinion in this regard has not changed. Just recently, she published an article on this topic with similar sentiments.[6] "I love being a researcher: . . . I often think about women of similar passion and personality, but facing circumstances that make it extremely hard or impossible to be successful as a scientist. Where are the problems, what can be done to solve them?"[7] To facilitate a solution, she started a foundation to give grants to young, talented women who have children, to pay for household help.

SIGRID PEYERIMHOFF

Theoretical chemist

Sigrid Peyerimhoff in 1999 in Bonn. (photo by M. Hargittai)

Sigrid Peyerimhoff is Professor Emerita at the Mulliken Center for Theoretical Chemistry at Bonn University in Germany. She is well known for her major contribution to computational quantum chemistry. She had been involved with developing ab initio quantum chemical methods, which are used to compute the structure, energetics, and many other properties of molecules starting from scratch. The expression "ab initio" ("from the beginning") refers to such computations that utilize only fundamental principles. These methods have become an essential tool in modern chemical research, augmenting or even replacing laboratory experiments. One of the signs of the importance of the field was the 1998 Nobel Prize in Chemistry for John Pople and Walter Kohn for their pioneering discoveries.

Sigrid was born in 1937, in Rotteil, Germany. During World War II, her father, who held a relatively high position in the tax office, had difficulties because he refused to join the Nazi Party. Her mother was reported on by some neighbors, and the police took her away.

Peyerimhoff's interest in science appeared very early, and when she graduated from high school, she knew that she wanted to do science but not yet whether it should be physics, chemistry, or mathematics. She thought that it was more likely to get an industry job in physics than in chemistry or in mathematics. She was sure that she did not want to become a teacher. Although she already leaned toward theoretical work, she did not want to do her diploma work in theory, fearing that it would make finding a job difficult, so she chose experimental physics. She received her diploma (MSc degree equivalent) from the Justus Liebig University in Giessen in 1961, and her doctorate (Dr. rer. nat., PhD equivalent) and higher doctorate (Dr. habil.) from the same university in 1963 and 1967, respectively. Dr. habil. is a prerequisite in Germany and a few other European countries to obtain a professorial appointment.

Sigrid started to use computers in the early 1960s. She spent a few years at the University of Giessen and the University of Mainz, and gained postdoctoral experience in Chicago, Seattle, and Princeton. For many years, she regularly spent months at the University of Nebraska to use its excellent computational facilities. In 1972, she accepted a position at the University of Bonn and has stayed there ever since.

From the early 1960s, computational quantum chemistry made great advances. Before that, chemistry was an experimental science. Even though the basics of quantum mechanics were developed as early as the 1920s, for a long time there were no computers to do tasks as complex as the calculation of the structure of a molecule. There is a famous saying by Paul Dirac, one of the pioneers of modern physics: "The fundamental laws necessary for the mathematical treatment of a large part of physics and the whole of chemistry are thus completely known, and the difficulty lies only in the fact that application of these laws leads to equations that are too complex to be solved."[1] In the early 1960s, computers capable of handling chemical problems appeared. Many scientists jumped at this opportunity and started to develop programs to calculate structures and other molecular properties. Peyerimhoff was an active participant in this exciting new development.

She and her coworkers were interested in calculating not only the structure but also the spectroscopic properties of molecules, for which they had to calculate not only their so-called ground state structure but also the excited state structures, which correspond to a higher energy condition of the molecule. "When we could not calculate them reliably, we developed new methods to calculate them. We had a lot of interaction with experimentalists in this work. The chemists did vibrational spectroscopy, some of the physicists did electronic spectroscopy, and so on."[2] Later on, their range of interest broadened; they became involved with the study of carbon clusters, molecules that exist in the higher atmosphere, and with photochemistry.

She realized gradually that computational chemistry was not exactly the typical profession for a woman; women used to be about 1 or 2 percent in this area of chemistry. For many years, she was the only woman on their science faculty. Lately, the situation has improved and academic jobs have been opening up for women. At the time of our meeting, in 1999, her university introduced a new rule stipulating at least one woman on all search committees. She observed a change in the general attitude:

> Whereas it used to be taken for granted that women stayed home to do the housework, I often see now with my younger colleagues that the husband takes care of the children, changes their diapers, and puts them to bed.
>
> I married five years ago. Before that I was single all my life and I would not have made it this far if I had been married and if I had had children. My husband is a retired professor of physical chemistry. For a long time we could spend only our weekends together. Now he lives here with me. We even wrote a paper together, which just appeared. He used to be an experimentalist but he has no equipment to do experiments any more. I bring him lots of material to read, more than he can manage. He does some of the reading for me. [As for relaxation,] sailing is absolutely important to me. During the fall we always go for two weeks in the Mediterranean. It is total relaxation. Then sometimes during the summer we go sailing in a smaller boat in a lake in Bavaria. If there is no wind we do hiking in the mountains. When I am at home, I like gardening. I also like down-hill skiing.

Peyerimhoff received many awards and recognitions; among them, the Gottfried-Wilhelm-Leibniz Prize of the Deutsche Forschungsgemeinschaft is the most prestigious, being the highest recognition in Germany. It means three million German marks of research support for five years. She is a member of several German academies of sciences, including the prestigious Academy of Göttingen and the German Academy of Sciences-Leopoldina. She is also a member of the Academia Europaea (London).

She has held positions in many professional organizations. I asked her whether she considered this as recognition of her accomplishments or as a means to improve statistics. "If I am on a committee of eight, they have 12.5% woman participation. If I leave this committee, suddenly, this percentage goes down to zero. I see that often they just want me for the quota. However, if I see that I can do something useful even if I am invited only because of the quota, I accept it."

Not long before our meeting she was invited to be in a committee that looked into the possibility of hiring women for higher positions in science. She was concerned that the officially requested support for woman scientists sometimes backfires.

> When there is a competition for an academic job, for example, the university has to submit a list of the three top candidates to the ministry. Some of the states then chose the woman candidate even if she was number two over the number one male candidate. As soon as this practice had become known, those compiling the list either put the woman candidate into the number one slot or saw to it that the women disappeared from the short list altogether. These are games that people play and a woman candidate may end up in a worse position than she might have without any positive or negative discrimination. Some other ministries request

extra evaluation from the outside. This means an additional three or four months and people try to avoid it. On the other hand, I just got another letter from the Max Planck Society that we should try to get more women in high positions lest our support will dwindle. Another way may be to identify individuals worthy of support regardless of the area of research. The situation for women has improved but it is still far from where it should be.

MIRIAM ROTHSCHILD

Entomologist

Miriam Rothschild in 2002 in Ashton Wold. (photo by M. Hargittai)

When Miriam Rothschild passed away in January 2005, all major British newspapers carried obituaries about her. Her family was famous, but she was an exceptional person in her own right. A "high-spirited naturalist,"[1] "entomologist extraordinaire,"[2] sometimes called Queen Bee, other times Lady Flea, a lover of and fighter for all living things, she was like no other. Although she did not receive a formal education, she received honorary doctorates from eight universities, among them Oxford (in 1968) and Cambridge (in 1999). She was elected Fellow of the Royal Society in 1985, and Queen Elisabeth knighted her for her achievements in science in 1999.

In 2002, my husband and I visited Dame Miriam on the Rothschild family estate, where she was born and lived most of her life. She was ninety-three years old, still full of energy and plans for the future. It was an exceptional experience to talk with her; she was a wonderful host, pleasant and interested, and she enjoyed talking about her life as if reliving and enjoying it all over again. She was speedily crisscrossing around in her wheelchair, but we hardly noticed her incapacitation, and were surprised afterward when we looked at the snapshots we took of her.

Miriam Louisa Rothschild (1908–2005) was born in Ashton Wold, Peterborough, in Northamptonshire, England, into the English branch of the Rothschilds as the first of four children of her parents. Her father was Nathaniel Charles Rothschild, the second son of Nathan Mayer Rothschild, the first Baron Rothschild. Charles was a banker and a passionate entomologist. In 1907 in Vienna, he married Rozsika von Wertheimstein, a beautiful Hungarian woman from Nagyvárad (then in Hungary; today Oradea, Romania). The Wertheimsteins were a prominent Jewish merchant family, ennobled already in the 1790s, the third among the European Jewish families to have that distinction. The Rothschilds were ennobled in 1818.

Among Miriam's earliest memories were the trips they made to her mother's birthplace, where she and her father were busy catching butterflies and ladybirds. Charles considered this area the most interesting place in Europe for ecological and natural history.[3] He even built a ranch there and furnished it with a small laboratory. They spent their summer holidays there until World War I began. These trips triggered Miriam's interest in entomology, an interest that shaped her life and stayed with her forever. As far back as she could remember, she was only interested in things that were alive. Once when she was asked what she wanted for Christmas, she asked for baby chickens. When Christmas came, and she got little velvet chickens, "I just screamed my head off," she recalled later.[4] Then, her father gave her a few live mice and that made her happy.

Miriam's parents put special emphasis on making their children feel that their family was not special. The children were educated at home, as her mother did not have much faith in public education. When Miriam was about thirteen years old, she participated in a newspaper contest by writing a short story. She got a silver medal, and it puzzled her when a reporter told her that she probably got the prize because she was a Rothschild. At dinner she asked her parents what this comment meant; they just looked at each other and said, "We have no idea."[5] Eventually, she understood that her family was special. Her parents told her that if you were a Rothschild you had to set a good example. She did not understand why—but this was what was expected of them.

Both Miriam's father and uncle were amateur zoologists and entomologists, and that influenced her. Charles Rothschild spent all his free time with bugs, butterflies, and fleas. He never missed a day in the bank, but when he came home at the end of the day, even if his wife hoped to go to a concert or some other entertainment, he told her to watch him arrange his butterflies. He wrote about 150 scientific articles about fleas and described about 500 new species. One of his important discoveries was that the bubonic plague was carried and spread by a rat flea, *Xenopsylla cheopis*.

Miriam's uncle, Lord Lionel Walter Rothschild, as head of the Rothschild family in England, had to join the the N. M. Rothschild & Sons Bank, but he left the bank soon since his major interest was zoology. According to the legend, when he was seven years old he told his parents that when he grew up he would establish a museum, which he did. It is still there, the Natural History Museum at Tring in Hertfordshire, where at that time the family lived. The child Miriam visited it often. The museum became unique, with over two million butterflies and many other insect and animal species. It used to house Charles's collection of fleas—the largest of its kind in the world. Later, the flea collection was moved to the Rothschild Collection at the British Museum.

Charles Rothschild committed suicide in 1923, when Miriam was only fifteen years old, and that was a tremendous blow to her. She suddenly lost all her interest in

The young Miriam Rothschild. (courtesy of the late M. Rothschild)

natural science and thought that she would become a writer. After two years passed, her brother came home for a holiday from school with homework to dissect a frog. He asked her to help, and she was thrilled by seeing the internal organs of this frog. "It was the most marvelous thing I have ever seen; all the blood vessels, circulations, so I was immediately captivated and went back to natural science."[6]

When she was seventeen, she enrolled at the University of London to study zoology and English literature. Although she was excellent at both, she did not want to get degrees, because she felt she was in an impossible situation: "You always wanted to hear somebody talk on Ruskin when it was time to dissect a sea urchin."[7] At the end of the 1920s and in the early 1930s, she studied zoology in evening classes at Chelsea Polytechnic and decided to become a marine biologist. She went to work at the Marine Biological Station in Plymouth and studied parasites. Then came World War II, and that changed everything.

Already before the war, she had tried to arrange with the British government to allow German Jews to come to England. Eventually, she brought forty-nine Jewish children between the ages nine and fourteen from Germany and gave them a home at Ashton Wold. She helped refugee Jewish scientists during the war and afterwards; they often lived in her house. During the first two years of the war, she worked at Bletchley Park with other biologists, mathematicians, and philosophers under Alan Turing on the famous Enigma project, decoding German communications.

During this period, she met her future husband, George Lane, born György Lányi in Hungary. He was a swimming and water polo champion who went to London to study at the University of London. He stayed in Britain and during the war became an officer in the British Army, engaged in cross-Channel intelligence-gathering operations. He and Miriam were married in 1943. As she commented: "Altogether we lived a very peculiar life . . . we never knew if we would see each other again. It was very hectic—I was expecting our first child. I don't know how I survived it."[8] They had four children and adopted two more; there were two boys and four girls. In 1957, Miriam and George divorced but remained friends.

In 1947, Collins Publisher asked Miriam to write a popular book on parasites. Decades later she described the story in an essay, "My First Book."[9] She was expecting her second child, and writing seemed an ideal occupation. However, she was doubtful how popular such an unappealing topic could possibly be. Would people find it pleasing to read about the life-cycle of a worm, *Halipegus,* that gets from the water of a pond to the liver of a snail, then to the cavity in a shrimp's body, to the gut of a dragonfly larva, and finally under the tongue of a frog? Would anyone find appealing the story of a worm that lives under the eyelid of a hippopotamus and feeds itself from its tears? How about to learn that fleas have the most complicated penis among animals? But she was enthusiastic and so was the publisher, so she wrote it.

The book appeared in 1952, and "It proved to be my only successful book."[10] Its title is *Fleas, Flukes & Cuckoos: A Study of Bird Parasites.* Her aim was to make laypeople understand parasites and symbiosis. Written in a lively style, it is appealing and readable. This is how she started the part about fleas: "Birds' fleas and feather lice do not sing. Nor do they fly about flashing brilliantly coloured wings in the sunshine. It is scarcely surprising that in Britain bird and butterfly enthusiasts number thousands, but the collectors of fleas and lice can be counted on the fingers of one hand."[11] In Miriam's opinion, people hate fleas only because they do not know anything about them and misunderstand them. Among their curious characteristics, she mentioned that "fleas breathe through holes in their sides, have a nerve cord below their stomachs and a heart in their backs; or that certain arthropods lay eggs through their heads and regularly practice virgin birth."[12]

She wanted her father's work in science to be preserved. In one of her most ambitious projects, she and George Hopkins catalogued the fleas that her father collected; it became a monumental five-volume treatise.[13] It took three decades to complete it, while she was raising six children. She usually worked on the book during the night—she was lucky that she had insomnia. The book catalogues 30,000 specimens, and it made her the world authority on fleas. She studied their behavior and made several discoveries. Perhaps the most remarkable of these concerned the jump of the flea. She and her colleagues studied the jumping mechanism of the rabbit flea with the help of high-speed photography. She asked us whether we knew—we did not—that the height of the jump of a flea compared to their body size is as if a human jumped as high as the top of the Empire State Building. Their acceleration is enormous—twenty times the acceleration of a moon rocket reentering the earth's atmosphere. Some fleas can jump as many as thirty thousand times without stopping for a moment.

She described the rabbit flea as an insect that flies with its legs. Their legs are the residue of the wings from their evolutionary ancestors. After all, they do not need wings; they hatch out in the nest of the rabbit, but then they have to find the rabbit. "So there is a big jump to do, this is why they are powered by jumps. Wings are no good to you if you live in fur, they just are in the way, so as a parasite living in the fur of the rabbit, they have lost their wings but they replaced them by this stupendous jump."[14]

She dedicated another book to her father and his interest in nature preservation, *Rothschild's Reserves: Time and Fragile Nature*.[15] Charles Rothschild started the movement for nature conservation in England at a meeting in 1912. He suggested that certain areas of England have to be saved for nature and the future. They formed the Society for the Promotion of Nature Reserves. He suggested about 200 potential reserves, and he himself donated land in Woodwalton Fen, Cambridgeshire, as its first reserve. In this book, his daughter, together with Peter Marren, looked into what happened to these areas.

After the death of Charles Rothschild, Miriam became close to her uncle Walter. After Walter's death, she wrote a book about his life, *Dear Lord Rothschild: Birds, Butterflies and History*.[16] The book pays tribute to the zoologist Rothschild who founded the Tring Museum and to his lifelong work. It is full of anecdotes, such as how he was able to harness wild zebras to draw his carriage. Eugene Garfield writes that "as the only book about a Rothschild written by a Rothschild, the work has special authority."[17]

I mention here a few of Miriam's astounding discoveries concerning fleas, butterflies, and many other species. During the war, she was asked to look into the problem of tuberculosis among cattle in England. She investigated hundreds of tissue sections by microscope and determined that the vectors of the disease were wood pigeons with darkened plumage that had tuberculosis of their adrenal glands. Another of her amazing finds was that the reproductive cycle of the rabbit flea is adjusted to that of its host. The rabbit flea can time its fertility so precisely that the baby fleas can drop right away onto the newborn rabbits. This was the first time it was shown that the reproductive cycle of an insect parasite depends on its host. She and her colleagues determined that the disease myxomatosis—of which there was an outbreak in Britain in the 1950s—was carried by the rabbit flea and not by mosquitoes as was originally thought.

Several of her papers discussed the defense mechanism of insects. She wrote an account together with the Nobel laureate Tadeus Reichstein (1897–1996), and this is one of her most-cited publications.[a] Plants of the milkweed family contain heart poisons. The monarch butterfly feeds on this plant, and it ingests and stores the poison in much larger concentration than is in the weed. The monarch butterfly has developed immunity to the poison, and it stores the poison for defense, so as to make itself unwanted food for birds. Similarly interesting are her papers on warning coloration of insects. These insects have bright colors that alert birds and other predators to their toxicity, to save them from predators that hunt at daylight. Other insects produce distinctive defensive odors to divert predators that hunt at night.

When we visited Miriam, she was still active in research. She had a well-equipped laboratory at Ashton Wold, and scientists from all over the world came there to

[a] Tadeusz Reichstein received the 1950 medicine Nobel Prize jointly with Edward C. Kendall and Philip S. Hench for their discoveries on hormones of the adrenaline cortex.

cooperate with her. One of her research interests was smell, especially involving pyrazine. Her colleagues were testing three-day-old chickens to see whether the smell of pyrazine had any effects on their "recall"; she avoided using the word "memory." In a related experiment, they showed that chickens breathing in air saturated with pyrazine laid larger eggs than the control chickens.

She participated in countless social and civic activities. She started a movement to save English wildflowers and cultivated them. She designed a strategy to do this, and many people, among them Prince Charles, followed her ideas in their gardens. She created a Schizophrenia Research Fund to promote treatment. As early as the 1950s, she stood up for the rights of homosexuals, saying that a woman bringing up six children in a happy family is an ideal person to take such a stand. She fought for animal rights, for better conditions in slaughterhouses, and against the misuse of animals.

Although she published over 350 scientific papers, she did not consider herself a scientist. Rather, she regarded herself as the last of the old naturalists, a leftover species from the nineteenth century. The segmentation of today's science saddened her, even if she understood the reason. She realized that one has to know such a tremendous amount of small details in order to get ahead in a particular field that it becomes difficult to see the complete picture. The general knowledge of scientists is getting narrower and narrower, and thus they can no longer find a common language with their colleagues in related fields. This is how she characterized the situation: "today you cannot just sit down and talk about insects in general. All you can do is talk about the hind leg of a bee."

Miriam Rothschild would have been famous just because of her family, but that kind of fame did not mean anything to her. She was an idealist bursting with practicality. She wanted to make a difference, whether it was for sentimental or practical reasons. She fought against all kinds of injustice and wrongdoing, whether against humans or against the earth's fauna and flora. "I was always a tilter at windmills," she characterized herself.

VERA C. RUBIN

Astronomer

Vera Rubin in 2000 at the Department of Terrestrial Magnetism of the Carnegie Institution in Washington, DC. (photo by M. Hargittai)

Already as a child, Vera was interested in the stars: "I had a bed that was under the window and I could see the sky and I just got more interested in watching the stars than in going to sleep."[1] She graduated from Vassar College, where the famous astronomer, Maria Mitchell, had become the first professor of astronomy in 1865. Vera's master's thesis, at Cornell University, created quite a stir. She suggested that over large distances galaxies might be moving en masse. Most astronomers rejected her idea, but not George Gamow (1904–1968), one of the most original minds of twentieth-century science, the initiator of the Big Bang theory of the origin of the universe. Gamow had posed the question "Is there a scale length in the distribution of galaxies?" and Vera decided that this problem would be a good topic for her PhD project; Gamow became her thesis advisor. Her conclusion was that the movement of galaxies is not random; rather, they were moving in large clusters. Years later, her results were considered to be the first proof that over 90 percent of the universe is invisible dark matter, but it took a long time for the scientific community to appreciate them.

Vera Cooper Rubin was born in 1928 in Philadelphia into an immigrant Jewish family. Her father, Philip Cooper, was an electrical engineer from Vilna, Russia (today Vilnius, Lithuania). Her mother, Rose Applebaum, originated from Bessarabia (today, much of it is Moldova, between Romania and the Ukraine).

When Vera received her BA degree at Vassar College, she was already determined to become an astronomer. There were not many universities where a woman could get into an astronomy program; Princeton University, for example, would not even send her its catalog, because, until 1975(!), Princeton did not allow women into its graduate program in astronomy. It was a different time from 2005, when Princeton University conferred on Vera Rubin its honorary doctorate in a ceremony officiated by Shirley M. Tilghman, the school's first woman president.

After Vassar, Vera continued her education at Cornell University, where her future husband, Robert (Bob) Rubin, was a graduate student. She had exceptional mentors, Philip Morrison, Richard Feynman, and Hans Bethe. She completed her work for the MA degree in 1951. She and her husband became close friends with Morrison and his wife, Phyllis. Vera once asked him about dark matter at the time when it was still little discussed, and he told her, "The vacuum has energy." This statement has puzzled Vera ever since, and she thinks that he might have meant what today is called "dark energy."[2]

Vera Rubin in the early 1970s, measuring spectra at the Department of Terrestrial Magnetism of the Carnegie Institution in Washington, DC. (courtesy of V. C. Rubin)

She did her doctoral studies at Georgetown University under George Gamow, who was at nearby George Washington University. The arrangement was ideal, because Gamow was interested only in the implications of Vera's results, not in the details of her research. She completed her dissertation in 1954, but left Georgetown University only in 1965, when she moved to the Department of Terrestrial Magnetism of the Carnegie Institution for Science in Washington, DC, where she has stayed ever since.

Vera Rubin's studies have had great appeal not only for her colleagues, but also for the general public. Her most famous result was finding evidence for the existence of dark matter. She has always been interested in uncovering the nature of galaxies, this time the outer part of spiral galaxies, which had been little studied. It was known that near the center of a galaxy, the stars orbit with high velocities, and by analogy with the planets in the solar system it was supposed that toward the peripheries of the galaxies, the stars moved more slowly. In a joint work, she and Ken Ford found that in the galaxies they considered, the distant stars moved as fast as the stars near the center. This was a puzzle, and they had to look for an explanation.

> The best explanation is that the bright [visible] matter is responding to the gravitational attraction of matter that we cannot see. The distribution of this dark matter is very different from the distribution of bright matter. The bright matter is highly concentrated in the center, and then falls off rapidly with increasing nuclear distance. The distribution of the dark matter is less near the nucleus, but it becomes more significant with increasing nuclear distance; it falls off much more slowly, and extends much farther than the bright matter. It composes about 95 percent of the galaxy mass. Thus, the distribution of bright matter in a galaxy is not a good indicator of the distribution of matter.[3]
>
> . . . Our current cosmological theories place a limit on the amount of "normal" matter that exists, that is, atoms and subatomic particles. And this limit is less than the amount of matter required by the observations. So the remainder must be an exotic form of "matter"—neutrinos for example. But now we have a constraint on the neutrino mass, and it looks like that is not enough. Particle physicists think that their next generation of accelerators will give us the answer. We'll see (rather "know"; we don't "see" dark matter) in a few years.
>
> There is still another possibility, which is much less likely, but I am surprised we haven't been able to rule it out yet: the possibility that Newtonian gravitational theory does not hold over distances as great as galaxies. That would be a shock, truly a revolution! It is easy enough to write down equations to describe what we see but that's not enough because we know that Newton's laws work in some domain and relativity works in a different domain and whatever changes are made, the theory still has to reduce to both of them in the domains where they are valid. Thus you have to invent a new cosmology, which is a daunting task. But some scientists are attempting this.[4]

Rubin's results were among the first of the recent revolutionary discoveries in astronomy. There have been tremendous technological advances to foster astronomical observations, paralleled by advances in computation to digest unprecedented amounts of data. On the other hand, "there are still many major things that we still don't understand. I think that every couple of years or maybe even more often we are going to learn something very, very new and important."[5]

Ever since Vera faced the problem of choosing college, she has been aware of women's problems in science. One of their most conspicuous indications is that the proportion of women decreases rapidly as we go from the undergraduate level to higher positions in academia.

> You can write books just on this subject! It is perceived as a women's problem and I believe it will never be solved until it is perceived as society's problem or an academic problem. So I am really more pessimistic now than I was 50 years ago. Then, with so many women entering college, it looked like a gradual evolution would take place. Yet in the US the number of women full professors in science departments of the top 50 or so colleges and universities, is 6% (or was in 1998) and that's really outrageously low! Twenty years ago, some 20% or more (depending upon field) of all PhD degrees went to women—these women should now be professors, but they are not. In science, women now get more college degrees than men. And even in science PhD degrees, numbers of women are now about equal to men. So I think the real problem lies in academia. Some women get a PhD degree in physics and never have studied physics under a woman.
>
> There are still meetings in science, too many, where the speakers are all, or almost all men. Organizing committees are all men—without women on committees, on faculty, it is too easy for men to "ask their friends." That's why there must be women on faculty, on committees, in Academies of Science, to see that the injustices do not continue to propagate. I still hope that this change will come about easily. There are now overwhelmingly large numbers of brilliant women receiving PhD degrees in science. Science faculties will have to work hard to NOT hire them. So we'll see.[6]

At this point, I could not help asking her whether only men could be blamed for the situation. Could it be that women just were not sufficiently interested in science to persevere? They might even not be willing to brave the unfriendly environments.

> I don't think it is bravery. If you are bright, and you are not welcome, you will have the brains to go elsewhere. Too many women enter college thinking that they want to be scientists, but they do not survive. In many cases this is due to a lack of welcome, a lack of support, and the poor treatment by their male professors or colleagues. They are ignored, they are not listened to. If the

community of scientists was more welcoming, many more women would succeed. I often visit universities because I like spending time with students and I always talk to the women. At universities with large graduate science departments the women can name the professors they are warned not to attempt to work with. Although the colleges deny that such exist, the women know that they would never succeed with these male faculty. Someone called this the Bluebeard syndrome. If you go in his laboratory you will never come out alive. So the universities are often a large part of the problem; I think it unfair to blame the women. Being a graduate student is hard, being a scientist is hard, but if you add more impediments along the way, the likelihood that women won't survive is very large. Women who are determined enough surely can survive, but even less tough women should be welcome into the fields of science. Science will lose too many of the most brilliant minds, if it is not more welcoming. That's the tragedy.[7]

Vera Rubin is a much-decorated scientist. Her memberships in learned societies include the National Academy of Sciences of the United States (1981) and the Pontifical Academy of Sciences (1996). In 1993, she received the National Medal of Science from President Bill Clinton and in 1996, the Gold Medal of the Royal Astronomical Society (London), the second woman so distinguished (Caroline Herschel was the first in 1828).

Vera has a large family with four children. Her husband was more than just supportive of her ambitions; he was most encouraging. When I asked Vera about the greatest challenge in her life, she responded promptly, "Finding good care for my children."[8] Eventually, all their children became scientists. The only daughter became an astronomer, two of the sons are geologists and the third is a mathematician. The comments her children augmented Vera's autobiography in *Annual Review of Astronomy and Astrophysics* give a hint about their extraordinary family[9]:

DAVE: "One evening, when I was a child about ten years old, my mother told me that she knew something about astronomy that no one else knew. To this day, I remember thinking that this was extraordinary. . . . what my mom alone then knew was the beginning of the story of dark matter."

JUDY: "We saw our parents working hard and having fun being scientists, but none of us knew at the time that we would all choose to follow their lead. . . . I feel truly blessed and deeply grateful to be able to say, 'Vera Rubin is my mother'."

KARL: "I'm not sure when I realized that growing up in a household headed by two scientists was unusual. As a young child, I just assumed that almost all adults were scientists and that astronomy was a job for women. . . . There was never ever pressure to become a scientist, but it did seem like the natural thing to do. . . . I've learned that . . . having parents, who understand and encourage such a life is an advantage most of my colleagues didn't have."

ALLAN: "I think it's no coincidence that the four children all ended up doing science. A pervasive early memory of mine is of my mother and father with their work spread out along the very long dining room table, . . . At some point I grew old enough to realize that if what they really wanted to do after dinner was the same thing they did all day at work, then they must have pretty good jobs."

My personal encounter with Vera Rubin as well as our later interactions left me with an impression of her inner harmony that no external recognition, distinctions, or award could create. This is how she formulated her *ars poetica*:

> My greatest pleasure has come from combining the roles wife/parent/astronomer. None would have given as much joy alone. I love science because I have an unending curiosity about how the universe works, and I could not be happy living on earth and not trying to learn more. For me, it is the daily internal satisfactions that make a life in science so wonderful. Cold dark nights at a telescope have been among the greatest treasures of my life.[10]

MARGARITA SALAS

Molecular biologist

Margarita Salas in 2007 in Madrid. (courtesy of M. Salas)

The husband and wife team of Eladio Viñuela and Margarita Salas did great service to the science of Spain. After having established themselves as successful researchers in the United States, they moved back to Spain in 1967 and created the field of molecular biology in their country. Their long-range effect was yet broader. When Viñuela died in 1999, this was written about them in *Nature*: "They brought to Spain a new mentality and a new way of doing science. . . . As researchers, Viñuela and Salas complemented each other well—her systematic thinking and his wide-ranging and restless intellect led them to achieve defined goals."[1]

Margarita Salas was born in 1938 in Canero (Asturias) in northwestern Spain. Her father was a physician and her mother a schoolteacher. When Margarita was about one year old, they moved to the city of Gijón, also on the coast of Asturias, where she spent her childhood. She wrote in her autobiography: "My father always told us that the only inheritance he was going to leave us was a university career. That was, indeed, the best legacy he left us."[2] She attended Complutense University of Madrid, and at first she studied chemistry. It was a decisive moment when she met the world-renowned scientist Severo Ochoa. He was about to receive the Nobel Prize in Medicine, jointly with Arthur Kornberg, for their work relating to the biological synthesis of ribonucleic acid (RNA) and deoxyribonucleic acid (DNA), respectively.

Ochoa was a friend and distant relative of Margarita's father, and under Ochoa's influence she decided to become a biochemist.

On Ochoa's advice, she did her doctoral work in the laboratory of Alberto Sols and received her PhD in 1963 from Complutense University. Sols had studied under the Coris (see elsewhere in this volume) at Washington University in St. Louis. While studying chemistry, Margarita met her future husband, Eladio Viñuela, who was also Sols's student. They married in 1963. In 1964, Margarita and Eladio, both as postdocs, joined Ochoa's laboratory at New York University School of Medicine. This is what Margarita told me about their work: "During our PhD thesis, both with Alberto Sols as director, each of us had its own research work, although we collaborated in many projects and we have several joint papers from that period. When we were at Severo Ochoa's lab, he put us in different groups and we carried out independent research, although we collaborated in one occasion and we have a joint paper."[3]

Margarita Salas with Eladio Viñuela in 1967 in Extremadura, Spain. (courtesy of M. Salas)

They stayed at Ochoa's laboratory for three years. "I have unforgettable memories of my stay in Ochoa's lab. He taught us (Eladio and me) not only molecular biology, . . . But also his experimental rigor, his dedication to and enthusiasm for research."[4]

In 1967, they decided to go back to Spain and establish molecular biology research at the Spanish National Research Council (CSIC). This was a new area in Spain and they could not be sure whether they would be able to get their research funded. The summer before, they attended a summer course on phage at Cold Spring Harbor Laboratory (CSHL) on Long Island, and that helped them to choose their first project, the study of bacteriophage ø29. Bacteriophage is often referred to just as phage; it is a virus that infects bacteria and replicates itself within its host. The phages consist of nucleic acids enveloped by proteins. The phage Margarita and Eladio chose was relatively small but had a rather complex structure. It had not been much studied yet, which was important for them; starting a new research project with all the logistics involved, they did not need competition. "In Spain the beginning was not easy. There was no funding at all to do research. We only had our salaries from the CSIC.

To be able to start our work here, we obtained funding from United States, from the Jane Coffin Childs Memorial Fund for Medical Research. Without this funding we would not have been able to start our research in Spain." They were enthusiastic, and soon, in 1971, they published a paper in *Nature*. The famous codiscoverer of the double helix and director of the CSHL, Jim Watson, invited Margarita to the next Cold Spring Harbor Symposium.

Margarita and Eladio were successful in their joint research. However, Eladio realized that if they kept working together, this might jeopardize Margarita's scientific future. In her words:

> We worked together on bacteriophage ø29 for some time. Within our group I did not have any problem, but for our colleagues outside the group I was "only" Eladio Viñuela's wife. I did not have a name of my own. Eladio was very generous and he wanted me to become an independent researcher. Thus, in 1970, he decided to start a new project—the study of African Swine Fever Virus—and he started to leave the work on phage ø29. For some time, he continued the collaboration with me on phage ø29, but only part time in a second position. Later on, he left completely the phage work and thus finally it was under my exclusive direction. After that, we did not have joint works. Of course, he always helped me. I often say that, in addition of being my husband, he was the best of my teachers.

Eventually, their research got funding in Spain. In 1977, they moved to the new Center of Molecular Biology named after Severo Ochoa of the CSIC and the Autonomous University of Madrid. "Until a few years ago the funding was not too bad, although Spain has never increased the funding for research above 1.3 percent of the Gross National Income. The recent budgetary restrictions have decreased greatly the funding for research. In the last four years funding has decreased about 40 percent. At this time, the funding situation is extremely bad, and many groups doing good research have lost their funding."

Margarita has been involved with the study of this particular bacteriophage for about forty-five years, during which she and her associates have obtained many important results.

> The work in my laboratory has dealt with phage ø29 DNA replication and the control of ø29 DNA transcription. In the control of transcription we have studied the mechanisms by which the ø29 genes are expressed (activated) or repressed along the phage development. We have described a key phage protein involved in this process.
>
> With respect to ø29 DNA replication one of the most important results was the finding of a phage protein that is covalently linked to the phage DNA that we later found that is involved in the initiation of ø29 DNA replication, acting as a primer. This represents a new mechanism to initiate replication.
>
> Another important result was the finding that the DNA polymerase encoded by ø29 has three properties that make it very useful for biotechnological

> applications, mainly for DNA amplification. One of these properties is its high "processivity,"—meaning that the enzyme is capable of catalyzing successive reactions without releasing its substrate. Another important property is its capacity to produce strand displacement (to open the double helix) as well as a high fidelity. We patented the ø29 DNA polymerase and kits for amplification of circular DNA and linear genomic DNA were commercialized by Amersham Biosciences with very good economic results.

She taught molecular genetics at Madrid Complutense University for twenty-four years, and stopped only when she became director of the Severo Ochoa Center for Molecular Biology in 1992. She has received many awards and recognitions, among them the Severo Ochoa Research Award of the Ferrer Foundation (1986), the King Jaime I Research Award (1994), and the 2000 L'Oréal-UNESCO Award for Women in Science. She is a member of the Spanish Royal Academy of Sciences, the Royal Spanish Academy, and several foreign academies, among them the National Academy of Sciences of the United States. In 2008, she was appointed Marquise of Canero by King Juan Carlos I for her dedication and accomplishments in molecular biology. This is a hereditary title, and Margarita has one daughter, Lucía, who will inherit it. "It was my decision, with the agreement of my husband, to delay my maternity until the time I had my group going on. My daughter was born when I was thirty-seven years old," said Margarita. Lucía works in communications.

Margarita studied during the mid-1950s, and I wondered how common it was at that time for a woman to study science at a Spanish university. She said that about one-third of the students were girls, and they did not experience any discrimination. The professors were all men, and women could be found only among the laboratory assistants. When she did her PhD work, it was different: "I was very much discriminated against by my thesis director. He did not believe that women were able to do research."

Since the time of Margarita's graduate studies, the number of women students and of those who work in science in Spain has increased dramatically. The number of women full professors on average is 17 percent and of associate professors 38 percent; these figures are approaching the average in the European Union.[5] Concerning membership of the national academies: "At the Spanish Academy of Sciences we are at this time three women and another one is being elected. This means 8 percent. There is the Spanish Academy of Pharmacy, where there are eight women of a total of 50, the Spanish Academy of Medicine has two women out of 51, and the Spanish Academy of Engineering has three women out of 50."

MYRIAM P. SARACHIK

Physicist

Myriam Sarachik in 2000 at City College of New York. (photo by M. Hargittai)

In 1950, Myriam Sarachik started at Barnard College, a women's school in New York City, and decided to study physics. She had to attend classes at Columbia University because, except for one introductory course, there was no physics taught at Barnard at that time. She liked mathematics and music, and even though she found physics to be the most difficult subject, she decided on becoming a physicist because she felt that was what her father would have done had he had the opportunity. Myriam persevered and graduated in 1954. Four decades later she was elected to the National Academy of Sciences of the United States (NAS), and another decade later she became president of the American Physical Society (APS), the third woman at the summit of this institution (founded in 1899). She has received numerous awards and recognitions for her results in experimental condensed matter physics, among them the Oliver E. Buckley Prize in Condensed Matter Physics and the L'Oréal-UNESCO Award for Women in Science for North America (both in 2005). She has been active in the human rights movement and is a member or board member of several committees concerned with the human rights of scientists.

Myriam Sarachik (née Morgenstein) was born in 1933 in Antwerp, Belgium. She was raised Orthodox Jewish, with all the attendant expectations of what women can and cannot do. When she was six years old, her family had to flee from the Germans, but the first time they did not succeed[1]:

> . . . we were walking towards Calais but by that time Calais was under siege by the Germans and within a few days we were behind the German lines instead of being ahead of the German lines, so we went back to Antwerp. This was in May of 1940. Later we tried again and were caught when we tried to smuggle from German-occupied to unoccupied Vichy France; we were interned in a camp, then another camp but eventually we succeeded in escaping and managed to get to Cuba. I was eight years old when we got there and we spent there the next six years, after which we emigrated to the United States in 1947. My older brother ended up in England during the bombing of Calais but he eventually joined us in Cuba.

By the time Myriam got to college, she had already attended several schools in three countries. She had gained sufficient experience that she was not afraid to choose a profession that was not thought to be appropriate at the time for a woman. In her first-year physics laboratory, she met Philip Sarachik, and they got married following their graduation. Her husband went on to get his master's at Columbia.

> At one point one of my husband's professors asked him if he might be interested in going on for a PhD degree, which was something that he never had thought to do. He also comes from a background where that is just not the expectation. But he was intrigued by the possibility and he thought about it and said yes. In the meantime, I was taking courses and working at the IBM Watson Laboratories at Columbia University and was just dying to do the same thing! But it was not something that women did. But I had a friend who was a physics major and was a year ahead of me at Barnard and she was getting her degree at Columbia, she never thought twice about it! So I thought; she is going for a degree, why not me? She never asked herself whether it was appropriate for her to do it, she is just doing it—why should I not do it!? And I decided to do it and I went back to school full time. I actually did my thesis at IBM Watson Laboratories; my advisor was Richard Garwin.

Her PhD thesis work was in superconductors, and she received her degree in 1960. At that time, her husband was changing jobs, and they took time off for a trip to Europe. This was a different experience for Myriam from her migrations with her parents. "Thirteen weeks is much too long for me; having done so much forced traveling during my earlier years, I felt unsettled and anxious, but he absolutely loved this trip!" When they returned home, she took a postdoc position at IBM Watson Labs, working with Garwin. She became pregnant, and that made the situation difficult, because IBM had certain policies that were not too favorable. So after their daughter was born, she left IBM and took another postdoc position at AT&T's Bell Laboratories in New Jersey.

> At that point I had an enormously hard time finding a position because I was a woman with a PhD in physics, with a baby. It was really difficult; nobody wanted to deal with me. One of my professors at Columbia, Polykarp Kusch,[a] was very supportive; although he really did not understand why I wanted a job and argued with me strenuously for half an hour against it—but then decided that I did, for whatever reason, want to continue to work full time in physics; he was suggesting part time. He helped me get a position. I hired help to care for the baby.

In 1964, Myriam was appointed assistant professor at City College, part of the City University of New York (CUNY) system, and she has stayed there ever since. Her research has focused on fundamental questions in physics. The projects she is still engaged with involve metal-insulator transitions, metallic behavior in two dimensions, and the so-called molecular magnets.

As is well known, metals conduct electricity, while insulators do not. In between metals and insulators are the so-called semiconductors; although they are insulators, they may be turned into conductors by adding to them certain impurities. Myriam and her colleagues conducted experiments to study the conditions under which an insulator turns into a conductor; these experiments have to be carried out at very low temperatures, close to absolute zero. Myriam's group has determined several important characteristics of such transitions in a variety of substances.

Metallic behavior in *two dimensions* used to be considered an impossibility. Myriam, however, was open to the idea:

> I was at a meeting where I heard Sergey Kravchenko giving a talk in which he reported the discovery of metallic behavior in a high mobility, two-dimensional electron system. I thought that it was marvelous. It turned out that he was having a great deal of difficulty having his results accepted because everybody knew that there could be no such thing. Soon after, he was looking for a position and I happened to have an opening. . . . This was an ideal combination. I had worked in the field for a very long time and my work in the area gave perhaps some legitimacy to Sergey's findings and the subsequent work we did together demonstrated the possibility of a transition in two dimensions.

The two-dimensional system under study is a thin layer of electrons confined to the interface of some solids, or that can be formed in specially designed heterostructures. Myriam and her colleagues made several important discoveries in these systems.

[a] Polykarp Kusch (1911–1993), 1955 Nobel laureate in physics for his precision determination of the magnetic moment of the electron.

Another of her interests is single-molecule magnets. These are clusters of atoms, such as manganese or iron atoms, that are coupled together to have large magnetic moments. Beside their theoretical interest, the molecular magnets have important application possibilities in information storage and quantum computing.

Myriam's human rights activities have been going on parallel with her physics. She has always been interested in individual rights: "I have a very strong sense of fairness." She has been a member of the National Board of the Committee of Concerned Scientists. When the Soviet Union still existed, Myriam traveled there to visit the "refuseniks," Jews who had declared their intention to emigrate but were refused permission to leave the Soviet Union. She is also a member of the Human Rights of Scientists Committee of the National Academy of Sciences. "My effectiveness in working as a human rights activist has been seriously compromised by what our own government has been doing during the early 2000s, which I find absolutely atrocious, unforgivable and inexcusable." She gave an interesting example. She was the president of the APS during the year of the Iraqi invasion.

> It was tough because I personally was not only opposed to what was happening but I was furiously opposed, and I felt helpless to do anything about it. But as president of a scientific society I felt that it was inappropriate for me to express political views because I was not speaking for myself, I was speaking as president of a society whose business was science. When it comes to things like evolution, I feel quite empowered to speak my piece but when it comes to political views—although I think that most of the members of the APS do agree with the position I have that was, first of all, not universal within the society but more importantly, it was not within the purview of the society. Americans were very unpopular for what we were doing. I actually represented the society at the 100th anniversary of the Spanish Physical Society. There was a Nobel Prize winner who, instead of speaking about science, devoted his whole hour to talking about how horrible the United States was in what it was doing. Everybody in the audience was applauding and stamping their feet in agreement. When my time came to speak, I had gone to the web and downloaded the precise wording of a number of statements that had been made by the society concerning overarching political differences and serving as bridges between people. I think that was an important thing to do.

At eighty, Myriam continues her research as well as her other activities. In 2008, she was elected to the governing council of the NAS, where basically all their important decisions are made. She is a member of many advisory boards, far too many to list. She is also concerned about the slow pace of women's "appearance" in the sciences, especially in physics. She was a member of the Committee of the Status of Women in Physics. As president of APS, she tried to take steps in order to make physics more attractive for young women. “A woman has to be tougher and more tenacious than a man. This was certainly true when I started and I believe it still is true. Today the possibilities are equal for men and women; what women have to deal with is peoples’ inadvertent biases (including those of women themselves) which have not changed that much.”

MARIT TRAETTEBERG

Chemist

Marit Traetteberg in 1996 in her office in Trondheim. (photo by M. Hargittai)

When I was embarking on my career in scientific research, Marit Traetteberg was already a well-known figure in our area of research, the determination of molecular structure by gas-phase electron diffraction. She was successful and respected, and she had earned recognition not only among her fellow scientists in her specialization but also in the broader scientific community of her home country, Norway.

Marit Traetteberg (1930–2009) was born Marit Krogstad in Ålesund, a small seaport town in western Norway. She had a strong intellectual family background. Her childhood was happy in spite of the war conditions. She knew from early age that she wanted to become a chemist; she had an inspiring high-school chemistry teacher. She decided to enroll at the Technical University in Trondheim, but for that she had to have one year of practical training, which she did at the Norwegian Wine Monopoly in Oslo. This was the first time in her life that she had to live on her own: "It was a good practical training, they took me seriously; I was placed in all the different divisions (unfortunately not in the workshop, where I had wanted to go; that was considered unfit for a girl) and I learned a lot about life. I had friendly relations with the people there until they learned that I was going to attend the technical university. Then, the atmosphere changed and became more formal, perhaps they thought that I would look down on them afterwards since they did not have any university training; this bothered me a lot."[1]

It was not too common for a girl to choose the Technical University rather than the University of Oslo, but Marit wanted this challenge. Everything went well, and she spent some time at Indiana University in Bloomington and on occasion at other places, but she spent her entire professional career at the Technical University. She earned all the higher degrees in the Norwegian system, and reached the pinnacle of recognition for a scientist in Norway when she was elected to membership in the Norwegian Academy of Science and Letters. In addition, she served in high administrative positions at the Technical University. Her main contribution to the gas-phase electron diffraction analysis was in the concerted analysis of the motion and geometry of molecules, which made it possible to determine molecular structures more accurately, with a deeper understanding of its various features.

Norway used to have one of the leading laboratories in the field of gas-phase electron diffraction and I asked Marit how this happened.

> When this new area of research was born, there were gifted and interested scientists in the right place and at the right time to develop this field in Norway. I am thinking primarily of Odd Hassel, Christen Finbak and Otto Bastiansen. Bastiansen was the driving force in developing this field in Norway, and he attracted a large number of Norwegian and foreign students to his group. Besides being a brilliant scientist it probably also mattered that he had a friendly and open personality. People liked to work with him. There are other things as well. The Norwegian Research Council was established at about the same time as they were building up the electron diffraction laboratory, and the council assisted by supporting the development of this new area in science. Otto Bastiansen had extensive international contacts. He had spent a year in Linus Pauling's lab in the late '40s. I started to work with him in the middle of the fifties and I remember that we always had famous visitors from abroad, really top scientists, who came and not only gave a lecture, but stayed for a while. This was very inspiring to all of us. I remember, Linus Pauling, Dorothy Hodgkin, Kozo Kuchitsu, Yonezo Morino, Verner Schomaker and many other scientists spent time together with us.

Marit was married to Jens Traetteberg, a physicist. He worked most of his life at SINTEF, the largest research organization in Scandinavia.

> He died 6 years ago. I am convinced that he was a better scientist than me, but I always got more credit for my scientific achievements. This cannot be that easy for a man to bear. But he was always very nice about it, and this also shows what a wonderful person he was. . . . We often used to talk about our work at the dinner table. Once, when our daughter [Kari] was about 15 years old she was asked what she wanted to become in life. She answered rather vehemently that she would not do what her parents were doing because they always bore her to death with their problems! This was the time when I realized that we probably talked more about the problems we had at work than about the joy of doing science.

Kari started to work as a gardener, but later decided to get a university education in science and mathematics and became a teacher.

Norway is a rather special country that involves women at every echelon of society much more than happens in other countries. They already had a woman prime minister, and at one point half of the members of the government were women. Marit was proud of this, and not only of the fact, but that this was considered quite natural by Norwegian society. The government has introduced various rules, for example, that in all official committees there must be at least 40 percent of both sexes. All this, of course, started only a few decades ago; when she was a young mother, it was still not customary for a young mother to work. But by now it is natural.

Marit was always very active, not only in science but also in other endeavors. When she was working for her doctoral degree, she had a small child. It was a problem to get proper care for her, but this was inevitable if she wanted to continue working. Thus, on the suggestion of Otto Bastiansen, she initiated a day-care center at the university. Together with other students with small children, they built a nursery school, actually built it with their own hands, and it is still standing after almost fifty years.

Another achievement of hers is connected with the fact that first her husband, and within a year Marit also, was diagnosed with cancer. She remembered: "This was a turning point in our lives. When I got cancer I felt awfully lonely, I did not know anything about it. Then I met another woman who was in the same situation and we talked about how important it would be if somebody could give us some advice and support. Then eventually we took the initiative and started a volunteer service. We got into contact with the Norwegian Cancer Society which was starting up a similar service in Oslo at that time. They supported our initiative, and I have been involved in this work ever since." Marit fought her cancer with great strength and dignity for many years, until in 2009 she finally succumbed to the disease.

CHIEN-SHIUNG WU

Experimental physicist

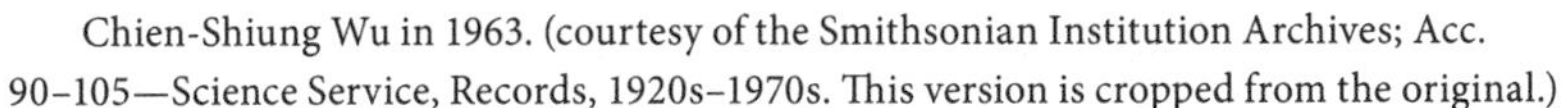

Chien-Shiung Wu in 1963. (courtesy of the Smithsonian Institution Archives; Acc. 90–105—Science Service, Records, 1920s–1970s. This version is cropped from the original.)

In my talks about prominent women scientists, sometimes I do not mention Dr. Wu, because I find others more relevant for the particular topic or audience. On such occasions, invariably someone asks why I did not mention her. Wu has become very well known as a great physicist, and rightly so. She has also become a symbol of Nobel-prize injustice, and in this, her story needs clarification.

Chien-Shiung Wu (1912–1997) was born in China, in a small town near Shanghai. Her engineer father founded the Ming De School, one of the first schools in China that admitted girls. He and his wife inculcated in their children the value of getting a good education. At the age of eleven, Wu moved to a boarding school for girls in a nearby town, Soochow, where she decided to study science and, in particular, physics, and she also learned English. After graduation in 1929, she spent one year at the Shanghai Public School, whose president, Professor Hu Shih, was a well-known scholar. Besides her father, Hu Shih had the most profound influence on her life.[1] From 1930, she studied in Nanking, at the National Central University, and in 1934 she received her BS degree at the top of her class.

After a year of teaching, she started doing research at the Academia Sinica. Her mentor was a woman professor who encouraged her to continue her studies in the United States. With the support of her parents, she enrolled at the University of California at Berkeley for her doctoral studies in physics under Ernest Lawrence, the director of the Radiation Laboratory and soon to be a Nobel laureate. Another future Nobel laureate, Emilio Segrè, was her mentor. Her PhD topic was the radioactive decay during the fission of uranium, and she learned how to devise experiments and carry out the measurements. She received her PhD in 1940 and was offered a research assistantship at Lawrence's Radiation Laboratory. Eventually, her work gained added significance in connection with reactor poisoning when it was recognized that the formation of xenon isotope ^{135}Xe was responsible for it.

During her Berkeley years, she met Luke Chia-Liu Yuan, a fellow physics graduate student, also a recent arrival from China. He was the grandson of the first president of the Republic of China. After graduating from Caltech, he received a job offer from Princeton. They married in 1942 and moved to the East Coast. She taught at Smith College, in Northampton, Massachusetts. In 1943, she was offered a teaching position at Princeton University. This was one of the several "firsts" in her career, as she was the first woman ever to hold a teaching assignment at Princeton, where at that time women were not even allowed to study! It was unprecedented for a young immigrant Chinese woman to teach one of the most difficult subjects, physics, to the Princeton male students. The reason it happened was the war; most physicists were engaged in war-related projects. In 1944, she was asked to join the Manhattan Project at Columbia University to work on radiation detectors.

Wu enjoyed Columbia and stayed at its physics department for the rest of her life. With the war over, she continued research in nuclear physics. Her primary interest was beta decay, a nuclear reaction that occurs when there are unusually more neutrons than protons in the nucleus, or vice versa. The resulting instability is resolved by emitting one or more beta particles—that is, electrons or positrons (positive electrons)—from the nucleus. The emission of an electron from the nucleus converts a neutron into a proton; the emission of a positron converts a proton into a neutron. Among the various forces in nature, beta decay is one of the so-called weak interactions. For a long time, physicists had a problem explaining beta decay. When Wolfgang Pauli discovered the neutrino, Enrico Fermi coined its name, and advanced a theory of beta decay by involving the neutrino. His explanation could not be verified by experiment until Wu devised an ingenious one that confirmed Fermi's theory. This made her well known in physics circles.

Her next experiment proving parity violation made her even more famous.[2] Parity is an inherent property of elementary particles. It refers to their behavior under reflection; in everyday terms, it means the relationship between a particle or process and its mirror image. The mirror image of a right-handed screw is a left-handed screw. Similarly, a particle spinning clockwise produces a mirror image that spins anticlockwise. Physicists used to assume that parity was conserved; that is, the parity of a particle could not change during the decay of the particle or during its production.

It was assumed that there existed a right-and-left equivalence in the world of elementary particles.

However, it had occurred to some physicists that there was no deeper reason for parity conservation. Two Chinese-American physicists, T. D. Lee at Columbia University and C. N. Yang at the Institute for Advanced Study at Princeton, noticed that while there were many cases when parity conservation was clearly observed in strong interactions, this question had never been considered in experiments on weak interactions.[3] Lee and Yang published their famous paper on the "Question of Parity Conservation in Weak Interactions" in the October 1, 1956, issue of *Physical Review.* In it, they briefly discussed the possibility that parity might be violated in weak interactions, and suggested some experiments that might test this possibility.

Lee and Wu worked in the same department, and already in the spring of 1956 they discussed the possibility of experiments to test parity conservation.[4] The possibility of an experiment using a cobalt isotope, Co-60, as the beta source came up. Wu recognized that this was a golden opportunity for a beta-decay physicist to perform a crucial test. She found the project urgent, and for its sake she even gave up a trip to a European physics conference and afterwards a visit to China, which she had left exactly twenty years before.

As it turned out, the challenge of experimental verification or disproval of parity conservation prompted two other groups to jump into similar work, though utilizing very different physical phenomena from what Wu's choice was. The main members of these groups were Richard Garwin and Leon Lederman at Columbia University and Jerome Friedman and Valentine Telegdi at the University of Chicago (Lederman and Friedman were two future Nobel laureates in unrelated areas, and Garwin and Telegdi were physicists of the same high caliber).

Getting back to Wu's story, she immediately started to plan her experiment, which was a complex task because two techniques that had never been combined before had to be used together. She was already an expert in beta-decay, but she also needed the ability to do the experiment at extremely low temperatures (close to absolute zero), which was not available at Columbia. She contacted a colleague, Ernest Ambler, at the National Bureau of Standards (NBS)[a] in Washington, DC, who had the right facility and expertise in nuclear orientation experiments.[5] In September 1956, Wu and Ambler invited three other NBS associates to participate in the project. On December 27, 1956, Wu's NBS colleagues saw the first signs of asymmetry in the experiment. As soon as she heard the news, she hurried to Washington. A few days later, back in New York City, she told Lee and Yang about the promising preliminary results.

Following a great deal of painstaking work, Wu and her NBS colleagues finally completed their report on January 10, 1957. In the meanwhile, in their famous "weekend experiment," Garwin and Lederman successfully showed that parity was violated during the decay of polarized muons in a cyclotron experiment. On January 15, the Department of Physics at Columbia University held a press conference to announce to the world that a basic law of physics, parity conservation in weak interactions,

[a] Today it is the National Institute of Standards and Technology.

had been overthrown. The papers describing the experiments of the Wu group and the Garwin-Lederman group appeared back to back in the February 1957 issue of *Physical Review*.[6,7] A third paper by Friedman and Telegdi about the experimental verification of parity violation was published shortly afterwards.[8]

Lee and Yang received the Nobel Prize in 1957, one of the fastest ever Nobel Prizes, considering that their paper had appeared only in October of the previous year. Their paper was a theoretical one, *suggesting* but not *proving* parity violation. The announcement at the Columbia University press conference of the experimental verification of parity violation on January 15, 1957, must have helped them get the prize, considering that the deadline of submission for the prize nomination is the end of January.[9]

It is an intriguing question: why was Wu not among the awardees of the 1957 Nobel Prize? After all, there was a "free slot," as according to the statutes of the Nobel Prize a maximum of three persons can share a Nobel Prize in each category. Wu's missing Nobel award has been the topic of frequent discussions, and from time to time it figures as an example of Nobel injustice toward a female scientist.

First of all, there was a legalistic reason why the experimentalists could not be considered for the 1957 prize. The Nobel Prize rules stipulate that the awarded work must have been published *before* the year of the prize, and all the reports of the experiments appeared in 1957. Had the Nobel Committee waited another year, it would have faced the daunting task of choosing one of the several experimentalist candidates. Of the experiment at NBS, which is often referred to as the "Wu experiment," Ambler could have been as credible a possibility as Wu. Then there were the participants of the other two experiments. Lee and Yang's preeminence in the considerations for the Nobel Prize is that they broke with the dogma of universal parity conservation and they were the ones to ask the question. Lederman noted: "the breakthrough was that [Lee and Yang] could consider that there are different forces and that different forces could have different symmetries. That was a tremendous insight."[10]

As to whether or not Wu could be singled out among the experimentalists, regardless of the Nobel Prize, there are many nuances to consider. These nuances are not without interest, and I was so intrigued by her story that I decided to delve into it. I examined the relevant literature and in 2012 contacted all the surviving physicist participants. Here, I briefly single out a few points from my findings, but I communicated a detailed account in a physics magazine.[11]

Wu was the first who suggested the beta-decay experiment in the early summer of 1956, but the actual experiments started only in September of that year. Telegdi and Friedman started their experiment in late summer 1956—not knowing of the other attempt—but then their progress was hampered.[12] Most probably it was the NBS team that had the first genuine signs of parity violation, on December 27, but they needed time to verify this under difficult experimental circumstances. Then, after hearing about their promising preliminary results, the Garwin-Lederman experiment took place in early January 1957—and it was ready in a flash; they completed the measurement at dawn of January 8. Theirs was the first conclusive measurement, and their article was written on the same day. The Wu et al. paper was ready on January 10. The two groups submitted their papers on the same day, and the papers were received at the journal on January 15. The Telegdi and Friedman paper was received two days later, on January 17. Certainly, all the participants of the three papers deserve credit for their hard work, their insight, and for embarking on a project that most physicists thought a waste of time, either because they did not believe in parity violation or because even if they accepted that there could be something in it, they anticipated the effect would be too small to measure.

Another question is about the role of Wu compared with the NBS team in the Co-60 experiment. I quote here Val Fitch's opinion, who shared the 1980 physics Nobel Prize for discovering charge-parity violation. He told me: "There were five people . . . who did the Co-60 experiment and they all contributed to it in a major way. Ms. Wu is often given the credit but I think that the most dispassionate view would be to recognize that those other guys were very important and it would not have happened without them."[13]

From my interactions with the surviving participants of the NBS experiment, I formed the impression that the roles of Wu and of Columbia University may have been overemphasized during the first euphoric days after the discovery. The press conference was held at Columbia University, where Wu was professor; she was the one who had suggested the experiment, and it was convenient to call it "the Wu experiment." There were some oversights in their short report. Significantly, it was not even mentioned that the experiment was conducted at NBS; even its terseness—the paper comprised a mere two pages—does not justify this omission. Thus, the impression could be formed that the experiment might have been carried out at Columbia.

There is something also to add to Wu's merits concerning the experiment at NBS. In the scientific and popular accounts, it is almost invariably mentioned that Wu decided to do the beta-decay experiment *after* Lee and Yang had suggested it in their famous paper. However, according to the records, it was Wu who had originally suggested to Lee this particular experiment as a possibility for the verification of Lee and Yang's ideas, well before the publication of the Lee and Yang paper.[14]

In conclusion, Wu had an outstanding contribution to bringing down the notion of parity conservation in weak interactions. But to say it was an injustice that she was not included in Lee and Yang's Nobel Prize is an oversimplification of a complex story.

Following the parity violation experiment, Wu continued her research with the same dedication and interest as before. Another important study by her and her associates at Columbia dealt with the symmetry properties of matter; they showed "the symmetry between the weak and electromagnetic currents, and set the cornerstone for the unification of those two basic forces into the electroweak force."[15] She continued her studies in particle physics by studying different subatomic particles and carried out research in condensed matter physics as well, using a variety of experimental techniques.

She was also a teacher, and she taught not only physics but also the importance of the high professional standards that she herself followed throughout her career. Wu's husband, Luke Chia-Liu Yuan, worked at the Brookhaven National

Laboratory. They had a son, Vincent, who is also a nuclear physicist. Both Wu and her husband were proud of their Chinese heritage. She was involved with helping the development of scientific infrastructure in China and Taiwan. When she passed away in 1997, her husband donated many of their possessions to Southeast University in Nanjing, China, where they established a Chien-Shiung Wu Memorial Hall. When Yuan retired, he helped funding the Synchrotron Radiation Research Center in Taiwan.

Chien-Shiung Wu was fortunate to have seen during her life how much her colleagues and even the general public appreciated her scientific achievements. Those who knew her say that she did not feel that she had ever experienced discrimination. She received numerous awards and got proper recognition. There were many "firsts" connected with her name. She was the first woman who taught at Princeton University (1943). At Columbia University, she was the first woman to get a tenured position at the physics department (1952), followed by the first woman professorship (1958) and the first woman to have the Michael I. Pupin Professorship of Physics (1972). She was the first woman who received an honorary doctorate from Princeton University (1958). In 1978, she received the first Wolf Prize in Physics in Israel. She was also the first woman elected to be president of the American Physical Society (1975). Beside these and other firsts, she received many awards and recognitions; she was elected to the US National Academy of Sciences (1958), received the National Medal of Science from President Ford (1976), and was—posthumously—inducted into the American National Women's Hall of Fame (1998). With her perseverance, her thirst for knowledge, her experimental skills and rigor, and her dedication to her students, she is a wonderful role model to all young women and men interested in science.

ROSALYN YALOW

Medicinal physicist

Rosalyn Yalow in 1998. (photo by I. Hargittai)

The highlight of the annual Nobel Prize celebration is the award ceremony, followed by the Nobel banquet. The banquet consists of a festive dinner and concludes with dancing. According to custom, one of the new laureates in each category gives a two-minute speech toward the end of the dinner. On December 10, 1977, from among the three winners in the category of physiology or medicine, Rosalyn Yalow delivered this speech. When the time comes for the speech, a student of Stockholm University goes to the table where the laureate sits and escorts him or her to the podium. On this particular occasion, the student saw on his seating chart that there were two Dr. Yalows, and he assumed that it had to be Dr. Aaron Yalow, Rosalyn's husband, whom he had to escort. He duly went to Aaron's seat. Rosalyn noticed the mistake, smiled, said something to the king of Sweden, sitting next to her, stood up, and walked unescorted toward the podium. The student at the other side of the table realized his mistake and caught up with Rosalyn at the end of the table. She kindly whispered something to him, went to the rostrum, and delivered her speech. The student's mistake was understandable—it was still very rare for women scientists to become Nobel laureates.

Rosalyn Yalow (1921–2011) was born Rosalyn Sussman in New York, into a poor Jewish family. "My mother was four when her family came to this country and my father was born in New York. They both came from poor East European immigrant families. She completed the sixth grade and he the fourth grade only, so they didn't even have the advantage of a high school education. But they were determined for their children to have a college education. I went to school in the Bronx. It was not a great school but we had good teachers and the pupils were very motivated."[1]

Rosalyn was an outstanding student whom science teachers took under their wings, and soon she decided to become a scientist. After high school, she entered Hunter College, part of the New York City college system, which was a women's college at the time. Her favorite subject was physics, and, to her luck, by Rosalyn's senior year, it became possible to major in it at Hunter: "Basically they started the physics major just for me," she said.[2] She wanted to go to graduate school, but the start was not very promising. Purdue University, where she applied for graduate assistantship, wrote back to her professor, "She is from New York. She is Jewish. She is a woman. If you can guarantee her a job afterward, we'll give her an assistantship."[3] Hence, she tried the "back door." While still at Hunter, she found a secretarial position at Columbia University. There was a bonus: Columbia let its employees attend classes without tuition. She made plenty of use of this perk.

In 1941, she graduated summa cum laude as Hunter's first physics major. The wartime circumstances increased her chances for getting into graduate school. She noted many years later: "The war made it possible for me and for many other young Jewish students to enter graduate school. While in Europe the Jews were being killed, the war made all the difference for Jews and for women in America."[4] She went to the University of Illinois at Urbana. Among all the faculty plus teaching assistants at the College of Engineering, she was the only woman. One of the students in the entering group of physicists was Aaron Yalow, her future husband.

Both Rosalyn and Aaron became Maurice Goldhaber's graduate students. Goldhaber was a brilliant physicist, a refugee from Germany, working in nuclear and particle physics. Later he became director of the Brookhaven National Laboratory. Goldhaber's wife, Gertrude Goldhaber (see elsewhere in this book), was also a physicist, and Maurice knew only too well how hard it was for a woman to stay afloat in science. "I did my graduate research in nuclear physics. This was very much the thing to do at that time," Rosalyn Yalow noted.[5] In 1945, she received her PhD in nuclear physics.

In 1943, while still in graduate school, Rosalyn and Aaron got married. He was an Orthodox Jew and kept kosher; that was a new challenge for Rosalyn, but she accepted it. Aaron, on the other hand, understood that she wanted to be a professional, and he appreciated that. He did not have ambitions for research and became a college professor, but was always supportive of her ambitions; he died in 1992.

When Rosalyn finished her graduate studies and her husband had not yet completed his, she went back to New York to find a job. After a short-lived engineering position, she returned to Hunter College to teach physics, but did not find it satisfying; what she wanted to do was research. In late 1947, she got a part-time job at the Bronx Veterans Administration (VA) Hospital to set up a radioisotope service. A janitor's closet served as her first office, and there she designed and built an equipment for radiation detection. She realized that the technique of radioactive tracers was a most useful approach in medicine that vastly expanded the application of radioactive isotopes beyond treating cancer, which it was mostly limited to in those days. For the discovery of the technique, George de Hevesy received the Nobel Prize in Chemistry.

Rosalyn benefited a great deal from a book by Rudolf Schoenheimer, *The Dynamic State of Body Constituents*, published in 1942. Schoenheimer had committed suicide in 1941, but his book became very influential. Rosalyn soon realized that she needed a partner, a medical doctor, someone skilled in internal medicine. She found her partner in Solomon A. Berson (1918–1972), a new resident doctor at the VA hospital. This was the beginning of their twenty-two years of joint work, which led to discoveries that changed endocrinology forever. She would later characterize their interactions[6]:

> Sol and I made a great team. For many years we shared the same office. We had two desks in the office, and we were discussing things all the time. I never had any formal training in biology. He taught me everything I needed in biology and medicine, and I taught him some physics. He knew a lot of physics but he was an M.D. Compared with a university setting, we had the great advantage of having no competition between us.

From her young age, Rosalyn was rather aggressive, determined, and stubborn. There are stories about her fighting with her teacher in first grade because the teacher had been rude to Rosalyn's brother years before. People describe Solomon Berson also as a rather aggressive and demanding person. How could they tolerate each other? Rosalyn obviously recognized his talent, his abilities, and that the two of them together might have a chance to achieve what she had been dreaming about. She was set on doing great science and making discoveries, and although she saw signs of a male sense of superiority in him, she accepted this. She allowed him "to take the spotlight, . . . because Sol was certainly worthy of being first . . . He was a leader in whatever he did . . . And really, there was nothing to lose. Why not let him go first since it mattered to him?"[7]

This turned out to be a clever strategy. He may have been a "male chauvinist,"[8] but he was always fair to her in what really mattered. During all their joint work, even if he was in the spotlight, he also made sure that they got the credit equally. They alternated first authorships of their papers. They went to conferences together, and both participated in the discussions. They were tough; either of them could easily stand up and comment on a presentation they did not like and say that it was nonsense. He considered her an equal partner. At the same time, being an old-fashioned gentleman, at the social events of the conferences he would suggest she sit with the wives.

They were like a married couple in their work; they developed a special language, in which one could finish a sentence the other started. Although they planned the experiments together, she was the one to set them up and carry them out. She did much of what might be considered secretarial or "women's work": she made the plane reservations and did the necessary paperwork. One day she told a friend, "I forgot to make Sol's lunch."[9] As it is often the case with scientific couples, they wonderfully complemented each other. In their science, he was broad-minded and romantic, while she was logical, mathematical, precise, and practical. He was a genuine leader, and she accepted the "woman's role."

Solomon Berson and Rosalyn Yalow in 1957. (courtesy of the late R. Yalow)

The first important work they did together was the development of the Yalow-Berson method for the in vivo determination of human blood volume. They introduced a radioactive tracer into the bloodstream and monitored the decay rate of the radioactive signal. From the decay rate, they determined the amount of blood in the body. Their next major accomplishment was the use of radioactive tracing in the clinical diagnosis of thyroid diseases. They developed a method by which they injected radioactive iodine into the bloodstream and determined its clearance rate; the method is still the best one for this purpose.

Their most famous discovery came out of their investigation of the malfunction of insulin in type II diabetes. This is the kind of diabetes where the patient has plenty of insulin in the blood, yet the insulin cannot remove the sugar from the blood, so a high sugar level persists. The insulin they injected was prepared from pigs or cattle, and the human immune system fought it by producing antibodies. Berson and Yalow determined that these antibodies were large gamma globulin proteins. This was already a new discovery, since at that time it was believed that insulin was too small a molecule to induce antibody production.[10]

In their paper, they described for the first time how radioisotopes could be used to study the reaction between an antigen and an antibody. The antigen is a substance that evokes the production of antibodies; the word "antigen" comes from "*anti*body *gen*erator." The truly great idea was that if they could measure the amount of antibodies that are bound to the insulin, they might be able to determine the amount of insulin itself. This is what they called radioimmunoassay (RIA).

RIA is an unbelievably sensitive method, and that is important, since peptide hormones such as insulin are present in the blood in very, very small concentrations. "And the RIA method has all of the specificity inherent in the precise reaction of an antigen with its specific antibody. The insulin antibody will find and bind and measure only the insulin among the myriad substances, many in billionfold higher concentrations, which are present in your blood sample. It is also inexpensive to do, and it is nearly as easy and quick to assay thousands of samples as just one or two. RIA was a breakthrough approach."[11]

Yalow and Berson could have stopped here, but they understood that they had a method in their hands that was practically limitless in its uses. Indeed, they went on with the same energy as before and found numerous other applications, such as measuring vitamins, steroids, prostaglandins, tumor antigens, enzymes, and viruses in the blood. They also measured the hepatitis B virus, and by this work they brought RIA into the fight against infectious diseases. By measuring human growth hormones, they could determine whether or not children needed to get growth hormone treatments at an early age.

Underactive thyroid glands might cause mental retardation in children, and by the time its symptoms can be detected, the brain damage is impossible to repair. Just from a few drops of blood, newborn babies can be diagnosed by RIA and treated in time. RIA is also used to determine drug concentration in the blood, for example, in case of heroin abuse or steroid use by athletes. RIA has revolutionized endocrinology and clinical medicine. The unselfish attitude of Yalow and Berson in not taking out patents for their revolutionary method is remarkable. "When Sol and I discovered the radioimmunoassay technique, at the beginning it had a slow start, but we knew it would catch up quickly because it was a very sensitive and very useful tool. We were very determined not to take out a patent on it but to do everything to help it spread. We organized courses to teach physicians to use the technique."[12]

At the end of the 1960s, Berson decided to move on, and he suggested to a young physician, Eugene Straus, that he join Rosalyn, which he did. In 1972, Berson died of a heart attack. This was an awful blow for Rosalyn. Many outsiders and even close colleagues thought that Berson was the brain behind all their discoveries. They did not think Yalow would be able to continue the way they performed together. One of her former graduate students, Mildred Dresselhaus (see elsewhere in this book), said that "being a woman in a team of a woman and a man, it was assumed that she was his assistant. The external relationship was like that all the time although it wasn't like that internally."[13] Moreover, for years there had been talk about a possible Nobel Prize for their team; now this seemed no longer feasible—up till then there had never been a Nobel Prize given to a surviving partner of a team, and there is no posthumous Nobel Prize.

Yalow's stubborn nature came to her rescue; she decided to prove herself one more time. She named their laboratory after Berson, so that her later publications would still carry Berson's name; it was a beautiful gesture. She took over many of his engagements and continued their research with her usual determination.

Within the next four years, she and her young coworker, Eugene Straus, published about sixty papers and made several new discoveries. In 1975, she was elected to the US National Academy of Sciences, and in 1976 she received the prestigious Lasker Award.

In October of 1977, the call finally came from Stockholm—she was awarded half of the Nobel Prize in Physiology or Medicine "for the development of radioimmunoassays of peptide hormones." She was the second woman to receive the medicine prize, after Gerty Cori (in 1947), and the first American-born woman to receive a science Nobel Prize. Eugene Straus writes in his book about her that the prize changed her life—perhaps even changed her. She said: "Before the Nobel, nobody heard of me. After I got the Nobel I was in the spotlight, people listened to what I had to say."[14] She received a large number of honorary degrees and awards; the National Medal of Science came eleven years after the Nobel Prize. The distinctions further enhanced her self-confidence, and this had negative side effects: she "has not been universally admired, and that's an understatement."[15] She became even more critical and opinionated than before, and her criticism of other scientists was sarcastic and unpleasant. She never forgot the rejection of their major paper back in 1955 by the journals. She felt that the occasion of her Nobel lecture was the right moment to show this; the printed version of her lecture has a figure with the relevant part of the letter from the editor of the *Journal of Clinical Investigation*.[16]

In the mid-1990s, Yalow's health deteriorated. Her struggles to overcome the consequences of devastating strokes is described beautifully by Eugene Straus.[17] Her last position was Senior Medical Investigator and Director of the Solomon A. Berson Research Laboratory. Although she retired in 1991, she kept going to the laboratory. We recorded our conversation with her in 1998 in her office at the laboratory in the VA hospital. She said then: "I've had three strokes, I have difficulties in moving my right hand, and my right leg is partially paralyzed. I come to my office regularly though, read my mail, keep up with things around me."[18] She continued to give talks about the importance of science and on issues related to women in science. She talked about the importance of day-care centers at universities so that young women should not need to be away from their research for too long.

When Rosalyn married Aaron Yalow, she knew that she wanted not only to do big science but also to have a family, to be a wife and mother. She waited to have children until she felt confident that she would be irreplaceable in the laboratory. The Yalows had two children, Benjamin (1952) and Elanna (1954). This is how in 1998 she remembered this period of her life[19]:

> My husband died five years ago, but we had a wonderful marriage. He was always very supportive. He was Professor of Physics at Cooper Union College here in New York. He was engaged in teaching, not research. We have two children. When they were small, in the 1950s, my mother used to come to help us with the children. We also had maids, first a sleep-in maid, later a part-time maid. They

were wonderful, bright black women from the South. They came to New York, and they couldn't go to study, so they found work as maids. This made it possible for me to carry on with my job. For me, this was a very fortunate situation. Today this would be impossible.

Our daughter, Elanna, studied educational psychology and has her PhD. She lives in San Francisco, has two children, and works all over the country, setting up day-care centers. Our son, Benjamin, used to do computer work. He doesn't have a formal job currently. I live together with him.

When the children grew older and the Yalows had only part-time help, she did all the shopping and cooking. She was old-fashioned in this; she felt that "running a house is a wife's responsibility."[20] On weekends, she took the children to the laboratory, where they played with the animals. If there were requests for help in field trips from the schools, she was always ready. She may have been a dedicated scientist and in this respect unconventional and demanding on herself as well as on her family, but in her private life she was a traditional woman.

At work, she was the same motherly type with her students. Mildred Dresselhaus told me about her[21]:

> She had a strong personality. She helped me become a scientist. Her academic expectations were very high and she may have frightened away some, whom she did not think should be there. . . . There are sides of Rosalyn that the public doesn't see but I've seen. She can be very motherly. When I was just starting my career, she and her husband would come to my 10-minute APS [American Physical Society] talks, and she would bring a shopping bag full of stuff, a little like a housewife. She always dragged her husband along too. . . . Whenever I needed something, she was always there.

An intriguing aspect of Yalow's life was her relationship with her husband versus her relationship with Solomon Berson. According to all accounts, Rosalyn and Aaron had a wonderful marriage. He was very proud of her achievements and helped her in whatever way he could. He understood her relationship with Berson and was not concerned about it. How about that relationship? Outsiders often thought that they were married; at conferences they were inseparable, they traveled together, and spent all their long working hours with each other for two decades. According to Straus, they "had an intellectual and scientific marriage, never a love affair. Neither one wanted that. . . . what she had with Berson was more significant, more stable, and even more exciting to her than a sexual relationship. Just . . . to share the work, to make their way in the world together, to make his lunch, it was more than enough, it was more than anything."[22]

Rosalyn Yalow was a talented, ambitious, and aggressive woman who needed all these traits to overcome the barriers she faced at different stages of her life. These traits made her stronger. Her ambition to become a successful physicist was rare and unconventional at her time, and she wanted even more. She also wanted to be a wife and a mother, in the traditional meaning of these words. She wanted it all, and she could not imagine compromise in either of her goals. Even if she paid a price in human relations for her success, she succeeded spectacularly. This is why she felt justified in encouraging her student audience about the opportunities for women in science: "You can have it all"!

ADA YONATH

Crystallographer

Ada Yonath in 2002 in Budapest. (by M. Hargittai)

The BBC announced the 2009 Nobel Prize in Chemistry as the "Nobel Prize for the chemistry of life," awarded jointly to Venkatraman Ramakrishnan, Thomas A. Steitz, and Ada E. Yonath. They received it for their results in the elucidation of the structure and function of the ribosome. The ribosome is a giant system of molecules that can be best described as the cell's "protein factory." The ribosome translates the genetic information carried by the DNA with the help of another nucleic acid, the so-called messenger RNA, and thus the proteins can be produced. The historical overview in the prize announcement went back to Darwin's general theory of evolution and stated that the 2009 Nobel Prize was the third in the series of Nobel awards for discoveries providing evidence in support of Darwin's ideas. The first was for the double-helix structure of DNA (James D. Watson, Francis Crick, and Maurice Wilkins, 1962); the second was for the understanding of how information is copied to the messenger RNA molecule (Roger Kornberg, 2006); and, finally, the one in 2009 was for demonstrating "how the simple DNA code can manifest itself not only as hearing, feeling and taste, or muscles, bone and skin, but also as thoughts and speech."[1]

Ada Yonath explained to me:[2]

> The biosynthesis of proteins happens with fantastic speed in the ribosome. When a chemist wants to create a peptide bond, it may take days and high-temperature and other extreme experimental conditions, whereas the ribosome can do this fast, within microseconds, and under mild conditions within the living cell. Also, scientists often make mistakes; the ribosome hardly ever makes mistakes.
>
> Since all cells of all living organisms starting from the simplest bacteria contain ribosomes, they can be an ideal and obvious target for drugs; understanding their structure and function should help in designing new antibiotics.
>
> In principle, any ribosome can read any genetic code. A ribosome in the human body can translate the genetic code of bacteria and vice versa. The ribosome is a factory for making proteins and can follow any genetic instructions. However, the ribosome of a higher organism—mammals and eukaryotes—is more complex than the ribosome of bacteria. The higher complexity is a consequence of additional tasks concerning regulations and selectivity, and it has to do with more interactions with the cell. The differences between bacterial and mammalian ribosomes are subtle. Even the active sites or areas near the active sites contain some differences. This is why ribosomal antibiotics can work. The antibiotics should impact the pathogenic bacteria only and not the patient, not even cause side-effects. Sometimes replacement of one single nucleotide can make the difference in the effects on bacterial and mammalian, that is, human ribosomes.

Ada is a strong-willed, dedicated woman; she could not have achieved what she did without that. She was born in Jerusalem in 1939 as Ada Livshitz; her father was a rabbi, and her parents immigrated to Palestine from Germany just after Hitler came to power in 1933. She was eleven years old when her father died, and from that moment on she had to help her mother in supporting the family; she tutored younger children and did other chores. Looking back, she feels that there was never time for anything. She had an insatiable thirst for knowledge: "As long as I remember back, I always felt that I wanted to learn more. I was never satisfied with what the curriculum would give me. In the high school I always went to the school library and I read a lot and I enjoyed reading and learning enormously."[3]

During her university studies, she was mostly interested in biochemistry and biophysics, and she received her MSc degree in the latter. She did her work for her PhD degree at the Weizmann Institute, where she studied the structure of collagen. This was followed by postdoctoral studies in the United States. First she was at the Mellon Institute in Pittsburgh, then, in order to do protein crystallography, she joined[4]

> . . . the group of F. Albert Cotton at the Massachusetts Institute of Technology in Cambridge. This meant a major turn in my professional career.

> Following the two years in America, I returned to Israel, and started my own group in protein crystallography [at the Weizmann Institute]. I was alone in the entire country. I had an instrument and some limited lab space, and it took almost half a decade before things started working.
>
> From the beginning of the 1970s, protein crystallography in Israel was slowly moving to the frontier of science. It was also at that time that I developed collaboration with Professor Michel Ravel at the Weizmann Institute. He had a technique to prepare what he called large amounts of the initiation factors, which initiate ribosome function in protein formation. The work involved an intensive collaboration with the late Professor Paul Sigler, who came to Israel for a year and a half. We had good interactions with the Chicago group; later I had spent a sabbatical year there. We were trying to grow the necessary crystals, but failed. During this year, at a meeting in Canada, Professor H. G. Wittmann of the Max Planck Institute in Berlin talked about the ribosome and reported the sequence of the initiation factors. We talked and he was interested in collaboration, and eventually I went to Berlin.
>
> It was just a few months before I was supposed to go to Berlin with one of my students that I was riding my bicycle to the beach—it was in February, and the weather can be beautiful in February in Israel—and I fell down in the middle of the street. I had a brain concussion. I was brought to the hospital. My brain concussion was more or less over within two weeks, but there were side effects, which prevented me from flying. Also, I needed an operation. After everything was done, I went to Berlin about five months after the original plan.
>
> When I arrived in Berlin, in November 1979, the initiation factors were almost ready, and in my "free time" I discovered that they had very active pure ribosomes in huge amounts from several bacteria and I suggested using them for crystallization. The people were very supportive. I knew that many prominent scientists had failed before in attempts to crystallize ribosomes, and if I would fail too, I would be joining a distinguished group of luminaries, including Francis Crick, Jim Watson, Aaron Klug, Alex Rich, and others. I knew though that this was my big chance. I went about the project very carefully; since I assumed that the difficulties in ribosome crystallization stem from their heterogeneity as well as their tendency to deteriorate.

To some people, the task Yonath addressed herself to appeared impossible. The ribosome is a huge and complicated RNA-protein complex. In case of the human ribosome, the so-called large subunit consists of two RNA molecules and about thirty-five to forty proteins, while the so-called small subunit consists of one RNA and about twenty to thirty proteins. This means that there are thousands of nucleotides and amino acids in a ribosome; and the positions of hundreds of thousands of atoms had to be determined to get the complete structure. No wonder she met with so much skepticism. But this did not deter her[5]:

> First of all, I went back to the old literature and studied everything there was written about the ribosome, especially techniques developed for maintaining their

> integrity for relatively long periods, required for crystallization. I took advantage of procedures developed in the sixties by A. Zamir and D. Elson. I spent only two months in Berlin, but after I had returned to Israel, they kept sending me almost every week pictures taken by a light microscope (neither fax nor internet were available in those days). In three or four months, we had micro-crystals, which were much too small to be studied as single crystals, but gave a promising weak powder pattern. Then it took about four years to get the first diffraction patterns that didn't look like garbage. Our paper about the micro-crystals came out in 1980, and for the past quarter century I have been involved in elucidating the structures of ribosome.

During long years, Yonath and her coworkers tried all sorts of methods to stabilize the ribosome in order to produce good-quality X-ray diffraction patterns from them. They examined the ribosomes of bacteria that live under hard conditions, supposing that their ribosomes might be more sturdy so that they could stand the "hardship" of all the manipulations. They took bacteria from hot springs and from the oversalted water of the Dead Sea. By the early 1990s, they managed to prepare good samples. They recorded nice diffraction pictures, but due to the size of the ribosome, the interpretation was extremely difficult. To make things more stressful, more and more research groups entered this area of research, and they found themselves participants in a race. Thomas Steitz's group at Yale University published the first X-ray structure of the ribosome in 1998, but without the atomic positions, because of the low resolution of the data.

The intensity of the race must have been almost unbearable. Finally, three groups published the most important results, and the three leaders of these groups shared the 2009 Nobel award. Steitz and his colleagues determined the high-resolution structure of the large subunit. Yonath and her colleagues and Venkatraman Ramakrishnan and his colleagues (of Cambridge, UK) determined that of the small subunit. After this, it gradually also became clear how the ribosome operates; they determined that the proteins are produced in the large subunit, and this happens extremely fast. Ramakrishnan determined that the small subunit translates the information that the RNA brings from the DNA into the "language of proteins."

Ada spent her entire career at the Weizmann Institute; in the late 1980s she became director of the Mazer Center for Structural Biology and of the Kimmelman Center for Biomolecular Assemblies. She has been recognized with memberships in science academies, honorary doctorates, and prestigious awards, including the Israel Prize (2002), the Wolf Prize (2007) and the L'Oréal-UNESCO Award for European Woman in Life Science (2008).

Our conversation took place five years before her Nobel Prize, but it was already in the air that she might receive it. This is what she told me in 2004, when I asked her about the possibility of the Nobel[6]:

> May I not answer this question? It is embarrassing. When I got the first microcrystal, I met with a Swedish professor, who is not alive anymore, who was one of

the founding fathers of structural biology. At that time I was working very hard, it was an exciting period of my life, I hardly ever slept. He noticed that I looked pale and haggard and asked me why? I told him that I might have crystals of the ribosome. He looked at me and said that this was a Nobel Prize project. This was right at the beginning of the work, in the middle of the 1980s. I never talked about it, but it has stayed with me. When we got the first high-resolution results, there was a scientific advisory committee meeting at the Weizmann Institute, and there, again, some people expressed the same opinion. Such impression tends to leak out, and people very often ask me about the Nobel Prize, a question which I do not like. But I know that this is a project that is very much in the center of attention, so it would be useless to deny that I am aware of the possibility.

To the question about her state of mind when October and the prize announcement is approaching[7]:

I hardly think about it except when people like you are asking me. It was never the prizes; it was always the intellectual stimulus that drove me. With this I am not saying that I am not happy by getting recognition.

Concerning the question of discrimination against women[8]:

The only thing I can think of is that sometimes I had the impression that they expected more from me than they would from a man in my position. Even this was not a very strong feeling. I was making very slow progress in the beginning because I was inexperienced in crystallography, but this was not because I was a woman. There was one occasion, when we were still together with my husband, the Weizmann Institute did not promote me, and one of the professors told me that they were not worried about my leaving them because my husband was there, too. This I didn't like, of course. I can't recall any other negative experience. Actually, I was offered some jobs, since I am a woman . . .

Ada has been divorced for quite some time and has one daughter, who she believes must have felt at times that her mother was not there when she needed her. Her daughter became independent and responsible at a very early age.

Since Ada received the Nobel Prize, she has often been asked about the situation of women in science; for example, whether it was more difficult for women doing science than for men. She responded that doing science is difficult not because you are a woman: "What it is difficult to be a woman about, is the same for a scientist, for a business woman, for a reporter."[9] In her opinion, the problem is with the society that does not encourage women to go into science as a profession.

During the 2013 Lindau meeting of Nobel laureates with young students, she gave a talk. A reporter noticed that many young women attended her lecture and was contemplating how to ask Ada about women in science, realizing that it was her scientific results and not her gender that was important. Even before the reporter could make up her mind, Ada concluded her talk by addressing this issue: "Young girls ask me: 'Should I stay in science or not?' They have fears about being a good scientist and having a family." On the next slide, she showed women of her laboratory and a huge ribosome-shaped chocolate cake that they baked. It was a gentle hint to show that you do not have to choose . . .[10]

Ada Yonath set for herself a goal that many of her peers thought was impossible—and she achieved it. When asked about the greatest challenge she encountered in her life, and about what she considered her greatest success, she said: "Scientifically, [the greatest challenge] was the crystallization of ribosome. Personally, it was the period when my father died. [And about her greatest success:] I have an eight-year-old granddaughter. Her kindergarten teacher asked me once to give a lecture about ribosomes to the kindergarten. I consider my greatest success that I kept them spellbound for an hour."

WOMEN SCIENTISTS IN RUSSIA

In spite of the magnitude and importance of Russia and the appreciation for science in the country, the question of women's participation in it has not been much discussed. There are books on women scientists in the late nineteenth and early twentieth century.[1,2] However, sources are conspicuously scarce about women scientists from the Bolshevik Revolution in 1917 through the Soviet era, let alone since the collapse of the Soviet Union. The scholar Olga Valkova noted the probable reason: "there was no 'woman's question' in the USSR, there was nothing to talk about."[3] Theoretically, women had the same rights as men, and all professions, even the ones considered most unfeminine, were open to them. Problems did not exist, or rather they could not be recognized as existing. Recently, Svetlana Sycheva published a book about contemporary women scientists using the example of soil researchers and geographers.[4] This is one of the signs that at least this issue is losing its taboo status.

Russian women's participation in the sciences was already strong in the second part of the nineteenth century; women of the aristocracy and intelligentsia became interested in studying at universities at home and abroad as institutions of higher education were slowly opening up for them. Most of these women recognized the need for social change in their country and that science would have an important role in attaining a better life for all their countrymen. These Russian ladies belonged to the first group of university-educated women in the world. The mathematician and author Sofia V. Kovalevskaya and the biochemist and physiologist Lina Stern were among them.

Sophie Kovalevsky—Sofia V. Kovalevskaya (mathematician)

Sophie Kovalevsky (1850–1891) was born as Sofia Vasilevna Korvin-Krukovskaya in Moscow into a noble family of Polish-German stock. She is the most famous among the first Russian female scientists. Mathematics started to interest her very early, and she was adamant in wanting to learn more of it. In her time, this was impossible for a woman in Russia, and to go abroad she needed her father's permission, which he denied. Her way out was a marriage of convenience at the age of sixteen. She and her paleontologist husband, Vladimir Kovalevsky, left Russia in 1867. After her marriage, her name became Sofia Kovalevskaya, but in time, she changed her surname from the Russian feminine form, Kovalevskaya, to Kovalevsky (in her publications, Kowalevski), and her first name to Sophie (or Sonia/Sonya).

She studied in Heidelberg, but women at that time were not allowed to graduate; the university merely let her audit the lectures. Eventually, she wrote three research papers on three different topics: partial differential equations, Abelian integrals, and the rings of the planet Saturn. According to her professor, Karl Weierstrass, in Berlin, each of these papers would have sufficed for a doctorate. Years later, in 1874, she received her doctorate, summa cum laude, from Göttingen University. She tried to get a university position, but was unable to. She left for Sweden, where she was appointed to the honorary position of Privatdozent, more or less at the level of today's associate professor. Soon, however, she received an extraordinary professorship from Stockholm University—the first time for a woman in Northern Europe (there had been two women professors in Italy).

Sonia edited a new journal, *Acta Mathematica*, and she returned to one of her earlier occupations, writing. As recognition of her activities, she received several prizes and became a corresponding member of the Imperial Russian Academy of Sciences in St. Petersburg, although in Russia she still could not obtain a university position. She died at the age of forty-one from influenza and pneumonia.

Sophie Kovalevsky was already famous during her lifetime, and when she travelled she was treated as a celebrity. Even the greatest mathematicians of her time considered her to be an important member of the European mathematics community. Besides, she was regarded as a promising writer, and she was friends with many celebrated authors, such as George Eliot, Anton Chekhov, and Henrik Ibsen. She was also a political activist and a fighter for women's rights. Her fame did not fade away upon her death; books have been written, films produced, stamps issued, and busts made that memorialize her. Negative opinions also appeared, trying to defame her with all sorts of accusations, relating to both her mathematics and her private life. Ann Koblitz writes, "her life was so colorful and her achievements so outstanding"[5] that it is not surprising at all that this happened; society was not yet used to accepting so much from a woman—talent was supposed to be man's privilege.

Lina S. Stern (biochemist)

Lina Solomonovna Stern (1878–1968) was a biochemist-physiologist. She was born in Libau in Courland, then in the Russian Empire (today Liepāja, in westernmost Latvia). She was among the first women to study at the University of Geneva. After graduation, she stayed there and carried out research in biochemistry and neurology. She was the first female professor at the university. In 1925, at the invitation of the Soviet government, she moved to Russia, where she continued her active research involvement. She was head of a laboratory in a medical school in Moscow. In 1929, she became the first director of the Institute of Physiology of the Soviet Academy of Sciences. She was successful in her research, and her most important result was in the study of what is called today the brain-blood barrier. She was much valued both in the young Soviet state and in the West. She was elected to the Leopoldina Academy in Germany and in 1939 she became a full member of the Soviet Academy of Sciences—the pinnacle of a scientist's career in Russia; she was the first woman on whom such an honor was bestowed.

In 1939, she became a member of the Communist Party, and in 1943 she received the Stalin Prize. During World War II, many of her research achievements were turned to practical use. She introduced novel treatments for neurological disorders, and her procedures saved thousands of lives on the front. Eventually, she became a victim of Stalin's anti-Semitism as well as his distrust of science and scientists during the period between the end of the war and his death in 1953. First, she was stripped of all her positions. She was a member of the Jewish Anti-Fascist Committee and of the Soviet Women's Antifascist Committee. The former was an organization originally set up by the Soviet government to mobilize Jews for the struggle against Nazi Germany worldwide. At one point, all members of the Jewish Antifascist Committee were arrested, tried, and—with the exception of Stern—executed. She was incarcerated and later sent to internal exile. She was allowed to return to Moscow only after Stalin's death. She was exonerated, and her membership in the Academy of Sciences was reinstated. She continued her research activities and headed the Department of Physiology at the Biophysics Institute until her death in 1968.

In tsarist Russia, science was open for women only in a very limited segment of society. Soviet power brought about major changes, and not only for women. It introduced equal rights for all citizens, and thus it extended the opportunities of education and involvement in science to the vast majority of the population. Women could study and find jobs, and for scientists this meant jobs in the educational system and at research institutions. The Bolsheviks understood that women's participation was important because they could not afford to not utilize half of the society in productive work. There were propaganda campaigns showing that women were able to perform in traditionally male occupations, even as pilots, sea captains, driving heavy machinery, and, of course, in science. But even powerful propaganda cannot change traditions, and the task of taking care of the children, old parents, the husband, and doing household chores largely remained women's duty. They worked in their jobs and had to take care of their family and household during the rest of their time. Progress was especially slow in those republics with especially strong traditions of male-dominated societies.

My husband told me about his experience of a visit to Baku, the capital of the then Soviet republic of Azerbaijan. In the early 1980s, he visited the Institute of Crystallography and attended scientific discussions in which both male and female members of the Institute participated without any difference. The wife of one of the leading Azerbaijani crystallographers was one of the respected and active members of the Institute. In the evening, my husband was invited to their home for dinner. The dinner guests, the host, and the couple's teenage son sat around the ornate table, but the wife—the respected crystallographer—did not join in, not even after the dinner was over and the friendly conversation continued. My husband learned that women had no place in the company. There was thus a split attitude—equality at the work place and traditional restrictions at home.

Full gender equality did not happen even in science. The vast majority of women held lower positions, and not only because they had just recently joined the workforce. Some of them eventually learned how they could advance in science; there were a variety of possibilities. Some joined the Communist Party and manifested appreciated activities in the trade unions, which did not have the role of protecting employees' rights and were subservient to the regime. Others chose branches of science that were especially important for the new regime and badly needed experts; geology was such a branch, for example, and it often involved going on expeditions to faraway regions of the country. Yet another way of helping themselves advance was marrying a professor or a fellow student with a promising career. Actually, marrying a professor and thus ensuring participation in scientific work was a common phenomenon from the start of higher education of women in the late nineteenth century.[6] Even later, a well-positioned husband could do a great deal to advance his wife's career.

The Soviet system existed for seven long decades, and it would be unjustified to consider it as uniform in its handling science and scientists. A good example is the situation of Jewish scientists. Under the czars, there were severe restrictions for Jewish participation in higher education, which were often unadvertised but ruthlessly observed. The first period of Soviet power lifted these restrictions. The situation markedly changed in the period between the end of World War II and Stalin's death, characterized by drastic anti-science and anti-Semitic actions. Covert, but rigorously observed, limitations on the number of Jewish students and Jewish associates of scientific research institutions remained in effect throughout the existence of the Soviet Union. Whereas the restrictions under the czars did not extend to women, and there were hardly any in higher education, the restrictions in the Soviet system applied to women and men alike.

Information about the current situation of women scientists in Russia is scarce, but indications are that the diminishing number of female researchers along the academic ladder is similar to if not worse than it is elsewhere in the world. Full membership in the Russian Academy of Sciences is the highest point scientists can reach in science in Russia. According to recent data, of the 528 full members of the Academy, there are ten women, a mere 2 percent, probably the lowest among the industrialized nations. There are, however, indications that the problem manifested in this disproportion is getting attention, and change may be expected in the near future. Unfortunately, whatever change in this respect there will be, there is an unmistakable tendency in present-day Russia that science is rapidly losing its lure and respect, and, significantly, its financial support is sliding as well.

Women scientists in Imperial Russia traveled widely and thus had the ability to learn about new research results in the West and interact with their colleagues there. Isolation became one of the most serious problems in the Soviet Union—not just for women, but for everybody. Isolation of scientists in Russia has eased since the collapse of the Soviet Union, but it seems that science in Russia has a long way to go to achieve full integration into international interactions.

Ada S. Kotelnikova (chemist)

In Soviet times, isolation often hindered promising scientific areas from developing fully, especially if they were not strategically important for the military. The story of Ada Kotelnikova (1927–1990) comes to mind. In the 1950s, the group of V. G. Tronev had an ambitious research project producing new rhenium compounds and determining their properties at the Institute of General and Inorganic Chemistry of the Soviet

Academy of Sciences in Moscow. The element rhenium, a metal, was discovered by Ida and Walter Noddack in Germany (see a separate chapter about them). The Noddacks produced some compounds of rhenium, but they could not isolate and reliably identify them. Tronev was a recognized leader in the field in the Soviet Union and headed a productive laboratory. In the mid-1950s, in the course of these studies, a young researcher, Ada S. Kotelnikova, produced several new compounds in which she supposed that there was a direct rhenium-rhenium bond. Such direct metal-metal bonds were considered quite unusual at that time. These direct rhenium-rhenium contacts turned out to be strong linkages. Kotelnikova published her findings in a paper jointly with Tronev.[7] After her discovery of these rhenium compounds, Kotelnikova moved on to other areas of inorganic chemistry.

Somewhat later, the internationally renowned American F. Albert Cotton became involved in the chemistry of rhenium compounds, and he discovered that the rhenium-rhenium interaction—discovered but never fully understood by Kotelnikova—was exceptionally strong. It was about four times stronger than the usual "single" bond, and Cotton named it "quadruple bonds." Had Kotelnikova been in close contact with international colleagues, she might have herself produced further important results in this unusual chemistry. In his turn, Cotton added a tremendous number of new results to the field, but this did not change the fact that the virtually unknown Russian scientist, Ada Kotelnikova, had made the pioneering steps in this area of research. She was a young researcher when she made the discovery and then faded into oblivion.[8] When, in 1968, her institute celebrated its fiftieth anniversary, the Soviet post office issued a stamp to honor the event. The stamp prominently displayed one of Kotelnikova's substances with the direct rhenium-rhenium linkage. By then nobody remembered her pioneering contribution, but everybody was aware of F. Albert Cotton. My husband and I met her accidentally, stumbled onto her story, and have tried to keep alive the memory of her contribution.

Five Russian scientists are introduced in detail in this section: a physicist, three chemists, and a mechanical engineer.

IRINA P. BELETSKAYA

Chemist

Irina P. Beletskaya in 2004 in Budapest. (photo by M. Hargittai)

Irina Petrovna Beletskaya was a rising star in Soviet chemistry in the 1950s and 1960s. The 1950s, especially the early years of the decade, were a difficult period for Soviet chemists. In 1951, a national meeting condemned the theory of resonance and for a long time turned away gifted young people from theoretical chemistry. This was the time when Beletskaya was a budding scientist. Although she apparently stayed out of these affairs, her superiors must have been deeply involved in them. Beletskaya succeeded in staying apolitical and focusing on her own career, which took off spectacularly.

She was born in 1933, in Leningrad (today St. Petersburg), and must have had a hard childhood, which included the Siege of Leningrad in World War II. When I posed a set of questions to Irina in 2004, her responses were laconic, and to some of my questions she did not respond at all. Among the unanswered questions was the one about her childhood experience.

In school, she liked mathematics and literature, but she did not consider these as possible professions. Rather, she entered the chemistry faculty at Moscow State University. After graduation, she stayed there for her candidate of science studies for the PhD-equivalent degree that she received in 1958—very young, but not extraordinarily young for this degree. It was more unusual that in a mere five years, in 1963, she earned her higher doctorate, the doctor of science degree. In the Soviet—and now Russian—system, this degree is a prerequisite for a professorial appointment. Irina was eventually appointed professor of organic chemistry at Moscow State University, the flagship university of the Soviet and Russian system of higher education. She became active in both teaching and research, and her career did not experience any slowdown; she was elected a corresponding member of the Science Academy in 1974 and was made a full member, an academician, in 1992.

Beletskaya did a lot of work in uncovering the mechanism of organic reactions, and in this she worked together with her mentor, Oleg A. Reutov, who had had an even more spectacular rise on the academic ladder and had become an academician already at the age of forty-four. To understand the way chemical reactions happen opens up the possibilities of constructing new reactions for producing desired compounds with useful properties. This has been a long-standing goal of organic chemists. Beletskaya initiated new approaches in chemical syntheses; she involved rare earth metals and produced new compounds that could be expected to become catalysts in heretofore unknown chemical reactions. Some of her reactions have led to useful applications in industry.

Catalysts make chemical reactions easier to carry out. For example, a chemical reaction may happen at a very high temperature, but in the presence of a catalyst it may happen at a considerably lower temperature. When such an approach is applied in an industrial process, enormous amounts of energy may be saved. Usually, small amounts of catalysts suffice for such functions, and even that small amount can be recovered at the end of the reaction. Recently, Beletskaya's interests have expanded to so-called green chemistry. The label refers to environment-friendly chemical processes. Catalysts play a decisive role in green chemistry.

I corresponded with Beletskaya and met her in person as well, but never felt that I could penetrate an invisible wall that she seems to have built around her. According to others who have known her much longer than I, she must have built this wall from early youth. I have posed her questions and more questions, but received only very brief responses. In our personal encounter I found her friendly and seemingly communicative but never opening up to any extent. Some of her responses were puzzling, indicating depths that remained under a lid. What could I do, for example, with the response she gave me to my question about whether she was religious. She said: "I am not religious, which, of course, I regret."[1] But let's go through some of the questions and her responses in a variety of areas. The questions may be pruned, but her responses are quoted verbatim (in translation from the original Russian). The exchange took place in 2004.

What turned you to science?
That work would not be monotonous.

What made you become an organic chemist?
My interest in substances.

Has politics ever influenced your work or career at any time?
Never.

What do you expect of science in the 21st century?
I would not take upon myself to guess; everything is progressing so fast—who could have predicted cloning?

Asked about her childhood and about her memories of the war.
Life was difficult rather than interesting; I remember the war very well; father was a border guard.

About her family and about how family conditions impacted her career and work.
My son is 42 years old; my husband always helped me to the extent he could; we have been together for 43 years.

I asked a number of questions concerning her spectacular career, especially for a woman, and one of the questions was about the long period from having been elected corresponding member of the Academy in 1974 to full membership in 1992. (Here we need to remember that many corresponding members never reach the stage of full membership; the promotion is not at all automatic.) Further questions included her present ambitions, the situation of women scientists in the Soviet Union and in the new Russia, the impact of the political changes especially in view of her spectacular success under the Soviet system, the development of international interactions, foreign travel, and suchlike. She gave an umbrella response to this whole set of questions.

> The long period (from 1974 to 1992) was connected exclusively with my bad character; I could not have an official career because I was not a party member without which I could not have even become head of a laboratory. My ambitions were always connected first of all with my own evaluation of my results and with the opinion of my foreign colleagues (since this determined all the invitations to conferences); being a woman makes it more difficult under any circumstances. Now it is easier as they let us go abroad and I travel often. In Soviet times, contacts and western grants were impossible; now this is possible, however, currently we have no support from our government.

In her activities in the international community of chemists (see below) she was involved with the issue of eliminating chemical weapons: "It was one of the many projects I had been involved with, and it was quite successful. Chemical weapons—this is a ruthless thing, they have been produced to cause pain for people."

What was most important in your life?
The health of the people close to me and the results of my work.

Have you ever experienced discrimination for being a woman?
Just as all others, not more.

What would be your advice to young women concerning professions, careers in science, building up a family?
I never give advice.

How do you compare women scientists in Russia with women scientists abroad?
Life for women abroad is easier, that is, just everyday life; everything else is the same.

Do you think your early success in your career had anything to do with your being a woman?
Not at all; rather, the contrary.

Anything else you would like to communicate, but I did not ask?
You did not ask about what traits I value in people; my response would be, in addition to having a sense of humor and self-control, I value the ability to apply irony to oneself.

Beletskaya held important functions in the organization of IUPAC—the International Union of Pure and Applied Chemistry—from the 1980s on. Of course, Soviet representatives used to be delegated there by higher organs rather than elected by their peers. But she must have gained the trust of her colleagues, because by the early 1990s she had arisen to leadership of the Division of Organic Chemistry of IUPAC, and until 2001 she served on the IUPAC Committee on Chemical Weapons Destruction Technologies (CWDT). She has earned a number of awards and other distinctions, both in the Soviet Union/Russia and internationally.

RAKHIL Kh. FREIDLINA

Chemist

Rakhil Kh. Freidlina. (courtesy of Jan J. Kandror, Wiesbaden)

Born into an impoverished working-class Jewish family in the Russian Empire at the beginning of the twentieth century, growing up deaf for years, and becoming head of a large laboratory of the Soviet Academy of Sciences (SAS) was a long way to go. It needed luck, perseverance, and special circumstances.

Rakhil Khatskelevna Freidlina (1906–1986) was born in Samoteevichi, in the Mogilevskii Region of what is now Belarus. Her father was a workman and her mother a homemaker, and she had many siblings. Her family was poor, and her childhood coincided with World War I, the revolutions, and the civil war in Russia. In addition, she suffered from a special predicament, deafness, during her preschool years. Her hearing started slowly developing only after she had begun attending school. Her condition made her very sensitive and perceptive. It was a unique experience for her when she started not only hearing but also learned listening. This she developed to such an art that she could remember the lectures she attended without having taken notes. A friend of hers called the change a transition "from sound vacuum to sound symphony."[1] Even later in her life, if she heard a poem that she liked, the next day she could recite it.

Rakhil and most of her siblings went on to higher education and became scientists or medical doctors under Soviet rule. She studied at the chemistry department of Moscow State University, followed by her PhD-equivalent candidate of science degree in 1936, under the mentorship of the renowned chemist and science administrator Aleksandr N. Nesmeyanov. By then, she was already an associate of the Institute of Organic Chemistry of the SAS in Moscow. Soon after Nazi Germany attacked the Soviet Union on June 22, 1941, the institute moved to Kazan, and she spent the period 1941–1943 there. She received high awards for her achievements in defense-related activities.

In 1954, she moved to the newly founded Institute of Element-Organic Compounds (INEOS) of the SAS. INEOS was founded by Nesmeyanov, and he made sure that the institute had conditions conducive to successful research. He was in the best position to ensure such conditions, as between 1951 and 1961 he was president of the SAS. He was an outstanding chemist, innovative and knowledgeable, but his inflated list of publications would make it difficult to realistically evaluate his achievements. He served the Soviet system unconditionally, but was braver than most institute directors in hiring Jewish scientists. He was a seasoned politician. When Freidlina's husband was arrested—on trumped-up charges, as often happened—he immediately removed her from her position of being in charge of her laboratory. This could be interpreted as a move of expediency or as a move to protect her by diminishing her visibility. On her part, Freidlina was grateful to Nesmeyanov for making her invisible during this most trying period of Stalin's last years when his anti-Semitic terror reached its peak.

Rakhil's husband, Georgii E. Syroezhkin, fought on the front during World War II. In Kazan, she and their eleven-year-old son lived in a small room which they shared with another family. When they returned to Moscow, the three-member family lived in a room not larger than a hundred square feet in a communal apartment. This meant many rooms with one family living in one room each, all sharing a common hall, toilet, and kitchen. Freidlina spent most of her time in the institute working for her higher doctorate, the doctor of science degree, which she received in 1945. Eventually she became the head of the Laboratory of Synthesis of Element-Organic Compounds of INEOS.

Her highest recognition came in 1958, when she was elected a corresponding member of the Soviet Academy of Sciences. Not all corresponding members become academicians, and neither did Freidlina, but in their local environments a corresponding member still wields great authority. Like other members of the SAS, she was entitled to considerable perks and privileges. I hasten to add that her associates and her peers in her field respected Freidlina for her knowledge rather than her position. Nesmeyanov said of Freidlina, "Sometimes the extraordinary clarity of her thinking makes me feel terrible." Then, he added, "You must understand, she has a man's mind!"[2] Her life motif was her research. She was also devoted to the Soviet regime, which provided extraordinary opportunities for her, coming as she did from the most disadvantaged layer of Russian society.

Her devotion to the regime does not mean that she was not bothered by some of the actions she was expected to take, although she took them anyway. One of her former pupils told the following story. Some time in the early 1970s, when Freidlina arrived in the lab, she noticed two of her associates reading a newspaper. She had not seen the latest newspaper yet and looked through the pages intently. As it turned out, she was looking for a statement by scientists that she had been obliged to sign the previous day. She read the statement—it was condemning Israel for some military action—and said with a sigh that it was not as bad as she had expected. She made it obvious to her associates that she had had to sign the statement without having read it.

Freidlina started her research in metal-organic chemistry and gradually expanded her activities to the most diverse areas of element-organic chemistry. One of her most successful areas of research was the utilization of the telomerization reaction, a chain reaction of unsaturated organic molecules (that is, of molecules with one or more double bonds). One of the reactants participating in the reaction is called a telogen, and it acts as the carrier of the chain. In the process, the telogen splits and forms reactive radicals that attach to the ends of the unsaturated molecule. She did not invent this reaction, but she developed it in order to prepare large classes of new compounds. She was a prolific author. Her first publication appeared in 1934 and altogether, she had 740 publications and she had an enormous number of coauthors on her research reports. Over two hundred different names appeared on her papers, including excellent associates from the most diverse areas in the country, among them Uzbekistan, Kazakhstan, and Armenia.

Rakhil was a typical product of the harder Soviet times also in that she did not speak foreign languages, hardly ever traveled abroad, and had no international interactions. However, she read the scientific literature in German and in English without any difficulty. Her real home was her laboratory and her real family was the collective of her associates. She was not happy with either her son or her daughter-in-law, but she adored her grandson. For her coworkers, Freidlina was a real "Jewish mother." She was always ready to help. She was most circumspect in playing the role of the boss. When she wanted to praise a suggestion, she liked to say "genidea," that is, genial idea. She had a way to generate enthusiasm and make her coworkers work and think more. She did not teach her associates directly, but educated them by example. She was polite and respectful; she kept her promises, but made only promises that she could keep. If she had to make a critical remark to somebody, she never did it in the presence of others. Freidlina was much appreciated, not only by those who loved her but even by those who did not.

ELENA G. GALPERN

Computational chemist

Elena G. Galpern. (Courtesy of E. Galpern)

In 1985, the discovery of a new form of carbon, beside graphite and diamond, made the news. The molecule of this new form looked exactly like a soccer ball consisting of sixty carbon atoms at the apexes of a truncated icosahedral shape and was predicted to be "superstable." The discoverers in due course received the Nobel Prize.

There was a lot of excitement, and as time passed, we learned that more than a decade prior to the celebrated discovery, there were two publications in which the stability of such a C_{60} molecule was already predicted. Unfortunately, both of these publications appeared in not very accessible journals, one of them in Japan and the other in the Soviet Union, although the Soviet journal existed in full English translation in the West. The Japanese paper by Eiji Osawa reported a *suggestion* of the truncated icosahedral shape of a C_{60} molecule. The Russian paper, independent of the Japanese study, came to such a conclusion based on quantum chemical computations that were quite sophisticated for their time. Elena G. Galpern (1935–) was a coauthor of the Russian publication.

In the early 1970s, Elena was working for her PhD-equivalent degree at the Laboratory of Quantum Chemistry of the Institute of Element-Organic Compounds

(INEOS) of the Soviet Academy of Sciences; she was a research associate of Dmitrii A. Bochvar (1903–1990), who had established the laboratory years earlier and determined the general direction of its work. The director of the Institute, Aleksandr N. Nesmeyanov, at one time thought of the idea of producing cage-like molecules consisting of carbon atoms, with the cage accommodating an atom of a different element, or even small groups of atoms. Nesmeyanov envisioned a plethora of uses for such substances. The first task was to find the carbon cages that could house such "hetero" atoms. The emerging technique of quantum chemical calculations appeared to be a convenient means to see whether such an approach would be feasible, and it seemed to be an appropriate dissertational project for Elena.

The carbon cages were quite large systems for the then available computations, hence they had to start with the smallest cages and gradually advance toward the larger ones. This is how she arrived at systems consisting of sixty carbon atoms. She had to find a shape that would be sufficiently stable; in the course of her computations, she had already tested many shapes, but there were so many other possibilities that testing them all seemed to be a hopeless task. One day, her senior colleague, Ivan Stankevich, who had just returned from a soccer match, suggested testing the shape of the soccer ball. It was only a few years before that they had started producing soccer balls by sewing together pentagonal and hexagonal patches. The ball is, of course, a sphere, but the polyhedral analog of consisting of pentagonal and hexagonal sides is the truncated icosahedron possessing sixty apexes. Elena's computations showed that such a shape of sixty carbon atoms would indeed be stable. In soccer, of course, the twenty-four players of the two teams kick that shape for ninety minutes and it withstands the abuse—it has to be sturdy. The logic was that an all-carbon molecule of such shape should also be stable.[1]

When Galpern completed her computations, she prepared a manuscript for publication, and her boss, Bochvar, decided to bring it out in the most prestigious Soviet periodical, *Doklady Akademii nauk SSSR* (Proceedings of the Soviet Academy of Sciences).[2] It duly appeared in 1973, and although an English translation of the journal made the article accessible also to non-Russian readers, the discovery went unnoticed both in the Soviet Union and in the rest of the world.

The paper appeared under the names of Galpern and Bochvar. In 1985, a team of scientists, Harold Kroto of Sussex University, England, and Richard Smalley and Robert Curl of Rice University, Houston, Texas, and their students observed the stable C_{60} in an experiment at Rice University. When they suggested the truncated icosahedral shape for it, this was a turning point in the story. This beautiful molecule was no longer just in Osawa's dreams and Galpern's computations, but had been observed in an experiment. Soon, combing the literature, the researchers spotted Osawa's suggestion and Galpern's computations. Osawa and Galpern received considerable attention, though not as much as Kroto, Smalley, and Curl, who received the 1996 Nobel Prize in Chemistry for their observation.

Elena was somewhat bewildered by the sudden publicity, even if it was short-lived. She had had a quiet career, nothing spectacular; all her working years she spent at INEOS without realizing that she had published her most important result in the course of her PhD studies. One could speculate how Galpern's career might have turned out had she or her professor realized the importance of their C_{60} molecule. What might have happened had somebody noticed her report and tried to produce the C_{60} molecule and thus recognized the importance of her work years earlier? In any case, nobody at INEOS continued this line of research. It seems that Nesmeyanov might have lost interest in the carbon cages and directed the attention of his associates elsewhere. When fame finally reached Galpern, she did not delve much on what might have happened, but she was happy to see so much interest in her early work, even if belatedly.

IRINA G. GORYACHEVA

Mechanical engineer

Irina G. Goryacheva in 2012 in her office. (courtesy of I. G. Goryacheva)

Statistics show that among all professions, engineering most often gets the label "unfeminine," since it has the fewest women. Even within engineering, one of the branches that is considered unlikely for women is tribology, a branch of mechanical engineering and materials science that deals with the design, friction, wear, and lubrication of interacting surfaces in relative motion. Not only did Irina Goryacheva choose tribology for her specialization, but she has become very successful in it. She is a full member of the Russian Academy of Sciences an academician, the highest and rarest distinction in science in Russia, and she has received high awards in her field both at home and internationally. Her latest recognition was the 2009 Tribology Gold Medal, which she received from the Institute of Mechanical Engineers, London.[1]

Irina was born as Irina Georgievna Mitkevich in 1947 in Ekaterinburg (Sverdlovsk, in the Soviet period), a large city in the Ural Mountains region, over a thousand miles east of Moscow. This is how she described her early years[2]:

> My parents lived there already before the Second World War. My mother moved to Neviyansk (a small town near Ekaterinburg) after graduation from Stalingrad Polytechnic University (Traktornii Institute) in 1937, where she worked at the

Irina G. Goryacheva receiving the Tribology Gold Medal and certificate in 2010 at the UK Embassy in Moscow. (courtesy of I. G. Goryacheva)

metallurgical plant. Her specialty was the pressure treatment of metals. There she met my father who graduated from the Chemistry Department of Ural State University in Sverdlovsk and also worked at the plant. In 1940, they moved together to Sverdlovsk because my father entered postgraduate education at Ural State University. This was the same year, when my brother was born. Later, my father received his doctorate and worked as professor of chemistry at different universities.

When I was two years old, my family moved to the town Niznii Novgorod near the Volga River. During my school years my family lived in Tolyatti,[a] near Samara, also on the Volga River. My father was invited to be professor at the new Polytechnic University there. My mother was a teacher in metal treatment at the Polytechnic School. Nature is very beautiful around Tolyatti. I liked to swim in the artificial sea (Zigulevsky sea) near the Samara hydroelectric station, to hike in the Zigulevsky mountains, and skate and ride a bicycle. I especially liked skating; later, when I was a student at Lomonosov Moscow State University [in short, MSU] I continued to skate—I was in the skating team of the University and participated in different competitions.

During her school years, Irina liked mathematics, physics, and chemistry. She competed in Olympiads in these subjects, and often she was the winner. Mathematics was her favorite: "I learnt more than what was in the school program from the books in mathematics that my parents bought me. I liked to try solving the tough problems and was happy when I succeeded." It is no wonder then that she wanted to go to the top university in the country.

[a] The town was named after the Italian communist party leader, Palmiro Togliatti, in 1964.

MSU is the best university in Russia (and so it was in the USSR). After graduating, in 1965, from my school in Tolyatti with a gold medal, I decided to try to enter this university. There were many other young people thinking like me. There was a very strong competition, 13 candidates for one place. All of us had to pass four exams: two in mathematics (in written and oral forms), one in physics, and one in Russian composition. Based on my results I was accepted. I had the choice among pure mathematics (Division of Mathematics), mechanics (Division of Mechanics), and informatics (Division of Numerical analysis and Informatics). I decided to study in the mechanical division because I liked the applied problems related to new materials, industrial devices, etc. My mother had told me about some of them and I was interested in this field. Already during my last school years, I visited some of the big plants that were built in Tolyatti and did some practical training there. In my university group, there were only a few women students.

After graduating from MSU, I stayed on for post-graduate education. My supervisor was Professor Lev A. Galin, a well-known specialist in mechanics. His first book in contact mechanics, published in 1953, was translated into English in 1961, and it was on the desk of most scientists in the field. I was fortunate to be his student and do my PhD thesis under his supervision. He made me interested in scientific research, and showed me the beauty of analytical solutions of mechanical problems. My thesis was devoted to the study of rolling contact of viscoelastic bodies and modeling of rolling friction. After my PhD in 1974, I joined the Institute for Problems in Mechanics of the Russian Academy of Sciences (RAS), where Professor Galin was the head of the laboratory.

I studied the roughness effects in contact interactions and developed a method to calculate the evolution of the stresses in contact due to wear process. This method is used to predict the lifetime of various junctions (bearings, gears, seals, piston rings, etc.). In 1979, I was awarded the highest prize for young scientists in the former Soviet Union—it was called the Lenin Komsomol Prize—for my achievements in science and technology.

In 1988, Irina received her higher degree, doctor of science, and in 1996 she was appointed Head of the Tribology Laboratory of the Institute. At the same time, she accepted a professorship at the Moscow Institute of Physical Technology (MFTI), where she had been giving courses in mechanics since 1979. In 1997, she was elected a corresponding member of the Russian Academy of Sciences and unusually soon, in 2003, she was elected a full member.

She is an expert in contact mechanics, the study of solids in contact with each other. It involves mathematical modeling and the investigation of what happens when two solids are in contact under pressure or when two solids in contact move relative to each other. The latter is the essence of tribology, and Irina is most interested in these aspects of mechanics due to their industrial significance. As I try to find examples of tribology, what comes to mind is the interaction of car tires with the surface of the road and the interaction of train wheels with the surface of the track. Thus, even if her actual work may sound above our heads, the applications side appears familiar. Railway companies, the tire industry, machine construction companies, and others have utilized her findings in choosing the most appropriate materials for their products and in deciding specific surface treatments for them.

Irina teaches courses at MFTI and at MSU, has graduate students, and performs functions at the Russian Academy of Sciences.

> I am the President of the Russian National Committee in Theoretical and Applied Mechanics (elected in 2011). This Committee was founded in 1956. The first president was the famous Academician N. Muskhelishvili. Now the Committee includes 450 members. We organize conferences in mechanics in Russia and connect with IUTAM (International Union of Theoretical and Applied Mechanics) in all their activities, including the organization of the IUTAM Congresses. I am also the Chair of the Scientific Council in Tribology of the RAS. The Council organizes collaboration between the tribologists in Russia and abroad, conferences in tribology, and so on. At this time, we are preparing the publication of the Russian version of the *Encyclopedia of Tribology*. Of course, I am very busy, and try to organize my work in the optimal way.

She is a member of the leadership of the Department of Energetics, Machinery, Mechanics and Control Processes of RAS. She is the only woman in the divisional leadership, but there is now one more woman member, newly elected, in the department. Irina has written four monographs in the field of contact mechanics and tribology and coauthored the first manual in the field, *Fundamentals in Tribology*.

A little about her family life. She met her future husband, Alexander Goryachev, at MSU, and they married in 1972, during their doctoral studies. His specialty is pure mathematics, and he is a professor at the Moscow State Engineering Physics Institute. This institute was founded in 1942 by Igor Kurchatov, the father of Soviet nuclear research; their current work is in nuclear and elementary particle physics. The Goryachevs have one daughter, Ekaterina, born in 1974. She graduated from the Faculty of Biology of MSU, and received her PhD in the field of biophysics. Now she works at the Shemyakin and Ovchinnikov Institute of Bioorganic Chemistry of the RAS.

We knew from our friends how difficult it was to raise a child in the former USSR, especially when both parents were dedicated scientists. I wondered how Irina and Alexander managed:

> When my daughter was young, my parents helped me very much in bringing her up. They lived in Kaliningrad near the Baltic sea, and every summer we moved there for our vacations. My daughter started to go to kindergarten when she was three years old. The director of my institute helped me to get the place for her. The kindergarten was near our home and I, or Alexander, took her there. We also got

our daughter to skate twice a week. She continued to learn figure skating during her first school years.

Ekaterina and her husband decided to have more children. They have four children, so I am very busy as a grandmother. Two of my grandchildren go to school now, one granddaughter goes to kindergarten, and the youngest one is at home with her other grandmother. Every weekend all the children are in my house and we spend Saturday and Sunday together. I like this time very much.

I asked Irina whether my impression is correct that scientists used to be held in much higher esteem in Soviet times than they are in today's Russia: "Yes, this is so. In the former Soviet Union, scientists were at a high-level position in the society. Young people felt that it was very important to get high-level education, and then to work in universities or research institutes. There was a strong competition between them and a strong desire to become a scientist or a good engineer to work at industrial development. Now young people choose to work in financial institutions, where the salary is much higher than in research institutes."

For young people who are interested in a science career, Irina Goryacheva can be a role model. She reached a position in Russian society that few do, either men or women. She has been a good teacher and advisor for students, and she has produced outstanding results in her chosen "unfeminine" field.

It is fitting to quote from the laudation of her recent award from the London Institute of Mechanical Engineers: "For clearness in expressing the basic concepts, as well as for her rigor in applying these concepts to overcome problems in engineering, Academician Goryacheva is a shining example of the academic teacher and researcher. Her contribution to science and engineering and her outstanding service, not only to Russian Science, but also to Science worldwide, makes her a worthy recipient of the world's highest award in tribology, the Tribology Gold Medal."[3]

ANTONINA F. PRIKHOTKO

Physicist

Antonina Prikhotko with her husband, physicist Aleksandr Leipunskii, in 1928 in Leningrad. (courtesy of Boris Gorobets, Moscow)

Within the borders of the Soviet Union, Antonina Fedorovna Prikhotko (1906–1995) was considered to be a great physicist, but she was hardly known to the outside world. She never ever set foot outside the Soviet Union. She refused to go through the humiliating application procedure, which would have included an investigation of her personal affairs by the organs of the Communist Party and state security. She once attended a party meeting in which she received firsthand experience of how unashamedly people's personal affairs were poked into. After that, she refused even to consider joining the Communist Party. As for foreign travel, she declared that she would travel to Paris when they, meaning the authorities, would bring her a prepared passport and visa. Needless to say, they never did.

Had she traveled to the West, she would have been the best advertisement for the Soviet Union, according to the biographer of her family.[1] She was the only female physicist and full member of the Ukrainian Academy of Sciences, and she was beautiful (this does not sound politically correct today, but it was always remarked on whenever she became the subject of conversation, and she lived in times before political correctness).

She was born into a Cossack family in Pyatigorsk. In 1923, she became a student at the Leningrad Institute of Technology, majoring in physics. There, she met a fellow student a few years her senior, Aleksandr Ilyich Leipunskii, and they married in 1926. He came from a Jewish family, and eventually that carried additional burdens on top of all the other difficulties in Soviet life. In 1930, she and her husband moved to Kharkov (then the capital of Ukraine), where they became associates of the Ukrainian Physical-Technical Institute (UFTI). Soon they had a daughter and Leipunskii was appointed director of UFTI. However, terrible times were coming. The years 1936–1938 have become known as the period of the Great Purges or Stalin's Terror, when people—including those in high military or civilian positions—were arrested on trumped-up charges, tried in show trials, and executed right away or exiled to slave labor camps for many years. Some of the brightest physicists of UFTI met such a fate, and Leipunskii was anticipating that his turn might also come. He was a leading physicist on a national scale, and under his directorship UFTI had become a leading institution in physics in the Soviet Union. His visibility certainly added to his being exposed to the severe danger represented by being labeled an "enemy of the people." Leipunskii was fully aware of the possibility, and he and Prikhotko discussed what they should do in case he was arrested. This much is known from their family members. What we cannot know, but can guess from their subsequent actions, was what their decision might have been.

Before anything happened, as a precaution they sent their daughter away to Prikhotko's relatives, far from Kharkov. Then, one day in 1938, the secret police did arrest Leipunskii. On the second day after his arrest, Prikhotko publicly condemned her husband for having lost vigilance in directing UFTI and letting German agents penetrate the institute. She categorically separated herself from him. Leipunskii might have anticipated the harshest sentence for his "crimes," but to his good fortune, his case happened during the subsiding period of the Great Purges. After two months of incarceration, he was let out of prison—to this day, nobody knows what really happened.[2]

We can only suppose that Prikhotko acted the way she did by their prior agreement, because after his liberation they continued their harmonious married life as if nothing had happened. When their daughter returned from the family visit months later, she did not notice anything different, and she learned only years later about what had happened in her absence. She did not even hear gossip about his father's "adventure," because people did not discuss such matters in those days.

Prikhotko started research projects as a student and found an excellent mentor in Ivan V. Obreimov, a future full member of the Soviet Academy of Sciences. The investigation of cryocrystals was her project. Cryocrystals are substances that solidify only under extremely low-temperature conditions. Solid oxygen became her favorite substance of inquiry. She soon acquired her PhD-equivalent scientific degree, and in 1943 she became the first female doctor of science (DSc) in the physical-mathematical sciences in the Soviet Union. The DSc degree has no strict equivalent in American academia, and the habilitation of the German system is not very close to it either. The DSc degree in the Soviet Union and now in Russia corresponds to a significant amount of research achievements and is a condition for a professorial appointment at a university or a laboratory head appointment at a research institute.

Prikhotko's feat was the more significant because she completed and defended her DSc dissertation while she and her laboratory were in evacuation during the war years. This was in the city of Ufa, 1,436 kilometers (892 miles) from Kharkov in the northeast direction. She and her colleagues were carrying out projects to help the defense efforts. In addition, Prikhotko made sure to bring her experimental material, on which she then based her DSc dissertation.

In 1944, Prikhotko, upon her return from evacuation, organized a division of crystal physics in the Institute of Physics in Kiev, which had in the meantime become the capital of Ukraine. She continued doing pioneering scientific work, and she is credited with the experimental discovery of molecular excitons, the phenomenon of the movement of excitation in crystal structures from one crystal cell to another. Subsequently, other physicists worked out the theory of molecular excitons on the basis of her experimental discovery. In 1966, the work on excitons earned the physicists working on this project, among them Prikhotko, the Lenin Prize, the Soviet Union's highest distinction, which had also the peculiarity that it could be received only once (the Stalin Prize—later, State Prize—could be received repeatedly, with no restriction).

She was not only an outstanding physicist; she was also a friendly human being. She was respected, and her rigor in scientific interactions earned for her the label of "absolutely iron lady." This, however, did not reduce her popularity among her colleagues, and when the institute was undergoing a crisis, the president of the Ukrainian Academy of Sciences invited her to serve as the director of the institute. The leadership of the Academy understood that only a bona fide scientist could have sufficient authority before all coworkers to have them follow directions. She accepted the challenge for a five-year period, between 1965 and 1970. For quite a while, after she was no longer director, she remained the principal scientist of the institute. Her approach to being in charge was simple, even transparent. She adhered to the principle that you promise only what you can deliver, and then you deliver what you promised.

Prikhotko had one weakness—she smoked a lot, which may have been her escape from the terrible tensions of her husband's arrest, the war, the evacuation, and the difficulties of rebuilding science after the war. Later, even in "normal" periods, doing scientific research—in competition with the West while also being isolated from it—meant existing under tremendous tension. She withstood all these tests, and, apart from her smoking, has remained a role model.

WOMEN SCIENTISTS IN INDIA

In the fall of 2011, I spent two weeks in India giving lectures on my research, structural chemistry, and on women in science. The lectures were presented at three major science hubs: the National Centre for Biological Studies in Bangalore, the Tata Institute for Fundamental Research in Mumbai, and the Indian Institute of Technology in New Delhi. The reception of my talk about women scientists, the large audience, the question-and-answer periods, and the vibrant discussions afterwards with professors and students alike left an unforgettable impression on me. I also had private meetings with women professors and talked with them about their life and about women's role in Indian society. This section is based on these conversations.

I have to start with a caveat. When we talk about Indian women in science, at all levels from students to professors, we are referring to a very small segment of India's population, and of course I sampled only a tiny portion of this segment. Nonetheless, I believe that I learned a great deal from my limited experience. Traditionally, Indian society has a complicated structure involving classes, castes, and the traditions of a deeply patriarchal society. According to the 2011 census, the country still faces major problems of fighting illiteracy[1]; only 65.5 percent of Indian women can read and write, compared with 82.1 percent of men. This places an enormous pressure on the government, and tremendous efforts are being made to establish adequate schools in the many remote regions of the country to eradicate illiteracy. My discussion does not deal with the social stratification of Indian society, because most women who study—and especially those at the professorial and equivalent level—come from professional middle-class families. They are atypical in Indian society.

The situation concerning women is intricate and complex. On the one hand, there is widespread illiteracy. On the other hand, India enacted universal adult franchise at the time of independence in 1947, and as early as the 1960s, there was already a popular woman prime minister, Indira Gandhi, at a time when there had been no such development in most Western countries.

Even if not many, there were well-known women scientists in India in the late nineteenth and early twentieth century. I will mention three of them, from different fields, based on their short biographies in the book *Lilavati's Daughters: The Women Scientists of India*[2]: Anandi Gopal, the first Indian woman physician; Janaki Ammal, a botanist; and Asima Chatterjee, a chemist.

Anandi Gopal (physician)[3]

Anandi Gopal (1865–1887) was the first Indian woman to receive education abroad and became the first woman doctor of India. Unfortunately, she did not live long enough to practice medicine. Born in a small town near Mumbai, she was married at the age of nine, and her husband, who believed in women's education, decided to teach her to read, write, and speak English. She was such a good student that he wanted her to study a profession, and he decided to send her to the United States. At that time, there were no female doctors in India, and women were often ashamed to go to a male doctor with their problems. This was an added reason why Anandi's husband thought that learning medicine would be a good idea for her. She studied at the Women's College of Pennsylvania, in Philadelphia. In spite of living and studying in a very different culture from her own, she succeeded in getting her degree; alas, she contracted tuberculosis. By the time she got back home, her condition had worsened. The doctor of the ship taking her back to India refused to treat a brown woman, and the specialist they contacted after arriving home did the same because "she crossed the boundaries of society." She passed away at the age of twenty-two. Her life has inspired young girls ever since, and there is a fellowship in her name for young women who work in women's health.

E. K. Janaki Ammal (botanist)[4]

Edavaleth Kakkat Janaki Ammal (1897–1984) was born into a cultured, well-to-do family in Kerala, a southern state. In some of its regions, matrilineality was practiced (meaning that descent was considered through the mother and maternal ancestors). In these families, women used to have much more freedom than in other parts of the country and girls were encouraged to study. Janaki Ammal started her studies in India, followed by a master's degree and then a PhD at the University of Michigan. She was a teacher for a while, then spent a considerable period in England as a cytologist (a specialist concerned with the structure and function of the cell). She investigated sugar cane, and produced several hybrids for it. She made discoveries in her studies of the chromosome. In 1945, she coauthored a book with a colleague, *The Chromosome Atlas of Cultivated Plants.*[5]

The first prime minister of independent India, Jawaharlal Nehru, invited her back to India to reorganize the botanical survey of the country. She studied a great variety of plants, including medicinal plants. She had various leading positions and different duties, always in the service of the government. For example, at one point, she was the head of the Central Botanical Laboratory in Jammu.

She helped the Nobel laureate physicist Chandrasekhar V. Raman and sixty-three other male professors to establish the Indian Academy of Sciences in Bangalore, and she was the first woman to become a fellow. She received many other awards and distinctions. Her award citation from the University of Michigan thus characterized her: "Blessed with the ability to make painstaking and accurate observation, she and her patient endeavors stand as a model for serious and dedicated scientific workers."[6]

Asima Chatterjee (chemist)[7]

Asima Chatterjee (1917–2006) was born in Calcutta (now Kolkata), in the Bengali region of East India. Her father was a physician and amateur botanist. Her interest was in chemistry; she received her MSc and then her DSc degree from Calcutta University—she was the first woman to receive a doctorate from a university in India. Upon completing her education, she founded the chemistry department at Lady Brabourne College in Calcutta and served as its head for seven years. After spending years at American and Swiss universities, she moved back to Calcutta University. In 1962, she became the first woman in India to receive a prestigious chair, the Khaira Professorship of Chemistry.

Her lifelong interest was medicinal chemistry, and her studies of Indian medicinal plants contributed significantly to this field. She patented two successful drugs, one against epilepsy and the other against malaria. She published extensively. She was the chief editor of a book series, *The Treatise on Indian Medicinal Plants,* with emphasis on medical uses.[8] The Indian National Science Academy elected her as a fellow in 1960, and in 1975 she was elected to be the first woman general president of the Indian Science Congress Association. It was yet another first among the "firsts" in her extraordinary life.

Following are stories from my personal encounters with present-day women scientists in Bangalore, Delhi, and Mumbai. They include some better-known women, members of the Indian Academy of Sciences, and some of lesser fame. All are dedicated to their science.

CHARUSITA CHAKRAVARTY

Chemist

Charusita Chakravarty in 2011 in New Delhi. (photo by M. Hargittai)

Charusita's parents came from large families, and there was considerable exposure to a wide range of academic, social, and political interests when she was growing up. "My father was a very formative influence on me because he had a broad intellectual interest and he had a genuine interest and encouragement for whatever I might be learning or reading. Being born in the United States, I had the option to live in America but I decided to come to India and that was probably due to that upbringing."[1]

Charusita Chakravarty was born in 1964 in Cambridge, Massachusetts, when her father was on the economics faculty of the Massachusetts Institute of Technology. Her paternal grandfather was a judge and came from a family of lawyers. Her maternal grandfather descended from a family of chemists who started one of the first chemical factories in Bengal, and he wanted his daughter to study chemistry and take the factory over. However, Charusita's mother was not interested in chemistry and switched to economics. She and her husband used to teach at Delhi University. They traveled extensively, either studying or working in Europe and the United States.

Charusita is married to a physicist, Ramakrishna Ramaswamy, and they have one daughter. To the question whether their marriage was arranged, she answered: "Oh, no, I think my parents would have been in a fix if they had to do that. I married someone who is older than me and, also, he was divorced, and so it was somewhat controversial." They have been married over twenty years. Until recently, they were lucky because their jobs were in the same city, Delhi. In 2012, however, her husband accepted the position of vice chancellor at the University of Hyderabad, a thousand kilometers (about 630 miles) away from Delhi, so now they have a commuting marriage; her husband can only visit a couple of times a month. Remembering and appreciating the nontraditional, cosmopolitan way she herself was brought up, she said: "I want my daughter to be able to decide for herself how she balances all these difficult questions that women have to face."

Charusita received her BSc degree in Delhi; her PhD is from Cambridge University in the United Kingdom. She spent her postdoctoral years at the University of California at Santa Barbara. Presently, she is a professor of chemistry at the Indian Institute of Technology, Delhi. Her interests are in theoretical chemistry and chemical physics. She uses both quantum chemical and classical computer simulations in order to understand the behavior and properties of liquids, particularly with regard to phase transitions and self-assembly. One of her topics is the study of the anomalous properties of water and the implications for hydration. She is also interested in understanding the relationships between structure, thermodynamics, and transport properties of water-like liquids which share the thermodynamic and kinetic anomalies of water, despite very different underlying interactions. Although her work is theoretical, it has a bearing on understanding important processes such as protein folding and nanoscale self-assembly.

Already many awards and recognitions have come her way; she has been elected a fellow of the Indian Academy of Sciences, Bangalore, and was awarded the S. S. Bhatnagar Award. She appreciates the difficulties women in science encounter, and we will return to her ideas at the conclusion of this section.

ROHINI GODBOLE

Particle physicist

Rohini Godbole in 2011 in Bangalore (photo by M. Hargittai)

The November 9, 2011, issue of *The Times of India* headlined: "IISc prof does India proud at CERN," where IISc stands for Indian Institute of Science in Bangalore. [1] It was about Professor Rohini Godbole of the Centre for High Energy Physics of IISc. Recently, she had nationwide press coverage due to her participation in a most grandiose experiment, the search for the Higgs particle, or Higgs-boson, at the Large Hadron Collider at CERN in Geneva.[2]

Rohini Godbole is a successful theoretical particle physicist who conducts her research in active international cooperation. She has been involved with studying elementary particles and has spent considerable periods of time abroad. One of her projects, carried out with a colleague, on how high-energy photons interact with one another has implications for designing high-energy electron-positron colliders. It has been often referred to as the "Drees-Godbole effect." With another colleague, she proposed a model that has been found very useful and has been referred to as the "Godbole-Pancheri model."

She has been involved with the search for the Higgs particle for the past thirty-five years. As a theoretical physicist, she was not involved with the actual experiment; rather, she was seeking ways to find the particle, and the experimentalists used some of her suggestions. The experimental discovery of the Higgs boson was announced on July 4, 2012, at CERN. She reflected: "When the announcement was being made around 9 am here, I was in the audience. I had goose bumps. For any particle physicist who has been Higgs-hunting for the last 30 years, this was like a dream come true. It is something we have been dreaming of and aiming for."[3] Incidentally, Rohini started her lecture series on the Standard Model at CERN on the day when the discovery of the Higgs particle was announced. The existence of the Higgs particle is essential for the validity of the Standard Model.

This is what Rohini told me about six months *before* this discovery took place[4]:

> What happens if we find it? What does it mean? Particle physics basically describes the interactions between particles and these have been known more or less during the past 50 years, understood within the standard model. Everything fits into place but there is a missing element, so this one discovery would be the apex. It is as if we have been building a house and this particle would be the top, the roof. Not finding the Higgs would mean that we have a house of cards that will just collapse, but if we do, it means that we have a house with strong foundations. But most particle physicists agree that it is only a question of time; we'll find it. In fact, I might even say that had we not found it, for theoretical physics it would be even more interesting because in that case we would need to build a completely new model to understand the interaction of particles and the origin of the universe. But as I said, I believe we'll find it. And, if we do, the minute we find it, its mass will give us some pointers about the physics that we still have not understood. It will help us to understand dark matter and the asymmetry of matter and antimatter. These are the two aspects of the Standard Model that we still cannot explain; the rest of physics can be explained with great accuracy within the model.

Today Rohini is as busy as ever.

Rohini Godbole was born in 1952 in Pune, a large cultural center about 150 km southeast of Mumbai. She had three sisters, and her parents—her mother was a teacher—and grandparents considered a good education essential. One of Rohini's sisters is a physician, and the other two are science teachers. Rohini went to a girls' school where they did not teach science, so eventually she had to learn these subjects on her own, since she wanted to pass a scholarship examination to continue science studies.

At that time, Rohini did not even know that there was such a profession as researcher. Her mathematics teacher was impressed by her abilities and offered to have her husband tutor her. She invited Rohini to their home, and those lessons and discussions had a major impact on her life. Rohini started to read magazines and books on math and science. First she wanted to become a mathematician, but then she got scared that it might not be possible to find a job, so she decided on physics. She received her BSc degree from the University of Pune, followed by a MSc degree in physics from the Indian Institute of Technology in Mumbai. She decided to go to the United States for her doctoral studies, and her parents were supportive. She received her PhD in particle physics from the State University of New York at Stony Brook.

Rohini has been active not only in research but also in organizing conferences and workshops, both in India and internationally. She is a member of the Scientific Advisory Committee of the Cabinet of India; chief editor of *Pramana—Journal of Physics*, published by the Indian Academy of Sciences; and a member of the editorial boards of several other journals. She has published over 250 scientific papers as well as a textbook on supersymmetry, and has been active in the popularization of science in schools. Many awards and distinctions have recognized her activities.

She has been involved with the issues of women in science in India. Her involvement began in 2002, when she was already a well-known physicist and she was invited to the 1st International Conference on Women in Physics, held in Paris. Only three women physicists attended the meeting from Asia: one from Japan, one from China, and Rohini. The Chinese delegate was a government minister. They were asked to talk about their life in science. This prompted Rohini to wonder if anything had happened in her career just because she was a woman. She had never before considered this question, and she had never noticed anything. Yet, she was not sure. Eventually she realized that she had known talented women who could not achieve what they should have because of society's negative attitude toward them.

After the meeting, she decided to help women in science in India. She was already a fellow of the Indian Academy of Sciences, the second woman ever to be elected among physicists. As a first step, she founded a panel, Women in Science (WiS), and the participants decided that their major goal would be to provide role models to young women. They wanted them to understand not only how fulfilling and exciting a career in science might be, but also that this was possible not only for men but for women as well.

Together with Ramakrishna Ramaswamy, currently vice chancellor of the University of Hyderabad, Rohini edited the book, *Lilavati's Daughters: The Women Scientists of India*. They asked two hundred successful women scientists from diverse fields to write about "what makes a successful career in science possible for a woman?" Short biographies of deceased women scientists were also included. The book was aimed at the general public, especially for young people, to make them interested in a science career. The title of the book came from a twelfth-century treatise, the *Lilavati*, written by the famous mathematician Bhaskara, in which he poses different mathematical problems to his daughter Lilavati. Besides the book, the panel organizes meetings where women scientists talk about their work to young people. The members of the panel know that it is not enough to talk to interested young girls; it is also important that they bring this problem to the attention of their fathers and the young men who may someday be their spouses. Without their understanding and cooperation, nothing can be accomplished.

SHOBHANA NARASIMHAN

Theoretical physicist

Shobhana Narasimhan in 2011 in Bangalore. (photo by M. Hargittai)

Shobhana Narasimhan was born in 1963 in Bangalore, into a highly intellectual family. Her father is Mudumbai S. Narasimhan, an internationally famous mathematician, and her mother is a journalist and a dedicated feminist. When Shobhana was young, she was interested in both writing and science. Initially, mathematics was her favorite subject, but she did not want to be in her father's shadow. She used to do radio programs for children. When she was about sixteen, she was asked to put together a program for the Einstein centenary. This made her read a great deal about Einstein, and this was her first exposure to the theory of relativity and quantum theory. She found them fascinating, and this made her choose physics. She realized that while being a physicist she could still write, but the other way around would be impossible.[1]

Shobhana's choice was considered to be rather unusual, as academically successful students were expected to go into engineering or medicine. For her master's degree she worked at the Indian Institute of Technology, and there were about 5 percent women on the campus. She received her PhD at Harvard University, where she was also active in the Women in Physics group. Sometimes she felt that she should focus completely on science and not get so involved in women's issues and other distractions. However, every now and then, when something unfair happened either to herself or to someone she knew, the experience made her stay involved. She knows that many of her peers, even women, think that she is crazy to be interested in these issues.

Shobhana was a member of a national task force in India whose goal was to help women in science. They suggested extending paternity leave, but it was argued that this was pointless because fathers almost never make use of this possibility. Most people, including women, were against it. There is a general attitude that if young men went home from work earlier, say around 6 p.m., because they wanted to help with domestic chores, they would not be taken seriously as researchers. Traditional Indian culture does not accept men sharing childcare with their wives. The general attitude may be changing, but—in Shobhana's opinion—much too slowly.

Currently, Shobhana is a professor of physics at the Jawaharlal Nehru Centre for Advanced Scientific Research (JNCASR) in Jakkur, close to Bangalore. What she does is just as much physics as chemistry; she and her colleagues study materials at the nanoscale level by computational methods. They model the changes of physical and chemical properties of materials at the nanoscale level and compare them to the changes at the macroscopic level. They aim at understanding these changes and, ultimately, at learning how to devise new materials with desired properties. They are increasingly interested in energy issues; for example, understanding how a catalyst converts carbon monoxide to carbon dioxide using nanoparticles. She combines her interest and curiosity in fundamental science with finding solutions to practical problems.

Shobhana has a large group of students working with her, about half of them women. She likes to collaborate with colleagues, mostly foreign scientists. She has the impression that in India men are reluctant to collaborate with her, so she feels scientifically isolated. She does not know whether this is because she is a woman or simply because of her personality. Her international collaborators are often women. She feels that being a physicist for a woman in India is hard and has often thought about moving abroad. Even her mother has encouraged her to do so, but Shobhana would find it too difficult to leave her home, her country, and her many friends.

Shobhana has not quite given up her original idea of becoming a writer. Besides her research papers, she writes articles about her experience in teaching hard sciences, and she writes about Chandrasekhar V. Raman. Sensing the need for it among young people, she has started a course on scientific writing at JNCASR.

In 2011, Shobhana Narasimhan was elected fellow of the National Academy of Sciences of India. She continues being involved with women's issues and was featured in the book *Lilavati's Daughters*.

SULABHA PATHAK

Immunologist

Sulabha Pathak in 2011 in Mumbai. (photo by M. Hargittai)

Sulabha Pathak was born in 1955 in the small town of Sangli, in Maharashtra, India, into a middle-class family in which learning was greatly valued. Her father was in government service, and he always treated her and her brother equally. From early on, she was interested in science and in many other things, such as travel and seeing the world. She wanted to teach, while her father thought that she ought to go into research. She liked most subjects, but what she really fell in love with was biology. Her parents thought that this would lead her to medical school, but Sulabha felt that the very busy and structured life of a doctor was not how she imagined her future. She chose microbiology for her master's project. After getting her degree, she did what she always wanted to do, she started to teach. She met the right man and they got married in spite of the wishes of her father, who told her "you can get married later; you should do your PhD first." Her father feared that she would waste her potential. This was very unusual in those days, exactly the opposite of how most Indian fathers would have thought. Sulabha's marriage was not by arrangement, but fortunately her in-laws not only accepted her but also accepted that she wanted to work. In Sulabha's own family, working women were not unusual; her grandmother had to work because her husband died early. Her grandmother was self-educated and eventually became the principal of a school. Sulabha's mother also worked for financial reasons. Sulabha's grandmother and mother did not have a choice, but for Sulabha, it was different: "I work because I do have a choice; what a big difference!"[1] Her husband and her father-in-law were supportive of her interests.

She taught for a few years, but eventually found it tedious, and when her chemical engineer husband got transferred to the Netherlands, she stopped teaching. They all went to the Netherlands—she had a daughter by then. They lived in Rotterdam, and she went to Erasmus University and did volunteer work in an immunology laboratory. Her daughter, three and a half years old, was Sulabha's main concern. Even though Sulabha worked only when her child was in school, she became so good at what she was doing that they started to pay her. After a year or so, her Dutch professor suggested she do a PhD project, and she became a part-time researcher working for her degree. Alas, in a couple of years her husband had to return to India. That was a major dilemma: what to do? She felt that their daughter was still too young and needed her. Then her mentor suggested that she should go back to India, work on her data, and then return to Holland for a few months to do more experiments, go home again, analyze the data, plan new experiments, and go back again. This is what happened. Since her husband also had to travel a lot, there was a question of who would take care of their daughter. They decided that they would share the responsibility and ensure that their daughter never felt insecure. He made sure that he was at home when she had to go, and vice versa. Eventually, she got her PhD at Erasmus University.

At this time, due to her husband's job, they lived in a small town over 500 kilometers (about 300 miles) away from Mumbai where there was no possibility of doing research. Thus, she wrote a book, with a friend, on immunology. The first edition was not very good, but later they rewrote it and published it in 2005, with a revised edition in 2011.[2] They have donated the proceeds from this book to children's education, "I feel that I have to give back something to the society that gave me so much," she said.

Soon after she finished the book, her husband was transferred to the United States, and she and their daughter accompanied him. By that time, their daughter was in her late teens, and she started her own studies in America. Sulabha found a place to continue her research as a postdoc, at Indiana University-Purdue University Indianapolis (IUPUI). Eventually she was accepted by MIT to work in immunology, with an opportunity to work at Harvard as well. When her husband had to go back to India, he and their daughter suggested that Sulabha stay in Massachusetts and finish her postdoc work. They told her, "you had always before supported the family; it is time for you to concentrate on your work and career." So she stayed there for another year and completed her project. People asked her if there was a problem with her marriage. Her answer was: "Oh, no, I am staying because of my science and for the fun of it!" She said, "Nobody believed me, of course, but I did not care. What was important that I did have a fantastic time doing science full time!"

After returning to India, with her experience in immunology she probably could have become an independent researcher, but that is not what she wanted. She started to work with Professor Sharma (also in this section) at the Tata Institute of Fundamental Research on malaria immunology, and this is what she is doing now. As she said: "This suits me better, since it gives me freedom to do other things I want to do. For example, I teach underprivileged children in my free time, I am a member of Jago India, an association involved with promoting democracy and good governance, and I am a member of the Residents Association where we live. This is the way I am balancing all the different fun-things I like to do."

Sulabha's daughter is pursuing a PhD in medical anthropology; she has followed the family tradition of professional women. She feels that the life she lived, with all her parents' comings and goings, was much to her benefit. Even as a little schoolgirl, she was proud to have a mother who worked outside the home.

RIDDHI SHAH

Mathematician

Riddhi Shah in 2011 in New Delhi. (photo by M. Hargittai)

Riddhi Shah was born in 1964 in Ahmedabad, Gujarat, in western India. Her mother was a homemaker and her father an engineer; they both put great emphasis on the education of their five children. When Riddhi's mother was young, she could not study because of financial difficulties, and she did not want this to happen to her children. All of them have graduated from university. The following is a quote from Riddhi's autobiographical note in *Lilavati's Daughters*.[1]

> My first school . . . was at one end of the street where I lived. In the beginning I loved going to the school, but soon I hated it and walked there as slowly as possible. The houses on our street were row-houses with common walls shared by the houses on either side. There were water taps just outside the houses and I chatted with people brushing their teeth, looked at the cows and buffaloes, still half-asleep on the street and managed to stretch the two minutes' walk to at least ten. As a result of reaching school late, I invariably got scolded and often punished by

> my class-teacher until one day I found her waiting for me eagerly, even beaming. It was my math result that made her happy. After that day I was punctual again, except on the day of the result, when I went late on purpose!

During Riddhi's school years and in college, she manifested excellent mathematical abilities. She wanted to become an engineer like her father, her role model, but her mother suggested she go for a PhD in mathematics. Riddhi took the advice without even knowing what she could do with such a degree. She first went to the Indian Institute of Technology in Mumbai, where she received her master's degree, followed by a PhD at the Tata Institute of Fundamental Research (TIFR) in Mumbai. Riddhi explained her research[2]:

> Mathematicians don't keep themselves busy by counting numbers. They like to prove theorems, which are logical consequences of the axioms of mathematics and are statements with some interesting fact. One theorem we all know is the Pythagoras theorem concerning the sides of a right-angled triangle. A related result was proposed by Fermat and came to be known as Fermat's last theorem, although it didn't become a theorem until Andrew Wiles proved it in 1994. My research in mathematics is in the area of group theory, a subject that arose in the study of symmetries. People usually associate symmetry with reflection; for example, the two sides of the human face look identical; this is mirror symmetry. There are also other kinds of symmetry. Imagine that you are in a very large open field and you look around, and in all directions it looks the same: that is rotational symmetry. Also, if you can move along any line and nothing changes: that is translational symmetry. Group theory is the study of such symmetries, but in a more general and more abstract way.
>
> One of the problems that I have worked on is the problem of random walk. In the case of the field mentioned above, this is precisely what the name suggests: an aimless walk, with a few steps this way and then that way, changing directions frequently and randomly. However, the essence of this can be generalized and studied in an abstract way. Some common questions are: How far does the random walker get from the starting point after a time? How long (on an average) does it take the walker to come back to the starting point? The answers to these questions can become theorems of the kind that I study.

Riddhi met her future husband, Debashis Ghoshal, a theoretical physicist, at the Tata Institute. Their parents on both sides accepted their wishes, even though they came from very different cultural and religious backgrounds, not to mention their different languages and dietary habits. Her family is staunchly vegetarian, while his Bengali family loves fish. But they accept each other's habits—Riddhi is still a vegetarian, while her husband and their one son are not.

With taking the interests of both Riddhi and her husband into consideration, they had a rather complicated life. After getting her PhD, she stayed at TIFR for years while her husband worked in a research institute in Allahabad, 1,500 kilometers (almost a thousand miles) away from Mumbai. Both of them worked abroad for long periods of time; she took her toddler son with her to these assignments. After fourteen years of separation, they both found positions at Jawaharlal Nehru University in Delhi in 2007; finally, the family could live together. As she noted in *Lilavati's Daughters*: "It has been a long way from the blissful ignorance of a small-town girl to the fascinating and cosmopolitan world of mathematics."

SHOBHONA SHARMA

Molecular biologist

Shobhona Sharma in 2011 in Mumbai. (photo by M. Hargittai)

Shobhona Sharma was born in 1953 in Calcutta into a Bengali family. Her father was an engineer who encouraged his three daughters to get a good education. He wanted them to go into quantitative sciences—physics, engineering, or mathematics. Shobhona, however, decided on chemistry, which her father considered rather irrelevant, but he did not object. Of course, he also felt the traditional pressure that his daughters should get married in time.

Shobhona wanted to study in the Molecular Biology Unit at the Tata Institute in Mumbai. Her father thought that Mumbai was a corrupt city and he wanted her to stay with the family in Calcutta. However, she was a rebellious young woman, and she had her way. Eventually, Shobhona met her future husband, a physicist, studying solid-state electronics, at the Tata Institute. That she arranged the marriage for herself at first upset her parents, but when they checked out the groom's background and met his family, their resistance disappeared. It helped that the two families belonged to the same social class. Their mother tongues were not the same, but they all spoke Hindi and English. Shobhona and her fiancé got married in 1978. Later, when Shobhona's younger sister married a man out of caste, her parents accepted that as well.

Shobhona received her master's degree in chemistry, but molecular biology fascinated her, and she obtained her PhD in it from Bombay University. She went for her postdoctoral work to New York University Medical Center. There she started her research on malaria, which has been one of the major health-related problems in India. To her, "The biology of the unicellular protozoan parasite *Plasmodium* appeared a challenging area of research."[1]

For a married woman to have a successful career, much depends on the support and understanding of her husband. In this, Shobhona was lucky, since her husband always encouraged her drive for advancement. Much against the established Indian social norms, from day one they shared household duties. He often opted to do the cooking while she performed other chores. They wanted children, but there were also career considerations. After seven years of marriage, when she was thirty-two, they realized that they could not wait any longer. He had just received an invitation from Research Triangle Park (RTP) in North Carolina, and she was still in New York. She could not find a place in North Carolina where she could continue her research on malaria. Her professor came to her rescue and found a place for her at Duke University, near RTP, in an immunology group. Their first child was born in North Carolina.

After their return to India, Shobhona was offered a job at the Tata Institute. Although she was anxious to get back to her experiments, she stayed at home for some months, as there were problems with child care. Eventually, they found a family on campus that agreed to take care of their daughter, and this worked out beautifully. The two families have stayed in touch ever since.

Shobhona is now senior professor at the Tata Institute of Fundamental Research (TIFR) in Mumbai and chair of the Department of Biological Sciences. She has received awards and has been elected fellow of the Indian Academy of Sciences, Bangalore. Her group explores various aspects of biology of the malaria parasite (*Plasmodium* species). A major focus of her lab has been the molecular dissection of acquired immunity to malaria. The group has identified several novel protective malarial proteins, of which certain ribosomal P-proteins exhibit fascinating "moonlighting properties," meaning that they can perform more than one function in the body.

Apart from the investigation of acquired immunity, she has set up extensive collaborations to examine metabolic changes; flexibility and dynamics; and the nanolipid carrier-mediated delivery of antimalarials in plasmodium-infected red blood cells. Recently, her group has determined that the parasite-infected red blood cells selectively engulf the nanolipid carriers, which results in a blockage of nutrient supply to the parasites. Thus the nanolipid vehicle, even in the absence of antimalarial medication, could be an efficient remedy against those parasites.

Shobhona and her husband were lucky that for a while both of them worked at TIFR in Mumbai. But they were not spared the usual worry that families with two scientists have in India, the danger of long-distance marriage. At one point, it seemed that there were greater opportunities for him at Bangalore than at TIFR for the research he wanted to do. By that time, Shobhona had already set up a malaria lab in Mumbai, so a move would have set her back. Fortunately, her husband found a suitable job in Mumbai at the Indian Institute of Technology, so their problem was solved and they could stay together.

Shobhona and the author (sixth and seventh from the right) with members of the Women's Cell in 2011 at TIFR, Mumbai (courtesy of Tata Institute of Fundamental Research, Mumbai)

VIDITA VAIDYA

Neuroscientist

Vidita Vaidya in 2011 in Mumbai (photo by M. Hargittai)

Vidita Vaidya is a neuroscientist working in molecular psychiatry, which is a research area aiming at understanding the molecular and cellular changes that contribute to the generation and treatment of psychiatric disorders. She was born in 1970 in Mumbai. Her parents are physicians; her father is a fourth-generation medical doctor in his family. After Vidita's undergraduate studies, she took a year off to teach children, explore her passion for dance, and start applying to graduate programs abroad. It was during this year that she met her future husband, but they did not marry right away.

They both wanted to complete their studies abroad and decided to go to the United States; she studied neuroscience at Yale University, and her husband went to Carnegie Mellon University for his business degree. They got married during the time she was working on her doctoral thesis. Theirs was not an arranged marriage, but, according to her, "both sets of parents were very happy that we had spared them the pain of having to 'arrange' a marriage!" Vidita and her husband had to stay separated for a couple of years just at the beginning of their marriage because of their different career paths. After finishing her PhD project, she worked in Sweden for a year before moving to Oxford for her second postdoc. Her husband joined her in England, and finally they could live together. After two years, they went back to India, where she is now an associate professor at the Department of Biological Sciences of the Tata Institute of Fundamental Research in Mumbai. Her husband works in the same city.

Vidita's specific interests are in understanding the neurocircuitry (connections in the brain) of emotion and its modulation by life experiences and psychotropic drugs. Using animal models of depression, some of which are based on perturbations of early life experience, her group studies the molecular, epigenetic, and cellular changes that contribute to persistent alterations in behavior. She also investigates the adaptations that arise in response to sustained antidepressant treatment.

Vidita and her husband have a seven-year-old daughter, and Vidita considers it very fortunate that the families on both sides support them, so that she can continue her research in addition to being a mother and wife. She is especially grateful to her in-laws for their help. She was not the only woman scientist in India who acknowledged such support, which seemed quite unique, as I have never heard similar stories in my encounters elsewhere in the world. I asked Vidita about it.[1]

> . . . it is a lucky few who end up with husbands and in-laws who place their daughter-in-law's hopes and aspirations on par with their sons. In traditional societies women's aspirations are often not even considered and so I realize how lucky and blessed I am in both my parents and in-laws that both mine and my husband's aspirations are considered and not prioritized one over the other. . . . My in-laws by rearranging their lives after my daughter was born have been invaluable in helping me successfully do the balancing act. It is in this that I am immensely grateful to them. While my parents are still active clinicians and scientists in their own right my in-laws have retired and have since retirement moved their entire life around, including moving their household so that they are within 10 minutes from us to help with childcare responsibilities. I would not have expected either my parents or in-laws to put away their own work interests to singularly focus on supporting us but it is in that regard that my in-laws have gone beyond anything expected to be remarkable sources of daily support. I could not do half of what I get done without their support and the unstinting help from my parents, every time I need to travel—which is pretty often. My gratitude to both sets of parents is immense not because they need to "allow" me to do what I love to do but because they have bent over backwards to help facilitate my work by being there for me whenever I have needed them. So while as a child I took my parents support for granted—they are the ones who have always encouraged me—I find that I am lucky enough to also take my in-laws support for granted (believe me, that is a rare thing in India) who are equal champions at doing everything and more to help me balance my work and home.

FURTHER COMMENTS

The above examples from India are probably the exception rather than the rule, because the woman's primary role in Indian society is in family responsibilities.[1] Tradition is deeply rooted, and from a very early age, expectations for boys and girls are different. For girls, the number-one objective is to marry and take care of her husband and the children. The societal pressure is very strong; a career might be considered, but only as something secondary. For men, the career is the first consideration. Women, if they still decide to have a job, are easily exploited in the workplace, because they want to perform well in all their roles. There is also the sentiment, especially among the older generation, that it does not look good for a man if his wife works because it suggests that he is not able to provide for his family.

In addition to looking after their husbands and children, at least in some segments of Indian society, women are expected to be intelligent and well educated. They often perform well in examinations, and they do go to college. In middle-class families, the parents want their daughters get higher education before they get married because it is considered that smart women make good mothers. There is a great demand for good schools, and there is increasing competition for a good education, even at the level of elementary school. In order to get in, the children have to pass demanding entrance examinations. The school interviews the parents as well. The father is expected to be highly educated with a good job. The ideal mother is also expected to be highly educated, *but* she is not supposed to have a job. The idea is that the mother should spend all her time at home rearing the children. It is not only the schools that look for such ideal mothers, but many men look for such women to marry them. It is quite common at universities that a male professor is married to a well-educated woman who is either at home or works in a job for which she is overqualified.

According to statistics, the percentage of women up through the PhD level in the sciences is quite high, about 36 percent, depending on the particular institution and field. It is at the postdoc and faculty level where there is a sharp drop.[2] This is in contrast to the pattern in Western countries, where there is a more gradual drop at every step on the academic ladder, which is described as the "leaky pipeline." The drastic change in India occurs at the postdoctoral level, which is the time when women may get married and have children. Even if later they would like to go back to work, the job should not be as demanding as scientific research.

Today it is being realized that the absence of women in the higher levels of science is an awful waste of talent and resources. The Indian government has started programs to attract back those women who seem to have been lost for science. The Department of Science and Technology and the Department of Biotechnology of the government have programs for qualified women who had been out of research for years. Now they get funding for five years to work in a laboratory, they get mentors, and after this period they are supposed to find themselves a regular position. There are also alternative options for work related to science, for example in science publishing.

I noticed in my conversations in India with successful women scientists that those who had families always mentioned how grateful they were to their in-laws. Having an understanding extended family makes a great difference for working women. When the child is ill or the woman has to attend a meeting, the in-laws may be as ready to step in as her own parents. Another advantage in India is that it is rather easy to get domestic help, and such help is affordable. Nonetheless, as Charusita Chakravarty said: "I have to still emphasize that this is a very traditional society so the way you think of yourself as a woman and the degrees of freedom and choices that you have as a woman can be severely constrained. But if you go beyond the traditional roles, and have some imagination and empathy in understanding and shaping your environments, there can be a lot of emotional and logistical support that can prove to be of great importance, especially with child care and support during various crises."[3]

Further, she added, regarding the women in science issue[4]:

> What interests me is what we might call "the sociology of science" and how it plays a role in determining how an individual scientist performs. An important aspect about doing science is that what you produce must be validated by your peers. That gives it a certain kind of robustness. That is the accepted method. But the process by which the peer group validation happens is through the members of the group where individuals and their personalities as well as their sociological predilections do play a role. When we talk about women in science and go beyond the level of holding a job and ask what it takes to be an integrated member of the professional community as a whole, then it is important to make this distinction. I have heard people sometimes saying in all seriousness that "when I look at a paper I do not look at where the writers are, whether they are from India or China; I only look at the work as a whole." One can say the same thing about women. But behind the fact whether you are able to develop your ideas and whether you are able to make your ideas taken seriously, there is an issue how well you interact with the peer group, and there gender, class, caste, nationality, all these things play a role. Social marginization does make a difference. But if you want to understand it and combat it, you have to somehow see which are the core professional values that you have to be seen as delivering on.

I interrupted her to say that these core values should not depend on gender. "Of course not. Very few people would say blatantly that they do. And still, we see what we see, that there are fewer and fewer women as you go up in the academic ladder. And one must ask why this happens. And part of the answer must lie in the extent to which socially marginal groups can participate in critical professional activities. I don't believe that the answers are very simple, and I believe that introspection on both sides of the divide is called for."[5]

WOMEN SCIENTISTS IN TURKEY

In 2008, I was invited to talk at a conference on Women in Science at Istanbul Technical University. I noticed with surprise that it is not uncommon in Turkey for a woman to hold the position of rector (president) of large universities. I talked with two such rectors, and these conversations and another with a chemistry professor of Istanbul University serve as the basis for this section.

The statistical data are quite astonishing. A study in 1970 showed that while women's participation in professions such as engineering, architecture, law, dentistry, and medicine was only about 5.7 percent in the developed countries, in Turkey it was 25 percent[1]—and this was more than forty years ago! The latest data show that the percentage of women full professors in Turkey's universities is 28 percent, compared, e.g., with Germany (15 percent) or the Netherlands (13 percent); the EU-27 average is 20 percent.[2] This, together with the relatively large number of women in high administrative positions in science and education, made me wonder what the reason might be.

Turkey is a predominantly Muslim country with a history of centuries of polygyny[a] back in the Ottoman Empire. Considering the patriarchal cultural heritage of the empire, one would not expect women to occupy important positions in society. The reasons for this surprising fact are rather complex.[3]

The major change in women's status in Turkish society happened along with the major political change of the country. The year 1923, under the leadership of Mustafa Kemal Atatürk, brought about the end of the Ottoman Empire and the foundation of the secular Turkish Republic. This event meant major changes in women's status. Primary school education became compulsory for all children, and soon it became coeducational. The Civil Code of 1926 outlawed polygyny and granted equal rights to women and men, at least in certain matters, such as divorce and child custody, although complete legal equality came about only in 2001. There were many other important changes, among them the lifting of bans on employing women outside the home. The wearing of the headscarf and full body covers for women was made illegal (until recently—see the discussion by Professor Ulubelen in this section). After the 1920s, women's position changed in a major way, and this was considered a measure of the modernization of the country.

Women who were interested in getting higher education and, later on, in entering the professions came from educated urban upper-class and middle-class families. With the rapid development of the country, there were jobs available for them; it was more reassuring for employers to hire women living close by than men living far away.[4] According to Professor Çiğdem Kağıtçıbaşı, in society's eyes there were no appropriate or inappropriate jobs for women, because earlier they had not been allowed to work outside the home in the first place. Thus, in this new world women could choose even professions that in the West were traditionally looked at

[a] "Polygyny" means one man having multiple wives. The better known word "polygamy" strictly speaking refers to one person having multiple spouses.

as "unfeminine." This is why there have been more women in engineering than in most Western countries. According to the latest statistics, for 2010, while the EU-27 average for full professors in engineering is 7.9 percent, in Turkey it is 19.1 percent. Even in the natural sciences there is a larger percentage of women then in the EU countries, 13.7 percent for EU-27 versus 25.7 percent in Turkey.[5]

There is, however, an important caveat. The relatively large ratio of women in academia does not translate to the general population. Looking at employment rates in the country, we find that Turkey has a lower rate for women in employment than any of the EU-27 states. While the average female employment rate is 62.4 percent (74.6 percent for men) in the EU member states, it is a mere 30.9 percent (75.0 percent for men) in Turkey; the data are for 2012.[6] One of the reasons is the difference between educated urban women and women living in rural areas. A large portion of Turkey's 75 million inhabitants live in faraway underdeveloped rural areas. Mustafa Kemal's ideal of abolishing the segregation of women meant that job opportunities should have opened up for women in all segments of employment. However, only a relatively small number of privileged women could benefit from the changing situation: young women belonging to the upper and middle classes, the intelligentsia, and bureaucracy—that is, mostly women living in the cities.

Furthermore, the traditional role of women in the family did not change, as is shown by the fact that in 89.6 percent of families with children younger than five years the mother does not have an outside job.[7] This state of affairs might also be a problem for women professionals who cannot afford outside help. However, because most professional women come from well-off families, they can usually afford help for child care and domestic work. Accordingly, their husbands do not need to feel that by "letting" their wives work, they would need to share household duties that might alienate them from the idea of a working wife.

The three women introduced below are successful scientists and equally successful science administrators. When I wondered about how women could reach such high positions, the general comments I heard were not too encouraging, in that positions in science and higher education may not be sufficiently attractive for men. However one-sided such opinion may be, it does not sound too unrealistic.

SEZER ŞENER KOMSUOĞLU

Neurologist

Sezer Komsuoğlu in 2002 (photo courtesy of S. Komsuoğlu)

After finishing her university studies in medicine, Sezer Komsuoğlu, together with other physicians, among them her husband, Baki Komsuoğlu, was sent to the Black Sea region of Turkey to participate in founding a new medical school in the Trabzon region. After fourteen years at the Karadeniz Teknik Üniversitesi (Karadeniz Technical University), in 1995 the Komsuoğlus moved to the Kocaeli region, between the Black Sea and the Sea of Marmara, to participate in setting up a new medical faculty. They were making great progress when, in August 1999, the Marmara earthquake destroyed all the buildings and equipment of Kocaeli University. The Komsuoğlus put all their energies into rebuilding the university in Umuttepe. They obtained help from abroad and from Turkish companies. New earthquake-resistant buildings were built, the equipment was replaced, and the new faculty was organized. Today, they have a modern university, hospital, and research center of international reputation. Kocaeli University has eleven faculties with over 60 thousand students. Since 2006, Sezer Komsuoğlu has been its rector.

Sezer Komsuoğlu with her husband, Baki Komsuoğlu in 2005 at Glasgow University. (courtesy of S. Komsuoğlu)

Sezer Şener was born in 1949 in Trabzon, on the Black Sea coast of northeastern Turkey. Her father, Ahmet Şener, was a politician, and in the 1970s he briefly served twice in Bülent Ecevit's cabinet. Her mother was a teacher. Education was considered of utmost importance in her family. During her high school years, there were only three hard-working, ambitious girls in her class, who found the sciences attractive. Each of them eventually chose science for her profession; the others chose chemistry and physics, and Sezer chose medicine. She attended the medical school at Atatürk University, after which she did a year of residency before passing her exams in neurology. In 1978, she went to England for her postdoctoral studies at Birmingham University and at the University of Aston in Birmingham, where she did research in neurophysiology for three years. She prepared her thesis on multiple sclerosis for her advanced degree.

Her experiences at Karadeniz University and after the earthquake at Kocaeli University made her experienced in administration and organization, and eventually she was elected rector. However, she has always been foremost a medical doctor and researcher. Her main research area is epilepsy and multiple sclerosis; she has also been involved with studying evoked potentials and hypertension. She is a clinician rather than a medical experimenter. She likes to see patients and has based her research on her clinical observations.

Sezer met Baki Komsuoğlu during her internship at Atatürk University; he was finishing his specialization in internal medicine and cardiology. They were married in 1975. They have two daughters; one of them is also a medical doctor and presently is an associate professor in pharmacology in Turkey, the other is a political scientist at the University of California at Berkeley and professor at Istanbul University. Sadly, Baki passed away in 2008 at the age of sixty-two. He was well known in Turkey and Europe as a cardiologist; he wrote ten books and over 400 research papers on cardiology. Sezer remembered: "He was always pushing me and our daughters to excel and to do good things. We were very good friends and we had a perfect partnership. It is very difficult; my life completely changed after he died but I try to teach myself that my life will be like this from now on and I have to face it. I don't know how it will work but I am trying. My husband is always there and I am happy that he is."[1]

Baki Komsuoğlu was rector of Kocaeli University; when he stepped down in 2006, she followed him in the position. Becoming rector meant that she gave up doing actual research, but she is still advising young faculty members. She sees patients and lectures to students on neurology. She also keeps up with the literature. It was different when they started to build their new medical school in Trabzon; there was no library there, so every now and then they drove to the central library in Ankara—a sixteen-hour trip—and collected the new results, and next day they drove back home. They shared the information with their colleagues, and the next month the others went. Today their medical library is excellent.

As a mother and homemaker, researcher, and clinical practitioner, not to mention her administrative duties, she has had to be very careful in organizing her life. "Every day I started very early and first did what I needed to do at home. Fortunately, we are a large family and we lived close, and I got a lot of help in this. Then I went to the hospital and got back home at 11 in the night—actually, this is still the case. If I say that I work very hard, that is true."

On the question of women in science, she believes that women are much keener on their work than men are. A scientist woman runs the family and takes care of the children; she works very hard. Sezer explains this by the left hemisphere of their brain being very keen on the details; hence women pay much more attention to details than men do. The female brain is different from men's—the hormones make it different—and women see many things that men do not. Women are better in multitasking, and they are gentler, more polite. Concerning science, Sezer believes that women are good in this field and the fact that they work hard is good because science needs hard work. Although she has no statistical data, she thinks that there should *not* be any correlation between doing science and not having children. If there is a correlation, then only society can be blamed for it.

Sezer loves to be a scientist and a medical doctor, and she has had a large number of graduate students. She opened the first clinical neurophysiology department at Karadeniz University in 1981, and she brought new methods there, such as electroencephalography and electromyography. Recently, she finished a book on epilepsy based on records of her own 210 patients over the past twenty years. At her university, she opened the epilepsy clinic run by one of her young associates. She has left her mark on the medical profession in Turkey.

GÜLSÜN SAĞLAMER

Architect

Gülsün Sağlamer in 2013. (courtesy of G. Sağlamer)

Istanbul Technical University (ITU) is a 240-year-old institution with over thirty thousand students. When Gülsün Sağlamer became its rector (president) in 1996, she immediately embarked on a comprehensive reform of the entire institution. This is how she characterized her eight-year tenure as rector: "Our research budget [increased] tenfold, our publications 2.5-fold, our state budget 3-fold; in all areas we managed to go beyond our expectations. We created new scholarships for students and we built dormitories for 3,000 students in one and a half years."[1]

She was born Gülsün Karakullukçu in 1945 in Trabzon, on the northeastern Black Sea coast of Turkey. Her father was a businessman and her mother was a housewife. Gülsün decided to become an architect when at the age of nine she read an article about various professions. "There was information about different professions. When I read a paragraph about architecture, it impressed me so much that then and there I decided to become an architect." Her brother studied mechanical engineering at ITU, and she followed him there. To get accepted was no trivial matter: "I was very happy; I did not know then that I was the first in the ranking. All my life I tried to do my best; the only person I felt that I have to compete with was myself. If you compete with others, you may lose your focus; you have to have your own goals and try to realize them."

On the encouragement of her professors, she stayed on after graduation for her PhD work: "I decided, all right, I will get the PhD; I can practice afterwards. I was dreaming about designing buildings throughout my study years." She had to postpone her dreams until she became an associate professor. She enjoyed her research project, which was quite new at that time: computer-aided architectural design. At ITU, she met Ahmet Sağlamer, who was a student in the Civil Engineering faculty. Eventually, they married and had a son. After getting their PhDs at ITU, they went to Cambridge University for their postdoctoral studies in 1975–1976. On their return, they received promotions to associate professor. In 1982, a new Higher Education Law was passed in Turkey according to which they were supposed to move to another university if they wanted to be promoted to full professorship. However, Gülsün and Ahmet decided to stay on. When, eventually, the rules changed, they both became full professors.

"Till the 1990s I focused on my original project, computer-aided architectural design. Gradually, from the 1980s, I also became interested in housing issues. I was teaching different theoretical courses, 'architectural design theory and methods' and 'present developments in architecture' for the undergraduates and 'computer-aided architectural design' and 'logic models of design' at the graduate level. I was also supervising the architectural design studio for seniors. I carried out many research projects, published many articles, and supervised many master's and PhD students."

Over time, she has developed excellent people skills. She became vice dean of their faculty, and in 1992, vice rector. By that time, their son had graduated from high school and had gone to the United States to study. This made it easier for her to accept the position. During the next four years, she understood what it meant to be responsible for the whole university; therefore, at the end of the term she decided to run for rector, and she was elected. She proved to be highly successful and served for two terms:

> We upgraded the teaching and research environment, we restructured all the undergraduate programs, graduate programs, we developed very close links with international organizations, we made ourselves visible at national and international levels. We accredited all the engineering programs at ABET [Accreditation Board for Engineering and Technology, USA] and our architectural program received accreditation from the US National Architectural Accreditation Board. We got through the European University Association Institutional Evaluation Program as well, in 2004. It was not easy to convince the people at the university that we should do this; it took three years to convince them to go to the international accreditation processes. That was a quite difficult period, but we had been successful. Most of the academics were excited about the possible changes. When I was asking them to be a coordinator of a particular project or to do certain work for the university projects they were happy to take the responsibility and all were

ready to do the work. Thus, I did have the support of the faculty and alumni in all these big reforms. This was an outstanding period.

Enormous changes have taken place in Turkish society since the War of Independence and the establishment of the Turkish Republic in 1923. These changes included the secularization of state, the separation of state and religion, and the establishment of equality of all citizens, including women, which in itself meant revolutionary changes for women. In 1934, women received the right to vote. A specific program served to improve the level of education in cities and in rural areas as well. A few decades after the establishment of the Republic, this is Gülsün's evaluation of what has been accomplished:

> The success of the Turkish revolutions can be observed in the cities rather than in rural areas. In the cities, you can feel the atmosphere. But as we were not able to establish enough schools in the rural areas in the early years of the republican period, were not able to penetrate into these areas to make the people feel that they own the result of the revolutions, the new lifestyle, as much as people living in the cities. After the Second World War as we started to establish democracy in the country, there was room for the opposition to use religion to revitalize the old traditional way of living again. This is my perception. In the cities, the republican revolutions were successful. I was brought up in a small city, Trabzon, in the Black Sea Region and there was a kind of atmosphere that looking back now at those years I feel that I was brought up in a European city with its physical, social, educational environment, all was perfect. I cannot believe how this was possible in such a remote place, to develop that atmosphere, that attitude. People tried to westernize their life, they wanted to send all their children, boys and girls alike, to universities; it was fantastic. The republican government created an atmosphere that suggested that girls should move forward, they should be given the opportunity to study at the university and to become academics. This was a wonderful period for us.
>
> After the 1960s, besides democratization of the country, we had to face a very intensive immigration problem from rural areas to cities and the big cities were not ready to absorb such an enormous influx of people. This relocation of masses (especially the young population) created a lot of infrastructure problems in cities and left many villages empty or with old people left alone. Squatter settlers started to invade the big cities, like Istanbul, Ankara, and Izmir, in illegal developments. These settlements provided a very good soil to exploit the people's weaknesses and nurture very intensively radical religious values instead of secular ones. Some political parties utilized this tool to become the government party and used religious symbols everywhere which was and still is against our constitution. Unfortunately, many religious foundations started to fund women students providing them accommodation and scholarships on certain conditions and universities themselves were banned to give scholarships to their own students under a law passed in 2003.

> On the other hand, we should not forget that our Republican reforms have achieved a lot. At the moment, Turkey is one of the leading countries in the World and Europe in terms of women representation in academia; 29% of full professors are female and the Turkish Higher Education sector has the lowest value of the so-called "Glass Ceiling index" in Europe, which shows that our pipeline during our academic promotions is less leaky then in most European countries. Especially in mathematics, computer science, engineering, and the health sciences, women's representation is higher than in other European countries.

Gülsün has been very active as a member or leader of a large number of national and international organizations, far too many to mention all of them; just three examples: she is an executive committee member (since 2003) of the International Association of University Presidents (IAUP), president of the Community of Mediterranean Universities (since 2012), and Chair of the European Women Rectors' Platform (since 2008). She has received plenty of recognition, including honorary fellowship in the American Institute of Architects and the Leonardo da Vinci Medal of the European Society for Engineering Education. She is a member of the European Academy of Sciences, Arts and Letters.

It is not difficult to imagine that she has also been involved with women's issues in science. "As I am the first female rector of ITU, I was invited to talk about being a woman leader in academia. The strange point was that I have never felt any segregation towards me during my study years and my academic life. But I understood the importance of this issue and I started to focus on it, in order to understand the dynamics of gender equality in academia. At the end, I found myself in the middle of many research projects on women in science, engineering, and technology and women leadership in academia."

AYHAN ULUBELEN

Chemist

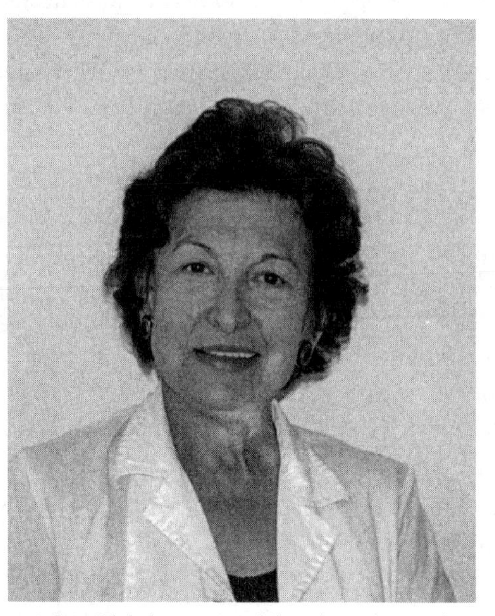

Ayhan Ulubelen in 1997. (photo by I. Hargittai)

Ayhan Ulubelen is "one of the pioneers in scientific research in Turkey and a worldwide recognized authority in Natural Products Chemistry," according to the guest editors in the special issue honoring her in the journal *Phytochemical Letters*.[1] In 2011, there was a conference in her honor celebrating her eightieth birthday and her sixty years in natural products chemistry. She has worked most of her life under much poorer conditions than most of her colleagues in the West.

Ayhan Ulubelen was born in 1931 in Istanbul. Her father was an army officer and her mother a homemaker. Originally, Ayhan planned to become a journalist, but during her high school years she saw a movie about Madame Curie, and that affected her so much that she decided to become a chemist. She said, "All the girls in my class did too, but only I went through with it."[2] She studied at Istanbul University and after graduation tried to get a job in industry, but at the time there was bias against employing women. Fortunately, the Faculty of Pharmacy of Istanbul University offered her a position. In a few years' time, she received her PhD there. She went to the United States for postdoc studies at the College of Pharmacy of the University of Minnesota. When she returned to Turkey, she became an assistant professor at Istanbul University, where she stayed as full professor until her retirement in 1998. She spent a few years in the United States, in Germany, and in Japan. Since her retirement, she has continued her research activities as professor emeritus.

There has been a strong interest in the plants of Turkey, especially the plants that have been used in folk medicine for centuries to treat various ailments. "The villagers use these plants extensively. We have special stores, even in Istanbul, called 'AKTAR,' that sell plants and plant extracts used as traditional medicine."[3] Ulubelen and her group have tried to identify ingredients that could be assigned to specific physiological effects. When they identify and isolate the active ingredients, they determine their structures using a variety of physical techniques such as nuclear magnetic resonance (NMR), infrared spectroscopy, and mass spectrometry.

Ayhan Ulubelen at her eightieth birthday in 2011; *from left to right:* Barbara Timmermann from Kansas University, and some of Ulubelen's former students, now all professors: Sevil Öksüz, Gülaçtı Topçu, Ayhan Ulubelen, Ufuk Kolak, Nezhun Gören, and Solmaz Doğanca. (courtesy of A. Ulubelen)

One of their studies in the late 1990s concerned a plant that used to be taken by pregnant women to cause spontaneous abortion. This particular plant had been used as an antifertility agent not only in Turkey but also in China. Both countries used to have high birth rates. Ulubelen and her colleagues reasoned that if successful, they could offer a natural remedy against unwanted pregnancy by causing spontaneous abortion without side effects. In Ulubelen's words[4]:

> Of course, I'm aware of the fact that researchers in many other places are involved in finding such abortive agents, and the World Health Organization also supports such research. The goal is to find something from natural sources that women can just drink.
>
> I happened to be in America when Carl Djerassi's birth control pill was being introduced, and it has prevented a lot of unwanted pregnancies. But it also contributed to the spread of free sex all over the world and that has seriously damaged family life. . . .
>
> The pill may also cause some cardiac problems and may also be carcinogenic. The agent causing spontaneous abortion may be much more advantageous, and it would be used only when truly needed. The women in our villages and in the villages of India and Pakistan, and many other countries, could use it when really

needed, and its use would be very easy. However, so far there is nothing like that has been found.

The plant they investigated was *Ruta chalepensis*; they checked both its roots and aerial parts. They isolated different types of compounds, and checked their effects with mice one by one, and they found several compounds that had the abortive activity. However, follow-up studies showed that some of the mice developed cysts in their ovaries and had other problems as well. Therefore, they could not suggest this plant as an abortive agent; on the contrary, women had to be warned of the hazards of its use, and the suggestion was to avoid using it.

This was about fifteen years ago, and eventually they dropped this particular research topic. As it happened, the Turkish Ministry of Health had some programs to reduce the birthrate and they apparently worked well—so much so that presently the country is facing the opposite problem, and the government is now trying to encourage women to have more babies.

Ulubelen and her colleagues have been involved with many other interesting studies to find out what compounds are the active ingredients of the plants used in folk medicine. Some of them have been used against cancer, others against HIV, and yet others in treating cardiovascular diseases, diabetes, and so on. The study of plants that are potential remedies against cancer has been their top priority. Ulubelen's group joined the plant screening program that had been started in the early 1960s by the US National Institutes of Health (NIH) to find cures against cancer. Within this program, about a hundred Turkish plants have been tested.

Ulubelen received a *Merendera* species, called *Merendera caucasica*, from eastern Turkey, where it has been used for its antitumor effect. She and her colleagues found that it is the alkaloids in the plant that have major activity, and they determined the structure of several alkaloids extracted from the plant. Furthermore, they identified different plants showing activity against different tumors.

Fifteen years earlier, they had been doing their research under much more modest circumstances than their colleagues in most other countries. Research funding was quite a problem. The university provided small amounts of money, and there was an additional source, called TÜBİTAK, the Scientific and Technological Research Council of Turkey. They helped, but only to purchase small things, not large instruments. At the same time, the Turkish government was keen on improving the country's science. Therefore, when researchers published their results in refereed international journals, they received a premium. This consisted of two parts. One was for personal use; this came from TÜBİTAK. The other was for their group to get minor items that they needed in their work.

The past fifteen years have brought further positive developments in science funding. Istanbul University and other universities formed their own Research Foundations, which are responsible for providing the necessary funds for research at their institutions. They themselves have to find the necessary money in some way, by doing commissions for small industry, by treating patients (in case of the Medical Faculty), or using other means. The University of Istanbul established a Central

Research Laboratory that by now has all the necessary instruments and equipment that the different research groups need, and when a new instrument becomes necessary, they can apply for it. The researchers have to pay from their own grants for the use of these instruments. There are several places to apply for grants; the State Planning Institute, the Turkish Science and Technology Institute, and the Turkish Academy of Sciences are the most important. Ulubelen feels that today it is relatively easy to do research at a high level.

The earlier financial difficulties notwithstanding, Ayhan Ulubelen has published about 300 research papers and received many awards and distinctions. She was a member of the NATO Scientific Committee for four years. She is a member of the Turkish Academy of Sciences—or, rather, she used to be. On November 2, 2011, 74 of the Academy's 137 members resigned, among them Ayhan Ulubelen. This was in protest against a government decision that, either directly or indirectly, the government would appoint two-thirds of the academy membership. A few weeks later, on November 25, 2011, seventeen former members of the Turkish Academy of Sciences—Ulubelen was one of them—founded the Science Academy Society as an independent, self-governing, civil-societal organization to promote scientific merit, freedom and integrity. Today, their members number well over one hundred. The new institution endorsed the time-honored traditions of integrity, independence, and social engagement, and aims to spread these traditions within the academic community.

Seeing the revival of religious dress for women in Islamic countries already at the time of our first meeting, we discussed the question of how women students dress at the university. This is what Ayhan said in 1997[5]:

> In the old days, that is, before the Turkish Republic, Turkish women used to have "çarşaf" [a garment that completely covers the body, together with most of the face] in the cities and baggy pants and a head scarf in the villages. Today, the women in the cities have modern clothes, and the outfits in the villages haven't changed much, except that the younger generation is switching toward modern outfits. During the last 10 years, increasing numbers of women started to wear "çarşaf," and many young girls wear uniformlike outfits, very long and loose coats and very long, large scarves that come down to their waistlines. There is nothing traditional about this way of dressing. And in a profession one should have proper clothing; a medical doctor, a nurse, a lawyer, etc., should dress whatever way is appropriate for the profession.

I asked Ayhan in 2013 whether there has been any change since then.

> Unfortunately, since 2002 we have a religiously inclined government. They, and some intelligentsia (newspaper men mostly and some university professors, some high bureaucrats) approved that this is more democratic, and free dressing was accepted by the university. Slowly but continuously they are increasing their influence on women; these women believe that by covering their head and dressing longer skirts, they are freer. In a way this is so because many of these girls

> are able to go out of their houses, free from their fathers but mostly from their brothers. They could freely flirt, walk around hand in hand with boys, and they could go out to the movies with boys; as long as they cover their head they are considered good girls by their families. Of course, this is only a group but I have to admit that the number of conservatives is increasing. At the moment at least 20% of the university girls are head-covered.[6]

Ayhan is not married. She adopted her nephew, who lives with her together with his daughter. Ayhan has been busy all her life with her science, and that has given her great satisfaction. Many of her past graduate students are now in important positions in chemical and pharmaceutical companies and in academia. Her former students, the three guest editors (all women) of the special issue in her honor, speak about her warmly. They emphasize her teaching abilities, her radiating enthusiasm about research, and her obvious joy at the bench in the laboratory. She is a role model for young women and men alike.

IN HIGH POSITIONS

For women scientists, it has not been a trivial matter getting to the helm of institutions of higher education. Women constitute an untapped pool of talent for such positions. There are women scientists who having succeeded in research and teaching are ready for additional challenges and possess ideas about what they would do in academic leadership positions. In the previous section, we have already seen two Turkish professors, Sezer Şener Komsuoğlu and Gülsün Sağlamer, who moved into high administrative positions in academia. There is a growing tendency to consider women scientists for such positions in the United States and Western Europe as well.

In 2002, just as we heard about a very successful scientist, Shirley Tilghman, becoming the president of Princeton University, I asked James D. Watson, one of the best-known scientists worldwide, whether such an appointment would not be a loss for science. He said: "No. You know that I also came here [from a Harvard professorship to be in charge of Cold Spring Harbor Laboratory]. It's kind of crazy that you should do the same thing all your life. It is important that some times good scientists run great institutions. It's good for science because there is then someone who understands science. You could be selfish and say, yes, it's good for us."[1]

The number of female university presidents in the United States has increased steadily if slowly for the past decades. In 2007, *Glamour* magazine chose the four female Ivy League presidents, Drew Faust of Harvard University, Amy Gutmann of the University of Pennsylvania, Ruth Simmons of Brown University, and Shirley Tilghman of Princeton University, as "Woman of the Year." As Lynn Harris wrote: "In the history of old boys' clubs, one is older and boy-ier than the rest: the Ivy League."[2] This old boys' club is a small one; altogether eight universities belong to it, and that means that these four women constituted half of the Ivy League presidents.[a] *Glamour* did a great service to the cause by increasing public awareness of this. According to a 2011 survey, about a quarter of US college presidents were female—this survey referred to all levels of higher education.[3] For PhD-granting universities, it was 22.3 percent, a great increase from 13.8 percent in 2006 and especially from 3.8 percent in 1986.

The European Union (EU) publishes information regularly about the representation of women in research and higher education. Their biannual report is titled *She Figures: Gender in Research and Innovation*. According to its 2012 edition,[4] 15.5 percent of the heads of the higher education sector were women overall in the

[a] As of 2014, their number is three, with Drew Faust at Harvard, Amy Gutmann at the University of Pennsylvania, and Christina Paxson at Brown University.

EU, though only 10 percent if considering only PhD-granting universities. Thus, Europe appears to be behind the United States in this respect, but the scatter is very broad among the European countries. Scandinavia is ahead of the rest, with Sweden 43 percent, Iceland 33 percent, Finland 31 percent, and Norway 25 percent—all these numbers refer to PhD-granting institutions. At the other end of the spectrum, in Cyprus and Hungary, no university had a female rector.[b] The United Kingdom was not included in the figures quoted above. According to 2013 data, 14 percent of university vice-chancellors (corresponding to rector and president in other countries) were women.[5] For India, among the Central Universities (the leading federally funded universities), there were four women vice-chancellors, making it 9 percent.[6] In Turkey, only 4 percent of all rectors of PhD-granting institutions were women.[7]

What follows below is a set of diverse examples, both for the origins and backgrounds of women leaders and the circumstances of their operations. Both the heads of universities and scientists who took up other kinds of leading positions are represented. As a general conclusion, we might hope that we are approaching a state when it will no longer be an outlier to have a woman at the summit of academic, educational, or other institutions.

[b] In 2014, in Hungary there is one female rector, heading the Liszt Academy of Music.

CATHERINE BRÉCHIGNAC

Physicist

Catherine Bréchignac in 2000 in Paris. (photo by M. Hargittai)

When in 2000 I visited Catherine Bréchignac, member of the French Academy, she had just completed her term as secretary general of the CNRS, the National Center for Scientific Research of France. I asked her whether her prestigious academy membership and the CNRS directorship were solely on merit, or whether being a woman may have had something to do with it. This was her answer[1]:

> Oh, sure, of course. More precisely, about the academic recognition, I don't know, I don't think so. But the appointment to be Director General of the CNRS definitely happened because I am a woman. Of course, I have a good reputation as a scientist and I also managed well the lab and the Department of Physics at CNRS earlier; but the politicians like to have pride in showing how considerate they are with minorities. Oh, I am sure. [Did you mind that?] No, not at all. I figured that

it was fine but after having been chosen, I simply had to prove that I could do the job better than anybody else, men and women alike.

Catherine Bréchignac was born Catherine Teillac in 1946 in Paris. Both her parents were academics; both were professors at the University of Paris. Her father, Jean Teillac, was a nuclear physicist, who succeeded Frédéric Joliot-Curie as director of the physical chemistry section of the Institut Curie; he was later high commissioner at the Center of Atomic Energy (CEA). Catherine's mother was a physician. Catherine was mostly brought up by her grandmother. With such a background, it is not surprising that for her to study for a profession was the logical path to pursue. Originally, she wanted to study mathematics or French literature, but later she oriented herself toward math and science. Why did physics win? "I think it was mostly because among the students I knew, I preferred the physicists; they were more social and provided a much more pleasant environment. I made this choice purely based on human interactions and not on the scientific disciplines themselves."[2] She received her PhD at the University of Paris-Sud in 1977; by that time, she was sure that she wanted to become a researcher.

From early on she worked at the Laboratoire Aimé Cotton in Orsay, and in 1989 she became its director. Her research topic is the study of clusters. As she explained, "Clusters are the component-precursors of the nano-world."[3] They may consist of just a few atoms or molecules—or as many as ten million units. They are important because they are the bridges between the gas phase and the solid phase. "However, because small is different from large, even if the properties of solids are usually known, the properties of these clusters are not. So I have decided to work with these clusters by entering a 'terra incognita', and it has been a real adventure."[4] She and her colleagues have made important discoveries in this field, and they always find new challenges. They study different-sized clusters and try to understand how their properties change with size.

Becoming the director of her laboratory was the first sign that she was well suited for administrative positions. After six years on this job, she became the director of the Department of Physics and Mathematics of CNRS, followed in two years by her appointment as general secretary of CNRS.

CNRS is government-funded and is the largest research organization in Europe for fundamental research. It has about 26,000 employees, of whom about 11,000 are researchers; they work with a budget of 3.3 billion euros (as of 2012). To become general secretary of such an enormous organization is a grave responsibility and a challenge. What they expected of her was no less than to rejuvenate this mammoth institution. Catherine turned out to be a good choice for the post. During her tenure (between 1997 and 2000), she introduced major changes in the organization in spite of occasional resistance and resentment. Her most important task was to turn the organization into an open network, and she succeeded in that. Openness refers to interactions and joint projects with the universities and industry, including those in other countries. By 2000, about 80–85 percent of their 1,700 laboratories were working jointly with universities, and they succeeded in establishing contacts with companies as well. She tried to boost the importance of social sciences and humanities and to strengthen interdisciplinary research. The British *Times Higher Education* wrote: "Bréchignac . . . has a formidable reputation for determination, decisiveness and an aptitude for analysing and clarifying complex matters."[5]

After finishing her stint as secretary general of CNRS, she returned to her laboratory to continue her research, which she never stopped completely while she was at CNRS. Friday was her research day, which she kept religiously. But how much of a born leader and administrator she is was proven by her further appointments. Between 2006 and 2010, she was president of CNRS, concerned more with policy making than actual directing. In the meanwhile, she also held other positions, not only in France but internationally. For example, she was president of the Institut d'Optique Graduate School and of the Palace of Discovery in France. She spent a term as the president of the International Council for Science (ICSU), whose goal is to strengthen international cooperation. Among her many awards and distinctions, beside her membership in the French Academy of Sciences (corresponding 1997, full 2005, and permanent secretary 2010), she is an officer of the Legion of Honour (2005) and commander of the National Order of Merit (2011).[6]

As to Catherine's private life, she met her husband, Philippe Bréchignac, during her university studies. He is also a physicist and professor at the Institut des Sciences Moléculaires d'Orsay of University of Paris-Sud. He is a successful scientist, but Catherine is the more successful and visible. He is not a member of the Academy. This is a rare situation, and I asked Catherine about it[7]:

> My husband has always helped me a lot and pushed me to succeed. He knows that I am most happy when I work, he knows that I like the research I am doing, and he is very open-minded and understanding. He does not complain if I come home late, if dinner is not ready; on the contrary, he helps to make everything easy. Of course, we have a paid help at home, because both of us are away from home very often. Sometimes he says that too much is too much and then we decide to do something together. . . . When I was elected, he was very proud; I could read it in his eyes. But our life together is not based on an academic success.

Catherine and Philippe have three children, two sons and one daughter. Kindergartens are not a problem in France, but Catherine's grandmother helped them out whenever the children were ill. Catherine understands now that it was difficult for the children that she spent so much time in the lab: "you never were there for us when we needed you,"[8] they told her. She also remembered a terrible day, when she had to take one of her sons to the lab in the evening because she had to finish an experiment. The boy was about four years old. "He was terribly bored just sitting there and looking at that machine, so he took a pair of scissors and 'bung,' he cut the wire and the current. That moment I understood the conflict." Fortunately, everything turned out to be fine, but neither of the children chose science as their profession. She finds it important though that they are not bitter now; whenever these old days come up in a conversation, the children mostly laugh about it. "Now when we talk about this with my children they all say that, at the end, they preferred having a lively working mother to an irritating mother at home."[9]

Catherine is very successful, both in her science and in her numerous leadership positions, and she has proved that even a mother of three can do it all. But it was not easy. She feels that her greatest challenge was to have parallel lives: "Private life, working life, having friends, travel—lives that do not always mix easily. I find it important not to focus only on one aspect of life because if something goes wrong then it is very difficult to survive the shock. You have to be happy in many aspects of your life and then when something happens in one, you still have other resources."[10]

When I asked her about what her message to young women would be, she said, "It is not easy but it is possible. You have to be ambitious. Do not fight against a big mountain, which you obviously cannot destroy, it is better to turn and go around the mountain than trying to go through it. It is important to have a goal and then to find the best way to achieve it. My advice is, do what you like, you have only one life."[11]

FRANCE A. CORDOVA

Astrophysicist

France Cordova in 2008 in Fort Lee, NJ. (photo by M. Hargittai)

France A. Cordova's biographical summary reflects an extraordinary career. She graduated as an English major from Stanford University and earned a PhD in physics at the California Institute of Technology. She served as head of the Department of Astronomy and Astrophysics at Pennsylvania State University at State College and was chief scientist of NASA, the youngest person and the first woman to hold this position. She was vice chancellor of the University of California, Santa Barbara and as the first Latina chancellor of the University of California (UC) system at UC Riverside. When we met, she was president of Purdue University. In March 2014, she became the Director of the US National Science Foundation.

Cordova was born in 1947 in Paris, France, where her Mexican-American father worked for a nonprofit organization after World War II. She was the first of her parents' twelve children. Science as a profession was not yet on her mind when she went to university. As a young graduate in English, after an archeological dig in Mexico, she wrote a short novel, "The Women of Santo Domingo." She entered it at a contest organized by *Mademoiselle* magazine, and finished among the top ten contestants. She then worked on the staff of the news service of the *Los Angeles Times*. At some point, she decided to change fields.

We met at a fundraiser for Purdue in a friend's home in Fort Lee, New Jersey. When I asked her the story of how these big changes in her life, from an interest in literature to majoring in physics, working at NASA, and eventually ending up in high-level university administration, happened, she told me that I would not have enough time to hear all the details[1]:

> It was just a long path, which I would not advocate to anybody. I grew up in a household of deep thinkers. My mother read all the works of the Church, about history and nature; we just grew up in a way that we talked about these things all the time around the table. Do these talks prepare you to be a scientist? Not necessarily, but they get you thinking about the big questions. Somehow, we always gravitated towards the questions of the origin of the universe. In my readings, I was also exposed to some of the great philosophers from a long time ago when science and the humanities were much closer to each other. My heroes in school were people like Albert Einstein. Physics in general was very highly regarded that time. We had science projects, and I knew that I wanted to do something that had real meaning, and stretched your mind to the deeper questions. But it was not before I grew older, in my twenties, when I started thinking seriously about what I was going to do when I am thirty? That time I was exposed to the space program, and it was just like a giant light-ball that went off and I thought, of course, that is just me, that is how I see myself!
>
> When I realized that I wanted to be a physicist, and I had only myself to fall back on, I started to explore how to get into the field, people were very helpful. I had excellent mentors—it is not the word that I would have used at the time because I wouldn't have known what the word mentor meant. But as I look back, they were people who just believed in me and my passion. They were very, very helpful. They were all men because that time only men were in the field. They gave me opportunities, each one was a small opportunity to do something well and I did that and moved to the next level. I did not have a roadmap to tell me how to get from here to there, I just kept doing the next thing and then people would lead me onto the right pass.

After she graduated from Caltech as one of the two women in her class, she spent ten years at the Los Alamos National Laboratory in New Mexico, involved with research in astrophysics. After this time, she moved to Pennsylvania, to her first appointment in higher education, at Penn State, followed by her move to NASA in Washington, DC.

Cordova's research involved various areas of astrophysics. She participated in experiments that measured the X-ray radiation from pulsars (see the Bell-Burnell chapter) and from so-called white dwarfs. White dwarfs are very compact, dense stars, similar to pulsars except that their mass is smaller. She was the leader of a project to study astrophysical processes in strong gravitational fields such as those of pulsars and X-ray binary stars; the latter refers to two stars that orbit around their common center of mass and emit X-rays.

Besides actual measurements, she contributed to developing space-born instrumentation. During her time at NASA, she became involved with the X-Ray Multi-Mirror Mission of the European Space Agency. The goal of this major program was to send powerful X-ray telescopes into space together with an optical monitor that had never before been part of an X-ray observatory. She was in charge of the optical monitor digital processing unit and acted as the US principal investigator. The Earth's atmosphere blocks the X-rays coming from space, and the goal of this mission was to get the X-ray telescopes beyond the atmosphere and detect the X-rays coming from distant objects in the universe. The mission was launched in 1999, and by that time France was already vice chancellor for research at UC Riverside. I asked her how she could manage these two very different jobs of high responsibility. "That was when I was in the final stages of building an experiment in a team for a flight. I just hired the right people to work on that; we also had a lot of students. Because it was a team effort, I was able to do both. I could not have done it if I was just a single investigator; I would not have the time to focus. But when you are in a team, you just make sure that everybody is doing what they are good at and you get the job done. This was in 1999."

In 2007, when she moved to Purdue, her NASA funding was winding down, her research associates were finding new jobs, and she stopped doing research. She felt that being the president of Purdue was too big to do anything else. "I know that scientists who go into administration try to have both for a while." She felt that even in this job, it was very good that she had the scientific background—she knew the language, she understood the problems, and she could ask the right questions when they were planning long-range research prospects with the faculty. "I like administration, I also believe that my modest skills, from English writing to the hard sciences, come all nicely together and help doing this job."

During our conversation, the topic of extraterrestrial intelligence came up. She is very much interested in this question but pointed out that it is important to distinguish between the possibility of life in the universe beyond the Earth and extraterrestrial intelligence. When she worked at NASA, on the search for life in the Universe, she understood that such projects might be easily misunderstood.

> People look for little green men. What is important is understanding life and how it came to be on our planet; and whether another planet could have the conditions for life. It was a breakthrough when they separated the search for life and the search for intelligence. I actually taught a class 'Life beyond Earth,' how life can exist at extreme environment. This opened up a lot of possibilities for having life in our solar system and beyond. The search for intelligence is another matter. It is rather improbable that we find it because the universe is such a huge place.

One of France's hobbies is rock climbing. During one such activity, she met Christian Foster, a high school science teacher. They fell in love and got married in

1983. They have a daughter and a son. I was wondering how she manages being a researcher, a high-level administrator, and a mom.

> I feel very good about the ways we raised our children. They did have a lot of moves but not as many as some in the military. I actually think that adaptability is a good thing, even though it is difficult. Our kids turned out to be very well; they are nice people, successful as students. . . . I tried to make it home for dinner no matter what and then I worked during late evening and the night. I don't know what my kids would say but I feel good about it. As far as my husband is concerned, I attribute a whole lot of the stability that has characterized my experience to him because he has always been so good about adapting to new situations. He makes friends easily and he has always been there for the children, more than 50% . . . he has just been completely supportive. He is also a very honest person; he is usually there when I give a talk and afterwards tells me if it was bad or I did not connect, or whatever else was wrong with my speech.

France Cordova's achievements have been broadly recognized. In 1984, she was chosen by *Science Digest* as one of "America's 100 Brightest Scientists Under 40."[2] In 1996, during her time at NASA, she was featured in a program by PBS called "Breakthrough: The Changing Face of Science in America," which introduced twenty Native American, African American, and Latino scientists from different scientific fields.[3] The following year, when she was already at Penn State, she was chosen as one of the "100 Most Influential Hispanics" by *Hispanics Business* magazine, and recently was among the "101 Top Influential Leaders in Hispanic U.S." in *Latino Leaders* magazine.[2] She was also among the recipients of the annual Kilby award that is given each year "to honor unsung heroes and heroines who make significant contributions to Society through Science, Technology, Innovation, Invention, and Education."[4]

Her message to young people, both girls and boys:

> No matter what you want to do there will always be times when you doubt yourselves; the tendency to think that you are not good enough is just part of the human psyche; there is always someone who is better, brighter, smarter. But that just should not weigh; because if you have a vision of what you would like to be, you have to get past these little hurdles and have an inner strength to move on. You have to keep an eye on what it is that you are going for and realize that once you get there people don't look back. Nobody will care about what grades you had; getting a bad grade might feel an awfully bad thing but later on, nobody would remember them. Of course, I realize that this is easier said than done.

MARYE ANNE FOX

Chemist

Marye Anne Fox in 2000 in Raleigh, NC. (photo by M. Hargittai)

As the first woman to be elected chancellor of the University of California in San Diego (UCSD), Marye Anne Fox contributed to an unprecedented growth of the university. When after eight years she stepped down, the president of the University of California system wrote, "During her tenure as Chancellor at UC San Diego, Marye Anne Fox has added striking breadth and depth to the university's already sterling reputation."[1] All this happened after she had already achieved similar successes as chancellor of North Carolina State University (NCSU)—besides never giving up her research career in organic chemistry.

She was born as Marye Anne Payne in 1947 in Canton, Ohio. Her father was a manager at a steel plant, and her mother was a teacher in grade school and high school. Although her father had a little bit of background in metallurgy, he was not a scholar, and her mother's field was journalism and Latin. Thus, there was no scientific background in the family. She believes that her interest in science was due to the time when she was growing up; this was right after the launch of Sputnik, which energized many youngsters in the United States to study science. She chose chemistry because "it had some of the cleanliness, the symmetry, and the beauty of the mathematical parts of science, with the experimental parts of chemistry and biology."[2]

Marye Anne married early, right after her undergraduate studies. Her husband was a physician, and that determined where she continued her education:

> I followed him around; I took a master's degree at Cleveland State University in chemistry because he had a one-year slot in Cleveland finishing an internship and I followed him to Dartmouth and chose my grad school basically because he had a three year residency. So I had to finish my degree in three years because that's how much time he had. I chose a post doctoral fellowship because he was drafted into the Air Force. I followed him there and chose my postdoctoral position. Then when we moved to Texas that was essentially my choice and he went along with that one.

She joined the Chemistry Department of the University of Texas at Austin in 1974. Her background was in organic and physical organic chemistry.

> I've always been interested in the relationship between structure and activity and in particular the way that light causes reactivity to change from normal ground state behavior. I have worked a long time especially on oxidation reactions and photo-induced electron transfer. I would say that my most significant scientific achievements are in those related areas, that is, what I call organic photoelectric chemistry. It is using light to induce reactions driven by single electron transfer on the surface of semi-conductors. We are working out how the intermediates are stabilized by the surface and how the reactions proceed. And secondly, something about long range photo-induced electron transfer and how a scaffold can be created. This way, electron transfer can be controlled over very long distances and so you learn experimental information about electronic coupling that facilitates electron exchange.

After twenty years of a successful research and teaching career, Marye Anne was appointed vice president of Research at the University of Texas at Austin and served in this capacity for four years, until 1998. Then she became the first female chancellor of NCSU. I asked her what made her interested in administrative positions. Apparently, her involvement with science policy dates back much earlier and is at least in part due to the influence of one of her heroes and mentors, Professor Norman Hackerman:

> He's a person who had himself followed an administrative career after a distinguished career as a scientist. He'd been President of the University of Texas and of Rice University. He was the one who got me involved in committee work on national science policy at the National Academy of Sciences. I was appointed to

> one of the commissions there very young; I was in my mid-30s. As a function of service there, I apparently did well enough to be noticed and I was nominated to the National Science Board [1991–1996] and was for two years chairman of programs and plans which was part of the National Science Foundation program providing support for research. These programs and plans deal with fellowships and education programs, so being the chairman exposed me to a lot of different disciplines. For example, astronomy; I went to the South Pole to see the astrophysics work that was done there and to the telescopes being built in Hawaii and physics, material science, to social sciences and biological and life sciences. I ultimately became vice chairman and learned a lot about interaction with Congress. While I was on the National Science Board, I was invited to become vice president for research so I went from being a faculty member without any intermediate experience to being a vice president and apparently, I did well enough there, so I was recommended for this job [chancellor at NCSU]. Now I'm a chancellor which in our system is the top official on a campus. This campus is the largest university in North Carolina and it is the one dedicated to science and technology.

She also mentioned a colleague of hers at the University of Texas, the late Michael Dewar, a famous theoretical chemist, whom she considers one of her mentors. She taught the basics of quantum chemistry (advanced molecular orbital theory) together with Dewar. She appreciated his telling her when she did not explain something well enough. "I think it is a hallmark of effective mentoring not only to support you but to tell you when you are making a mistake."

After having served as chancellor at NCSU for six years, in 2004 she accepted the position of chancellor of UCSD and served in this capacity for eight years until her resignation in 2012. She then returned to full-time teaching and research. But even during the previous years, her connection with chemistry did not stop—and that was to a large extent due to her second husband, James Whitesell, a chemistry professor, with whom she has been doing joint research. When I asked her how she could keep up with having a relatively large research group, she said, "I am only able to run that size here because my husband is a chemist and helps with the day-to-day crisis management that goes along with a research group. I love science so much that it would be almost impossible to abandon it completely." The two of them wrote two chemistry textbooks during the past twenty years.

Marye Anne has three sons from her first marriage. When I asked her about when she faced the greatest challenge in her life, this is what she answered: "I think, probably, when I was a graduate student and I learned that I was pregnant. I had to decide how seriously I wanted to do science; and I was twenty-seven. At that stage, I made a commitment to an active intellectual life and to the idea of getting into an academic institution. Once that decision was made firmly in my mind many other things became easier." She believes that her children did not suffer from having a fully dedicated working mother; on the contrary. "The relationship is very much different when you grow up in an academic family. I remember most of our sons' friends remarking that the dinner conversation was at a different level than at their own homes." She also felt that having the responsibility of a family helped her in her professional life as well: "It certainly was a help to me in my career. It caused me to focus and I've seen in my colleagues that those who don't have children don't have to focus as much as those of us who have family obligations. My entire professional career I've had that focus. So it helps you to learn to multitask early on and it was a big help in my career."

She already had three children by the time she got tenured, and that certainly had to teach her how to focus. When the children were small, she had help at home; later, she had to accept the "clock" of the day-care center in organizing her professional life. "There are, of course, always compromises. I carried my briefcase to the baseball game when my kids were playing baseball—but I was there."

Her involvement with US national science policy goes back to the time when she became a member of the National Science Board at the beginning of the 1990s; she was also George W. Bush's science advisor during the time he was governor of Texas in the late 1990s. She did not stop this activity in the following years either; she was a member of President George W. Bush's Council of Advisors on Science and Technology. Moreover, she has served on the boards of many different organizations; her board memberships take several pages of her vita. Similarly long is the listing of her awards and other honors. She counts the most important among them her membership in the US National Academy of Sciences (1994), and another is, obviously, the National Medal of Science (2011) from President Barack Obama. I mention one more of her awards (in 1996), because "The one closest to my heart is the mentoring award I got from Sigma Xi, the Monie A. Ferst Award, because it acknowledges what I value the most, which is seeing my students come to scientific maturity and achieving on their own."

KERSTIN FREDGA

Astronomer

Kerstin Fredga in 2000 in Stockholm. (photo by M. Hargittai)

"I have always been interested in astronomy; when I was ten years old, I told my parents that I want to get to the moon. Astronomy always fascinated me." So says Kerstin Fredga, professor of astrophysical space research, former director of the Swedish National Space Board and former president of the Royal Swedish Academy of Sciences.[1]

She was born in 1935 in Stockholm into an academic family. Her father was professor of chemistry in Uppsala; her mother was a kindergarten teacher before marriage. Kerstin was the middle one of the family's five children. "It was a very good home, because everybody was allowed to pursue his or her own interest. I had a younger sister who was interested in textiles and she pursued that; an older brother of mine went to cut wood in the forest. My parents gave me books on astronomy."

Kerstin went to Uppsala University for her studies, and in 1962 she received her PhD in astronomy. Then she went to work for a Swedish solar station on the island of Capri, off Sorrento, not far from Naples, Italy: "I am not a nighttime astronomer, but a solar one. There we observed the sun and saw all the eruptions and flares developed on the sun. This station was part of the international stations that had been watching the sun." This research was important for practical purposes as well. During World War II, people noticed that pilots could lose contact with their bases if there were strong eruptions on the sun. Knowing in advance about such events could save lives. It is still not possible to correctly predict these eruptions, but at least it is possible to predict a probability of their occurrence.

After Capri, Kerstin spent a few years at NASA, at the Goddard Space Flight Center. She built an instrument and had it launched on a rocket that she also designed: "That was a spectacular moment. I was interested in looking at the Sun in the ultraviolet region—that is in a wavelength range that you cannot see from the Earth because it is covered by the atmosphere. I built an instrument that took pictures of the sun and we could compare them with the pictures that we take of the sun on Earth. During my stay in the US at NASA I launched four rockets and I got all the support for them."

Before returning to Sweden, she spent a short period at the Astronomical Institute and Space Research Laboratory at the University of Utrecht in the Netherlands. In 1973, she was appointed professor at Stockholm University. At the same time, she started to work part time at the Swedish National Space Board and gradually drifted more and more into an administrative job. "I do not regret that, I liked my scientific period but I like this one, too." First, she was responsible for the Board's science policy. In 1989, she became the chairman and director general of the Board and held that job for a decade.

> That is our counterpart of NASA. We act as a research council for research within the area of space science. We fund the scientists who want to use rockets for their science. Sweden is rather unique in that you can go very far north and you are still in temperate region because of the Gulf Stream. We have a launch area up far north in Kiruna [the northernmost city of Sweden, about 150 kilometers north of the Arctic Circle.]. To have a launch area there is an advantage because, for example, if you want to study the Northern Light (also called Aurora Borealis) you have to shoot into the Aurora and for that, you have to be far north. There are other advantages of being so far north. For example, when you want to have a satellite and look down on the Earth then the instrument goes around the Earth over the poles; the Earth turns around and you can then have one stripe of the Earth and the next one and so on. You cannot do this if you are over the equator. There are several reasons why we have a large activity up at Kiruna. We also represent Sweden in all international cooperation; we are members of the European Space Agency. We also had to make sure that we help Swedish scientists to get their experiments on board or to see that the Swedish industry get contracts and you get return on your money. The budget at my time was about 700 million Swedish kronor per year.

When I asked which of her jobs she enjoyed more, being a scientist or being a high-ranking administrator, this is what Kerstin answered: "When I was in the States launching rockets, it was the most enjoyable one. It also came natural to develop the feeling that you have to care for others and give them possibilities to do their science, so talking to politicians, getting the money for their research, representing Sweden abroad, all this was very interesting and a lot of fun. The right thing at the right time."

She was still head of the National Space Board when she was elected president of the Royal Swedish Academy of Sciences, of which she had become a member back in 1978. It is not a full-time job, but it is a highly respected one, at the summit of Swedish science. The Academy was founded in 1759 and it has always represented a free voice, independent of the government. Added recognition for the Academy comes from the Nobel Prize; this institution awards the physics and chemistry prizes and the Nobel memorial prize in economics. They work all year round scrutinizing and evaluating the achievements of scientists, and she believes that they have managed to do this very well for over a hundred years.

To the question of the effect of the Nobel Prize on Swedish science, she said:

> It has a big effect. We have a lot of scientists who are involved with it. They have to be up to date in the fields. My father was involved and I remember our summers when he was just reading and reading—preparing for that year's evaluation. Moreover, we get the best scientists in the world coming to Sweden; giving lectures, visiting our institutes, talk to the students. The Prize certainly inspired young people to go into science, but perhaps the most important effect of the Prize on Swedish science is that our scientists have to keep themselves up to date.

Fredga feels that it was hard being the director general of the Space Board and president of the Science Academy at the same time. Sometimes she felt it was too much; usually she did one in the daytime and the other at night. "But at the same time the two positions benefited from each other. Being the president of the Space Board, I always had contacts with the space communities all over the world while at the same time with the Swedish National Space Board. Being president of the academy had an advantage; after all it is a high-prestige job."

I was wondering if being a woman helped her to get elected to these high-prestige positions. She feels that at the time she became director general this question was not yet considered the way it is today; she had already worked there for a while, so

> maybe they thought I could do the job. They saw I was interested. Today it might be different. We had a minister who tried to help by offering special professorships for women, and that is not a good way to do it because there will be a backlash. Nowadays you might be caught in that backlash or at least you are accused of having gotten your position because you are a woman. It has not been that bad in Sweden but the tendency is there. Nobody is happy. They are trying to help by doing the wrong thing. In my opinion, the only thing you can do is strictly go on the scientific merit when appointing a professor. We have a rule that if you have

> two persons of equal quality then give the job to the underrepresented sex that could be either a woman or a man depending on the topic. In natural sciences, it is usually a woman but in humanities it is usually a man. That is fair enough, but you cannot go further than that. You cannot have quotas.

She has been married twice; she has one son from her first marriage. Her first husband was also an astronomer; that is how they met. Her second husband is in the humanities. He was a former rector of Stockholm University, and they met at the Science Academy. Concerning the difficulty of bringing up children when both parents work, she said:

> That is always a difficult question and you find the solutions on an ad hoc basis. The husband also has to take part and share the burden. I myself started a whole kindergarten—we called it the "penguin" because they are birds that share taking care of their infants. There were several families and we got together and made a daycare center; the husbands and wives together. It was a private enterprise but we did not call it private since it has a wrong connotation in Sweden where everything is so collectivistic and socialistic. We called it "parent-cooperative-test case." It has worked very well and is still functioning. You have to find solutions; you have to get involved.

Her message to working mothers:

> Give priority to what is important when it is needed and then later you can always catch up. Sometimes you have to set back a little and not be so competitive but you can never ever let your family and especially your children to suffer. It is difficult but it is possible.

She believes that she has not encountered discrimination during her career:

> If you ask women in general in Sweden about this, probably many of them would say yes. But in my case, I don't think that I have ever been discriminated against because I was a woman. On the other hand, it very much depends on how you interpret what you meet in your life and how people behave. In the States, I thought that it was tough but I think it would have been tough for everybody. I believe that this very much depends on your attitude; if you do not see something as a gender discrimination, it does not necessarily mean that others would not. I have often debated this with my female colleagues and they often say "but don't you see it?" And I say "No, I don't." And if I don't see it, I don't react like that and I don't get any more discrimination. It depends a lot on you. You can also ask, have you been positively discriminated? No, not really, I don't think so. Nobody has accused me of climbing too far. Perhaps this is easier in the academic world . . .

CLAUDIE HAIGNERÉ

Neuroscientist

Claudie Haigneré in 2003 in Paris. (photo by M. Hargittai)

First, she was a medical doctor and a PhD in neuroscience. Later, she was the French minister for research and new technologies and minister for European affairs. In between the two, she was the first French woman astronaut. She participated in joint French-Russian missions; was the first woman to qualify for a Soyuz Return commander; and, finally, was the first European woman to visit the International Space Station as part of the Andromeda mission. She is a recipient (commander) of the Legion of Honour, the highest award in France.

Claudie Haigneré, née André-Deshays, was born in 1957 in Le Creusot, a small town in eastern France. She attended medical school in Paris and specialized in sports medicine, space medicine, and rheumatology. She worked in a hospital in Paris when one day she read on the notice board that the French Space Agency (Centre National d'Etudes Spatiales, CNES) was looking for scientists to participate in their experiments on microgravity. Space travel has fascinated her since childhood, when she watched Neil Armstrong walking on the moon. The CNES call seemed to her like a dream come true. Out of about a thousand candidates, she was among the very few selected for training in the mid-1980s. The agency was particularly interested in having scientists for doing experiments at the space station. She was eminently qualified[1]:

> Beside specializing in rheumatology, I also had a certificate in sports medicine. At this point in my life I understood the importance of this, so I decided to change my career and become a scientist. I started to take different courses, for example, in biomechanics and in the physiology of movement and also decided to get a PhD in science. I worked in the Neurosensory Physiology Laboratory of the CNRS[a] for about six years. During these years, I worked part time in the hospital and part time in this laboratory. I received my PhD in neuroscience. I had a very interesting experience in this laboratory because what I was working on here was in close relation to my training as a rheumatologist. We were working on gaze control at movement and we needed to look at the pathology of the neck and neuro-scientific basis of the movement of the head and gaze. This question is important in microgravity; it was a specific topic related to the organization of the cognitive mapping in the brain depending on the environment. This was an opportunity for me to get first-hand knowledge on the rules of scientific approach. This was also the opening for me to the microgravity experiments and to the international scientific cooperation in space research.

Eventually, Claudie joined the CNES, and was appointed to be in charge of the program that investigated the adaptation of cognitive and motor skills under microgravity conditions. Soon, she went on her first training for space flight. She met Jeanne-Pierre Haigneré there, who became her husband. "The first long training when we shared all that together was fantastic; I believe that we are a very special pair." Originally, he was a professional pilot. "There were two categories for astronauts, one was operational and in charge of human control of the space lab, and the other scientific, in charge of the experimental programs. He was in the first, I in the second."

Microgravity means almost zero gravity. "Microgravity science is performed in a very specific laboratory at the space station. On Earth, we cannot study the effect of gravity (or the lack of it) because you cannot extract gravity here; you can increase it by a centrifuge but cannot take it away." There are quite a few areas of science where there is great utility in research performed under microgravity conditions, such as materials science, combustion, fluid physics, nanotechnology, cell biology, and biotechnology, just to mention a few. However, there are only

[a] CNRS: Centre national de la recherche scientifique (National Center for Scientific Research)

limited opportunities to do so, so the research has to be well organized, the questions have to be relevant and well defined, and the actual investigations well prepared and organized. The most interesting questions concern phenomena that are masked by gravity on Earth. "I studied locomotion, movement control, with and without gravity; we studied how the brain adapts to this new environment. Our brain, our motor skills have to adjust to that very special—very different from our usual—environment."

During training for the space flight, she spent considerable time in Russia and learned the language. "It was essential that we understand each other perfectly. We participated in the same training and it was quite long but everybody wanted to bring the mission to success and we had a good relationship." It was an international group, Europeans and Americans alongside the Russians, and "we from Europe sometimes felt that we were a bridge between the Russians and the Americans."

In the early 1990s, she was responsible for the space physiology and medicine programs at the CNES. After years of training, in 1992 she was chosen to be the backup for Jeanne-Pierre in the flight that took place in 1993. During the flight, she monitored the biomedical experiments as part of the ground team. In 1994, she was selected for the "Cassiopée" mission that took place in 1996, when she spent sixteen days at the Russian "Mir" space station. She performed experiments in physiology and developmental biology as well as in fluid physics and technology.[2] Her most important flight was in the Andromeda Mission, in 2001. She was one of three in the crew, and the mission lasted ten days. They performed a wide range of experiments, in microgravity physics, life sciences, ionosphere studies, and Earth observation. They studied how the brain uses gravity in the process of perception. Later, they compared their results with the same processes on the ground. This should help scientists to understand how the brain adapts to a gravitation-free environment. At that time, she already had a three-year-old daughter, who went with her to "Star City" (the Russian cosmonaut training center) and stayed there for several months. There was a Russian nanny to take care of her.

I talked with Claudie about women in the space program. At the time of our conversation, in 2003, there were forty astronauts altogether in the world, among them four women, three Americans and Claudie.[b] Because of the long, rigorous, and physically demanding training, it is not easy for young women to take up this challenge. Claudie stresses that it is worth it, and it is possible to cope with the added difficulties: "My daughter is five years old and since my husband is also an astronaut I could only manage because my family was very supportive. They knew how much this meant to me. Here, now, I have someone who takes care of my daughter in the morning and after school. For me it is sometimes frustrating that I cannot be more with her—but she is happy.

[b] The first ever woman in space was the Russian Valentina Tereshkova, in 1963.

As the first European woman, Claudie Haigneré flew to the International Space Station in 2001. Copyright ESA/CNES. (courtesy of ESA and Claudie Haigneré)

Claudie retired from the European Space Agency in 2002. At the time of our meeting, she was minister of research and new technologies in Jean-Pierre Raffarin's government. I wondered how this happened.

> As the only French woman astronaut, I am quite well known in France and I know that I am a role model for many young women. I consider it important to share my experiences, my knowledge, and the whole adventure that I was lucky enough to be involved with. I also try to encourage young women to go into science; and let them know that if they have a dream, they should be brave enough to go for it. The Prime Minister called me and asked if I would give a chance and accept this position, with my experiences behind me. This was a new challenge and I accepted it. Again, I had to learn a lot, especially about the political side of the work. I hope that I will be able with the help of the scientific community to reorganize and rationalize the system. It is difficult. I never thought that it would be easy but it is much more difficult than I thought.

She stayed in the above position for two years, and then, between 2004 and 2005, she served as minister for European affairs and secretary-general for Franco-German cooperation. In November 2005, the European Space Agency chose her to be adviser to the director general.

In 2008, a few days before Christmas, she was hospitalized because she overdosed on pills, and there were rumors in the papers suggesting that it was a suicide attempt. According to an article in the newspaper *Libération*, she described the incident as a sign of "burnout." "One no longer closes one's eyes, one is empty on the inside, the emotions become dull and disappear, one feels useless, zero, insignificant. Above all, she wants to sleep, unplug the robot that she had become. Five or six tablets (pills) of something strong, or even more . . ."[3] Did she want to achieve too much, performing in such different high-level and demanding capacities? Did she feel that beyond all her achievements there was nothing to live for anymore? We do not know. What we know is that, fortunately, she has recovered and has since been busy again in numerous positions, being on the board of directors of major companies and foundations. In 2009, she became the president of the Cité des Sciences and the Palais de la Découverte, two science museums in Paris that she joined together under the name of *Universcience.* This new insititution was meant to encourage young people to study science and technology to enable them to creatively participate in the ongoing debate about science and society. Apparently, she has found a new niche for herself.

HELENA ILLNEROVÁ

Biochemist

Helena Illnerová in 2001 in Prague. (photo by M. Hargittai)

The story of why Helena Illnerová in Czechoslovakia (as it was then) decided to study chemistry is perhaps typical of the 1950s and 1960s in Eastern Europe. She was a child with broad interests in both the sciences and the humanities but she wanted to study something that was far from politics and ideology. Her lawyer father was fired from several jobs because he was not considered trustworthy. Eventually, he found a job in a glass factory and did research for them. This is how he got the idea that Helena should choose chemistry as a profession and work in the field of inorganic chemistry, which seemed interesting enough and ideologically neutral enough and thus not "dangerous." She did not mind, since although she was more inclined toward the humanities, she also liked the sciences. This is how she became a chemist, and she has never regretted it.

Helena Illnerová (née Lagusová) was born in 1937 in Prague into a Jewish intellectual family. Her mother did not have a profession but knew several languages, as did

Helena's father. Helena's maternal grandfather was a university professor and rector of Masaryk University in Brno. He was murdered by the Nazis during World War II. Helena's father and his family members were taken to a concentration camp; he came back at the end of the war, but the rest of his family perished. After the communist takeover in 1948, Helena's family went through hard times. She remembers: "My sister could not go to the gymnasium although she had the best marks, but we were considered intelligentsia and that was not good. Finally, both of us managed to get to the university. During our last year, at the end of the 1950s, our mother was put on trial, which was against members of the former intelligentsia, and she killed herself. That was a terrible end to our childhood."[1]

Helena graduated from Charles University in Prague in biochemistry and started graduate studies in the Institute of Physiology. She soon found out that she was not well prepared for the work, since during her undergraduate chemistry studies genetics was still taboo, so that anything related to it was not taught properly. When she graduated in 1961, the double helix structure of DNA had long since been discovered (in 1953), but it still was not part of the curriculum. She could only read about it in popular magazines. The situation in Czechoslovakia was similar to that in the Soviet Union, and Lysenko's unscientific teachings could not be challenged. This eventually started to change, and by the time of her doctoral research, genetics was no longer a forbidden field.

Helena's principal area of research became the biological clock in mammals, also known as the circadian clock. This happened more or less by chance. She worked at the Institute of Physiology and started to study a small organ called the pineal gland, which got its name from its shape, which resembles a pinecone. It is a small gland in the center of our brain between the two hemispheres, and one of its functions is to produce melatonin, which regulates our responses to the changes of day and night; that is, being awake or asleep. In her words:

> Around 1970, the study of the movements of mammals was one of the research topics in our department. I was interested in the pineal gland because I understood that it was affected by light and I thought it would be interesting to look at the development of this gland from the biochemical point of view. I noticed that small rats open their eyes on the 14th day after they were born and I thought that perhaps the development of this small endocrine gland might have a connection with the light coming through their eyes. I always used only red light in the room because its effects are smaller than those of normal light, but one day I accidentally opened the door and the light from outside came in. I noticed that my results changed completely. The serotonin [a neurotransmitter molecule in our brain] level in this gland, which is very low during the night, jumped up within a few minutes.

She did not understand right away why this happened; it took several years to figure it out and see its connection to the biological clock. She found the topic fascinating, and from that moment on Helena has continued to study the biological clock. Most of her life she has studied how the production of melatonin in the body is regulated.

The biological clock is a biochemical mechanism that drives our circadian rhythms. The expression "circadian rhythms" refers to all the physical and behavioral changes of an organism that follow the twenty-four-hour cycle related to the change of night and day, darkness and light. These rhythms are present in all organisms, humans, animals, and plants alike.

Melatonin not only signals day and night but also the seasons. This relatively simple compound serves as a chemical marker for the functioning of the whole biological clock, like the hand of the clock. Melatonin production starts to increase in the evening and starts to decrease toward the morning.

> After we understood melatonin's role, we went to the clock itself, to the special cells in the brain to understand the process. Melatonin is produced partly in the pineal gland in the brain and partly in the eye. First, they make serotonin and that is the precursor of melatonin. The enzyme that transforms serotonin to melatonin has such a strong rhythmicity that its activity during the night is at least 100 times higher than during the daytime. Can you imagine an enzyme that has such amplitude in rythmicity? This is why we use it as a marker to show when the melatonin production goes up or down. It is programmed by the biological clock.

Some years ago melatonin was widely advertised as a natural product that can help overcome the adjustment difficulties caused by jetlag after long flights over several time zones. Of course, the commercially available melatonin was a synthetic product, but its chemical composition was the same as the one produced in our body. Helena, however, warned me that people have to be careful of when and how they use it. The body produces it at a special time of the day, that is, at night but not during the day. If somebody takes melatonin at an "improper" time, that is, in the morning, this might give the body a wrong signal that it is night. This would damage the timekeeping system of the organism, which is responsible for the daily program of the body. Taking it on or before long flights, especially eastward, may be all right, because it helps advance our old clock to the new time. It also might help people to fall asleep and sleep deeper, but in this respect there is a caveat: it is safe only for older people, because it might influence the reproductive function—at least, animal experiments have shown this.

There are all kinds of rhythms in our body. There are clocks in different peripheral organs in the skeletal muscles, the kidney, the liver, and the intestines. Under certain conditions, these organs can behave as clocks. There is a hierarchy of clocks in the body; the "master clock" is in the brain, and it synchronizes the rest. This is a group of nerve cells, called the suprachiasmatic nucleus, that controls all the circadian rhythms that give the time organization to the whole body.

It is an interesting question how our biological clock develops. Babies are born already with their biological clock ticking. It has many rhythms that we can observe, for example, the heart function and body temperature. However, the baby is not yet synchronized with the outside word. The newborn baby, which does not yet perceive light, does what is called "free-running." The baby's clock free-runs during the whole day, and it can have night when the parents have day, and vice versa. There are probably even more such rythmicities in the baby's bodies. Some of these cycles may have only eight or four hours. It takes about three to six weeks before the baby becomes synchronized to the twenty-four-hour cycle. After that, the baby becomes trained to the environment and it starts to distinguish days and nights. It may not be just a question of perceiving light. Melatonin also has a role. The baby when it is born does not produce enough melatonin. In the fetus stage, it takes melatonin from the mother, but when it is born, it loses this information. Helena and her colleagues showed that there is a beautiful rhythm of melatonin in mother's milk. During the night, it has a high concentration of melatonin, but during the day it does not have any. The baby can get the information from the mother's milk. Helena got the idea to study this when her daughter was pregnant; she collaborated with the doctor in the hospital. She got the first milk for this study from her daughter; later on, of course, they extended the investigation.

Studies with blind people show that light is the most important factor in setting the biological clock. Blind people also have the twenty-four-hour cycle, but they have to synchronize it with the rest of the population, and this happens socially. Another noteworthy case is people living in the far north in Scandinavian countries. Helena commented:

> A psychologist friend of mine in Norway moved from Oslo to a university up in North, and he told me that that place is just one big chronobiological laboratory. People are paid more there not because of the inconvenience but because it really is hard up there during the winter months. During the day they have to subject people to very bright light that is similar to the Sun. People may have seasonal affective disorder and it mostly is seasonal depression when the days are too short and the nights are too long. This disorder is the more severe the farther north you go. People getting circadian rhythms and light treatment are mostly from Scandinavia and Canada. In fact, some people in the United States, where many people stay in their cars, at their home and workplace and do not spend any time outside during the day, may also have some problems because daylight and especially the morning light outside is very important. This is, why it is very good to have a dog that you have to walk in the morning. In fact, most people have a cycle that is a little longer than 24 hours, therefore, we have to face-advance each day and for that we need the morning light that face-advances us.

Helena accepted the position of vice president of the Czech Academy of Sciences in 1993, soon after the political changes, and in 2001, she was elected president of the Academy. This is a very prestigious job with a large workload. I visited her soon after she started her presidency. I wondered why she decided to go into administration in the middle of a successful research career.

> I took the job because so many people asked me to take it. I did not want to tell them no; I was honored by their trust. Before I became president, I was vice president and I was in charge of biology and chemistry. The people in these fields, hopefully, have known me as being honest and doing the job well. They trusted me. Also, maybe people felt that it would be good for a change to have a woman for the first time in this position. There were four candidates; the other three were men. I very much hope that I was not elected *only* because I was a woman. I feel that I have ideas about what to do with the job. I believe that exactly because I am a woman I can create a harmonious environment.

Helena met Michal Illner in a tourist club to which both of them belonged; they got married in 1963. Michal is a sociologist by training and worked at the Institute of Sociology of the Czechoslovakian Academy of Sciences (as it was then). He became director of the institute soon after the political changes, and held that job for eight years. They have two children; their daughter is a physician, specializing in endocrinology, and their son studied mathematical engineering. Helena and Michal have five grandchildren.

I wondered if Michal was proud of Helena for becoming the president of the Czech Academy of Sciences. She responded: "We have discussed many times before I decided to run for the presidency whether it was a good idea; I don't know if he is proud of me but I know that he is not jealous of me. I also make sure that wherever we go together, I play the accepted role of a wife; when he starts talking, I stop talking. He is much better in foreign languages than I am. Then, being a sociologist, he knows a lot about society and politics."

For Helena, being a career woman and having small children was a great challenge. However, the family came first for her. She made sure that she was always home in the evening, prepared dinner, and discussed the day with the children when they were small. After they went to bed, she returned to the laboratory to continue her experiments. It was hard, but she managed.

Helena was president of the Czech Academy of Sciences till 2005. After stepping down, she served on various committees; for example, she was president of the Czech Commission for UNESCO and of the Commission of Ethics of Scientific Work at the Czech Academy of Sciences. She continues her interest in her field and is researcher emeritus at the Institute of Physiology. The Czech radio described her as "the leading lady of Czech science" and "one of the top minds of the Czech Republic."[2]

CHULABHORN MAHIDOL

Chemist

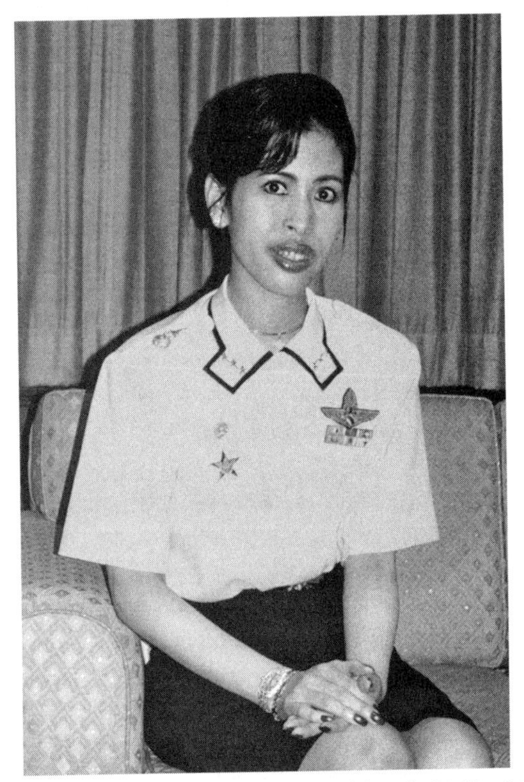

Princess Chulabhorn in 1999 in Bangkok. (photo by M. Hargittai)

The King and I with Deborah Kerr and Yul Brynner, and later *Anna and the King* with Jodie Foster and Chow Yun-Fat, are among my favorite movies. Somehow, I could not help feeling that parts of these movies came alive when in 1999 we visited Princess Chulabhorn Mahidol of Thailand in Bangkok.

Her proper title is Professor Dr. Her Royal Highness Princess Chulabhorn Mahidol; she is the youngest daughter of Their Majesties King Bhumibol Adulyadej and Queen Sirikit of Thailand. You might wonder how a royal princess gets into a book on women in science. The answer is simple: she *is* a woman in science, even if not a conventional one. She is the founding president of the Chulabhorn Research Institute, and a professor of chemistry at Mahidol University.

The author and her husband present the Princess with one of their books. (courtesy of the Chulabhorn Research Institute, Bangkok)

We visited her in her research institute, and our first obvious question to her was why chemistry? She is probably the only princess who is a chemist. When she was a little girl, she did not plan it this way, and wanted to become a concert pianist, but her father insisted that all his children learn a profession that would be useful for the future of such a developing country as Thailand. She only had the choice to take either physics or chemistry, and she chose chemistry. First, she studied at Kasetsart University and then earned her PhD at Mahidol University, both in Bangkok. After her chemistry studies, she was a postdoc in Germany at the University of Ulm in genetic engineering and in Japan at Tokyo University Medical School. Thus, she acquired a broad range of knowledge, "but it took me many, many years,"[1] she stressed.

After returning home, she founded the Chulabhorn Research Institute, whose goal according to its mission statement is to "improve the quality of life." According to the princess, this is her father's policy. Her interest, and the main research line of the institute, is natural products chemistry. This is a fitting topic for Thailand, where there is a large arsenal of plants that have been used for centuries to heal people. She told us: "I am always fascinated when old people are telling me about the curing and healing effects of various plants. . . . I have to see for myself what is in the plant and what the active ingredient is."[2] The introduction of one of her papers also explains why this research topic is so fitting for her country: "Thailand is uniquely located to represent the fauna and flora which characterizes the biogeographic province of Indo-Burma. A number of the eastern Himalaya temperate taxa penetrate south into the northern mountains of Thailand while the southern part is evergreen forest thus making this area one of the richest floristic regions of the world."[3] There are thousands and thousands of different species of plants in Thailand. The Institute is well equipped with modern instruments, the gifts of Germany and Japan.

Princess Chulabhorn teaches oncology, toxicology, and biochemistry. Besides, she is a vice marshal of the Thai Air Force, and she teaches chemical warfare to its personnel. She arrived to our meeting from one of her lectures and was wearing her air force uniform. To the question what she teaches in this course, she answered: "The purpose is for my students to know the chemicals, the biological agents, and to observe the circumstances around them, and thus protect their lives. I'm not teaching them how to kill somebody."[4]

Because of her many duties at the Institute and teaching at various universities, she is excused by her parents from attending most ceremonial functions. They understand that her role as a scientist is more important than her role as a princess. She is divorced; she has two children, two daughters, who were seventeen and fifteen at the time of our visit. They live with their father in the United States.

Even if Princess Chulabhorn does not have to attend all ceremonial functions, she has many other obligations beside teaching and research. She is actively involved with promoting scientific collaboration in Asia and the Pacific; she was awarded the Einstein Medal for this by UNESCO, being the third person in the world to get it. She is special advisor to the United Nations Environment Program and has a number of other functions. She was the first person from Asia to become an honorary fellow of the Royal Society of Chemistry in England.

When we were waiting for the princess to arrive for the interview, we were ushered into a room full of reporters and TV cameras. When she arrived, they all took their pictures, and after a few minutes, they left. In the evening in our hotel, we saw ourselves interviewing Her Royal Highness on TV. All channels start their evening news program with events concerning the life of the royal family. Our main impression on this visit was the enormous contradiction between a modern well-equipped research institute, which looks exactly like any such institute in the West, and the strong tradition apparent even in such an institute.

What we observed brought back the memories of nineteenth-century Thailand in the movies mentioned in the introduction. The elderly vice president of the research institute, a lady with a PhD in chemistry and an authority in natural products chemistry, was there with us for a while. This lady approached the princess crawling on her knees, and crawled retreating in such a way as to never show her back to the princess. We wondered how a scientific discussion would be taking place under such circumstances.

PAMELA MATSON

Ecologist

Pamela Matson in 2009 in Palo Alto. (photo by M. Hargittai)

Pamela Matson has been a dedicated environmentalist since her childhood. In her first job at one of the research centers of NASA, she studied the impact of deforestation and urban pollution on the atmosphere above the Brazilian Amazon rainforest. "Then suddenly I woke up one day and said: 'My gosh, this is bad!!! What can we do about this?' I began to focus more and more my research on not just understanding these things but also on trying to develop solutions that would reduce some of the problems that I was beginning to measure in the atmosphere and on land."[1]

She realized that although agricultural practices have a lot of negative impact on the environment, the growing population of Earth has to be fed. This led to her most ambitious project: the study in Yaqui Valley in Sonora, Mexico, which turned into a real success story of agricultural and economic development. The valley became a role model worldwide as "the birthplace of the green revolution."[2] It is also one of the seeds of the new research area called sustainability science.

"Sustainability is the goal of meeting the needs of people today and in the future while protecting the life support systems of the planet."[3] Pamela became a leading figure in this field. She is a born leader. After having served as director of several programs concerning the environment, she was elected to be dean of earth sciences at Stanford University in 2002, and has been in this position ever since. She became a member of the US National Academy of Sciences in 1994, has received numerous awards, and is member or chair of a number of organizations involved with environmental and climate issues.

Pamela Matson was born in 1953 in Eau Claire, Wisconsin. She grew up in Hudson, Wisconsin, at the border with Minnesota. Her father had college-level training in engineering but did not graduate; he worked for the Wisconsin Bell Telephone Company. Her mother was a homemaker, avid reader, and poet. They both loved nature; Pamela got her interest partly from them and partly from her paternal grandmother. The grandmother was a farmer and horsewoman who loved flowering plants. Pamela remembers picking flowers in the forest on her grandparents' farm.

Pamela attended the University of Wisconsin and graduated in biology. At the time she was not considering science for her career, but she knew that she wanted a profession that would involve the environment, perhaps working for the Environmental Protection Agency. She went for a master's degree to Indiana University. However, "doing my master's degree I realized that I love research and so I decided to go on for a PhD at that point." She went to Oregon State University, but for her experiments, she often had to use an apparatus at the University of North Carolina at Chapel Hill. There she met the ecologist Peter Vitousek, her future husband. After finishing his studies, Peter moved to Stanford University, while Pamela received a job offer from North Carolina State University—so they were facing a commuting marriage. But then, "the NASA/Ames Research Center in the Bay Area called and asked if I would be interested in applying for a new special position there in the emerging area of Earth System Science. I took the opportunity and I jumped at it! I never regretted that I worked at the first part of my career at NASA. After spending there ten years, I moved to academia," to the University of California at Berkeley.

One of Pamela and her husband's joint projects was the study of the ecosystems of the Hawaiian Islands.

> We were looking at developing ecosystems from brand new geological materials that were laid down by the volcanoes, all the way through to ecosystems that had been developing for five million years. We looked at how nutrient cycling changes and how the interaction between plants and nutrients changed over the five million year development in these places. It is very interesting just in terms of understanding how the world works but it also provides a very nice test system to ask questions about human-caused changes in the global system. We used the Hawaiian Islands as a model system. My husband is from Hawaii, he does all his research there and we have a house on the Big Island. Our children spend all their summers in Hawaii.

In 1997, Pamela moved to Stanford University, as professor of geological and environmental sciences. "My passion during the last ten years has been in managed ecosystems. I am interested in how we can manage agricultural ecosystems in such a way that we reduce the environmental consequences of agriculture, while still maintaining the production and yields and the ability of agricultural systems to feed the planet. This is a question where you have to be very pragmatic and this is a huge challenge for the twenty-first century." She started to organize the Yaqui Valley program; when they started it, the effect of pollution on the valley was very obvious, due to overuse of nitrogen containing fertilizers.

The field of ecology is interdisciplinary. It involves not only specialists from different branches of the sciences but other professionals as well. She has been involved in the work of economists, political scientists, agronomists, and representatives of a variety of other disciplines.

> We were interested in finding alternative practices that made sense for the people who were using them; i.e., for the farmers. It made sense economically and agronomically but also reduced environmental consequences. We were looking for win-win situations and the only way you could do that kind of work was by being in a very interdisciplinary team. My closest collaborator here at Stanford, beside my husband, is Rosamond Naylor, an economist. I was the science team leader and I brought the people together. But the real leader on the ground, the person who connected all of the research with the farming community was the agronomist Ivan Ortiz-Monasterio, who lives in Mexico and works very closely with farmers.
>
> An important part of our research was to understand how farmers made their decisions. First of all, they would have never accepted my advice; I am just this woman from the United States, and that is understandable. Ivan was the one who brought our suggestion to the farmers but he also wanted to make sure that we understood what the farmers were concerned about. So there was a two-way flow of information. One of the things that we learned in this work was that new knowledge helps the farmers to understand when and at what point in their crop management they are wasting money by buying too much fertilizers; they were just dumping it down the drain basically into the water system and into the ocean and into the atmosphere.
>
> Eventually we understood that it was not the farmers themselves who made the decisions but the credit union and the farmer's associations were telling them how to do it. Had we only talked to the farmers, we could have never achieved any change. Thus, we needed to approach the credit unions and farmer's associations as well and collaborate with them as well. This shows how important it is to step back and clearly understand how and where the decisions are being made. We developed new approaches that could make sense for the farmers and that would engage the credit unions. Finally, they have been using our approaches. The key of what we were suggesting was that if they apply fertilizers, they have to more carefully time the crop requirements; they should use them only when the

> crop needs it. They could save huge amounts, about 12–17% of after-tax profits and they would get great yields of crops. What they had to do was to apply the fertilizers later on the year rather than at the beginning of the year. We developed a hand-held device that measures the nitrogen content of the leaves, and by doing this they know when and how much fertilizer they need to use. It is a form of precision agriculture. In the USA there are very sophisticated approaches for precision agriculture, where tractors have remote sensing devices and computers on board and the farmers know when and how much fertilizer to give at different parts of the field. But in the developing countries of the world with such simple equipment that I mentioned they can still do precision agriculture and determine when and how much fertilizer to use. We developed the technology; the credit unions bought them for their farmer's associations. Probably the biggest lesson for me from this whole project was that if you are doing research, that is partly for the purpose of helping decision makers make their decisions, then you have to really understand all of the things that drives their decisions, you cannot just assume you know.

When I asked Pamela how she became interested in administration, she said she did not. She never applied for administrative jobs, only agreed to do them when asked, because she felt that she could be helpful in them. "I enjoy leadership—I don't call it administration, I call it leadership. I enjoy making sure that everyone's ideas are being heard, and developing a common strategy and working towards that together. I like teamwork and I very strongly believe that the universities of the world today have a critical role to play for the welfare of people and environment."

At the beginning of her career, she partly worked and published together with her husband. Such an arrangement often makes it more difficult for a woman to gain her own recognition. I wondered if it ever happened that people attributed her work to him.

> When I was very new, I remember at one time, in North Carolina, somebody said: "Oh, all of her work is his, anyway." I remember, I was very angry. That was the last time. I think this is because we do separate things as well and also, I am an articulate person and I interact with people so they can see that I know what I am talking about. Also, it was important for me at these early years that I became very active in the international community working on the biosphere-atmosphere interaction, and Peter did not do that. I had my own international community who saw me also just as me. It is wonderful to have a spouse that you can talk about your work with.

Pamela and Peter have a son and a daughter. At the time of our meeting, in 2009, their son was in graduate school in computer science. Their daughter had just begun college at the University of Vermont; her aim is to study natural resource planning. When the children were born, Pamela and Peter were already deep into research. She stayed at home for about six months with each baby, and later on always made sure to hire quality child care. Soon after her daughter was born, Pamela received the prestigious no-strings-attached MacArthur Fellowship, which means a considerable sum of money. Her decision about what to use part of the money for caught the attention of a reporter of *The Scientist,* whose article on the recipients includes the subheading: "Tag Along: Pamela Matson says she'll use part of her award money to bring her children with her to the field."[4] They hired a "third parent," a woman who traveled with them to their foreign destinations, and thus Pamela and Peter could do their research while the children were taken care of and were with them as well. Pamela mentioned that her husband had always been extremely helpful in every way. Very often, he stayed home with the children when needed, and the children learned that he was just as capable as she, so nobody worried when she was not at home.

Pamela feels that she has always been lucky; in all her jobs, at NASA, Berkeley, and Stanford, everybody was supportive, and she never experienced discrimination. But the most important thing has been—and this is her advice to all budding women scientists—"Find the right partner/spouse!"

KATHLEEN OLLERENSHAW

Mathematician (and politician)

Kathleen Ollerenshaw in 2003 in her home in Manchester. (photo by M. Hargittai)

She lost her hearing almost completely at the age of eight. She fell in love with mathematics, graduated from Oxford University, and solved long-standing mathematical problems. She was a passionate sportsman, playing hockey, and she won medals in figure skating. She fought for girls' schools and for improving education in general, and advised Margaret Thatcher's government on educational matters. She was appointed Dame Commander of the Order of the British Empire, and as a politician served as Lord Mayor of the City of Manchester, England. This is a brief summary of Dame Kathleen Ollerenshaw's rich life. She died on August 10, 2014.

Kathleen Timpson was born in 1912 in Manchester, England. Her parents came from very large families; her father was the seventh of twelve children and her mother was the eldest of eleven. "She was always kept in the apartment and had to look after the kids. The boys walked to the grammar school and the girls didn't, so she never went to school. She refused to have a big family because she was sick of having such a big family growing up—so she had only two children, my older sister and I."[1] Kathleen's father worked in the shoe business that his father started. "They were in retail trade and that was considered very lower-middle-class. The trouble with the Timpson family was that they had this terrible inherited deafness. At the age of about eight, I had a bad cold and suddenly realized that I couldn't hear at all. I can remember sitting in a room and I didn't know what anybody was saying. Most of my father's brothers and sisters became almost stone-deaf and I became deaf, too. That has made an enormous difference."

By this time Kathleen was at the Ladybarn House Montessori School, a small primary school, and the school arranged for her to learn lip-reading.

> My parents paid the enormous fees of 16 pounds per term; that was a lot of money. . . . This is where I met my husband-to-be, he also went there; we were the same age. . . . We had a very good teacher who taught us everything. Already that time I knew that I wanted to do mathematics. I have an older sister, who is four years older than I, so we never played together because she was always a stage earlier than me, so I was like an only child. I spent all my time in those days doing some things with pencil and paper because there was nothing else to do and I always used to draw patterns and I always played with numbers. After I was taught how to count, I realized that if I could count to 100 I could count as far as I wanted. I enjoyed this very much and by the age of six I knew my multiplication tables, not just up to 12×12, but up to 20×20. At the age of nine, the boys went off to their prep school; I had the very good fortune to have a new headmistress who was a mathematician. She was a graduate of Cambridge, Girton College. Of course, she didn't have a degree because women in those days were not allowed to have degrees. And she found this girl who was deaf and could not hear normally but absolutely had a thirst for anything that she did. She taught me and showed me challenging things, to see if I could deal with them. She taught me about infinity and difficult mathematical formulas, and many other things.

When Kathleen was thirteen years old, she went to a girls' boarding school, St. Leonards, in St. Andrews, Scotland. She was very much ahead of the others in mathematics but was behind in other subjects. She participated in various sports activities and joined in games that did not require hearing. "One of the reasons why I participated in these games was to take my mind off of the other difficulties in my life." Then, when she was fifteen and had all the credits for graduation, she took a year off. When she returned a year later, she was told that she did not get the proper classes in mathematics, so she could not move on, and, moreover, "there is nothing a girl can do with mathematics besides teaching but you cannot teach because you are deaf." They suggested to her to study something else that would help her to earn a living. She insisted that she wanted to go to university and study mathematics. Finally, the school gave up.

Remembering those times, she remarked: "I had no ambition to be a mathematician, . . . I had always planned to be a husband's wife and do what my mother had done; good work, keep the house and do mathematics for fun as I always had. I had no ambition for it, but I didn't want to do anything else, I wanted to do that all the time." Kathleen stayed at home for a year preparing for the university entrance exams. Her preparation included special studies with a mathematics professor at Manchester University, J. M. Child. Her exam at Cambridge went well, but when she mentioned that she was deaf, they said, "We won't be able to help you." When she went to the interview in Oxford, she tried not to show that she was deaf. They asked her what she did during her summer holiday. "Now by a freak chance, the summer before, they had the League of Nations' conference in Geneva [on disarmament], and I went there with a group from our school. And by chance, they asked me to write this up for the school magazine. . . . Of course, that was the perfect question. I said every name correctly with authority . . . they were impressed that I was a mathematician and that I was able to speak very clearly about a topic that had nothing to do with mathematics and because of this, I got the scholarship; there only was one."

She went to Somerville College in Oxford at the age of nineteen, and she received her BA degree in mathematics three years later. In the meanwhile, she got engaged to Robert, who had been studying medicine there. Unfortunately, there were no mathematicians at Somerville College, so she attended lectures at other colleges and spent a lot of time with hockey and other sports and socializing: "I did waste my undergraduate years at Oxford."

In 1936, Kathleen went to work at the Shirley Institute, related to the cotton industry. Her task was to do research on the efficiencies of different methods and ingredients used in weaving.[2,3] She studied statistical techniques and used advanced algebra, and was successful in problem-solving. She did not have any disadvantage due to her hearing difficulties, because there was so much noise from the machines that nobody could hear anything.

Kathleen and Robert got married in 1939; he soon left for war duties and was away for three and a half years. Their son was born in 1941. She left her job, but she missed mathematics. At the university, she met Kurt Mahler, a recent German refugee mathematician who suggested she work on an unsolved problem in connection with the so-called critical lattices. "Critical lattices relate to whole numbers in two or more dimensions and lead, by geometrical methods, to solutions concerned with 'close packing,' for example, how best to stack tins in a cupboard or oranges in a box."[4] She solved the problem in a few days' time. Mahler was so impressed that he persuaded her to go back to Oxford and study for a DPhil degree, which she did. She published five papers in two years, enough for her DPhil degree, which she received in 1945. She did part-time lecturing at Manchester University, but she also continued as wife and mother, and her daughter was born in 1946.

During the early 1950s, important things happened to her. The first hearing aids appeared, and this made an enormous difference to her. After living in complete silence for decades, she could hear again. The other important change was the start of her involvement in politics, especially in issues of education in general and girls' education in particular. It all started when she was asked to talk at the National Council of Women. She talked about the poor conditions of schools in Manchester. Her talk caught the attention of national newspapers and she decided to do a serious analysis of the situation of schools in the whole of England and Wales. As a mathematician, she understood that in order to be convincing, she had to provide numbers, based on facts. She presented a statistical analysis of the situation. With her background,

> not to mention an Oxford degree, which was so rare, almost impossible for a woman, I was the perfect candidate and I was fortunate enough to be elected to the city council. . . . I wasn't political, I wasn't a politician but I was made a member of the education committee. . . . I was in the minority party. I was the perfect female candidate to fill in on national committees; I was the only woman on almost every top committee there was that had to do with education, because I had an Oxford degree in mathematics and I know that was the passport that made it respectable.

She wrote articles and booklets on education, and they were widely read. The first little booklet, *Education for Girls*,[5] appeared in 1958; it discussed the current situation and suggested ways of improving it. She stressed the importance of proper education for girls in order for women not to be stuck at lower-level administrative and assistant positions because of their insufficient education. During the 1960s, she toured the United States and visited American schools; she wrote a series of articles about her experiences for the *Manchester Guardian*. She became well known for her work as an educator.

In 1971, her services to education were recognized. "I became Dame and at the same time, almost within a day, I got a letter saying that my daughter, Florence, had terminal cancer. In fact, she still went to Buckingham Palace with me. When you're young and you want to live, it takes a long time to die if you are fighting it." Soon after, Kathleen became part-time research fellow at the University of Lancaster. As her project, she tried to find out how women teachers who were married and had children could be attracted back to teaching after their children grew up.

During all these busy years, Kathleen did not give up mathematics. For her it was a way of thinking. Whenever she had a little free time, she thought of mathematical problems. She received inspiration from Hermann Bondi, the famous mathematician of Cambridge University. He encouraged her to take an interest in "magic squares." As is well known, the simplest magic square is a 4×4 array of numbers in which the sums of all rows, columns, and diagonals are the same. Her joint work with Bondi resulted in providing proofs for long-standing quandaries and led to scientific publications and two books.[6,7] Later in her life, she went back to another of her childhood interests, the wonder of the stars. She became an amateur astronomer and did that the way she did everything else, with full energy and vigor. At the age of seventy-nine, she climbed Mauna Kea in Hawaii to see a total eclipse of the sun.

During her life, she occupied numerous positions, such as director, chairman, and president of many societies. Her career in politics also continued; in 1975, she was elected Lord Mayor of Manchester for a one-year term. In 1978, she became the first woman president of the Institute of Mathematics and Applications, following Prince Charles in the position. She gave presentations on education and women's employment. She never gave up lecturing on mathematics, and she held the very prestigious Friday Evening Discourse at the Royal Institution in London in 1979 on soap bubbles, the honeycomb, and other examples of beautiful symmetrical shapes. Kathleen was the second woman ever to deliver a Friday Evening Discourse.

Kathleen said, "I had a wonderful double life—mathematics and public service—and meanwhile managing Robert and our children." Although mathematics stayed with her all her life, she was so popular in her public life that she became much better known as an educator than a mathematician. "I don't think that there is a school anywhere in Manchester where I have not given a speech, it became as it were quite a profession." In spite of this, she stressed that even in the busiest days she never stopped doing mathematics. "Mathematics is a way of thinking. It requires no tools or instruments or laboratories. . . . Archimedes managed very well with a stretch of smooth sand and a stick for his magnificent discoveries in geometry. . . . Mathematics is the one school subject not dependent on hearing. . . . I never aspired to being a professional mathematician or to being a professional anything for that matter. If you are deaf, you are glad to 'get by'."[8] She surely did much more than just "getting by" . . .

MARIANNE POPP

Botanist

Marianne Popp in 2001 in Vienna. (photo by M. Hargittai)

According to the website of the Department of Terrestrial Ecosystem Research of the University of Vienna, its mission is "to advance the fundamental understanding of metabolic and physiological responses of plants and microorganisms to their environment and of the role of plants and microbes, and their interactions, in the functioning of ecosystems."[1] The former head of this department is Marianne Popp, a member of the Austrian Academy of Sciences and former dean (2000–2002) of the Faculty of Science and Mathematics. Currently she is professor emerita. I visited her in 2001.

She was born in 1949 in Vienna. Her father was an orthopedic doctor. Her mother studied chemistry but did not finish it, because she got married and her first child was born; later on, she helped her husband in his practice. Marianne's elder sister studied medicine, and it was expected that Marianne would too. However, when she saw what her sister studied for her exams, she realized that medicine was not for her. She liked mathematics and biology in school, and eventually decided to study botany and zoology. At that time, those who studied biology usually became high school teachers, but Marianne did not want that either. When she realized that she did not want to deal with dead animals, after the first year she changed from zoology to biochemistry, so she graduated in botany and biochemistry. For her doctorate, she studied at the Institute of Plant Physiology of the University of Vienna and received her degree in 1975. She stayed on at the institute as a research assistant. For many years, she did research abroad, mostly in Australia and for about seven years at the University of Münster in Germany.

Marianne has been involved with a wide range of research topics. Most of them had a common thread, a particular molecular type, cyclitols, that was present in most of the plants she studied. The cyclitols are relatively small cyclic molecules, having several hydroxyl groups. They form in plants under extreme climatic conditions, such as in a lot of water or in a desert. One of the first systems she studied was some grasses belonging to the halophytes, plants that are adapted to a soil with a high concentration of salt. She and her colleague Roland Albert studied them at Neusiedlersee, a large lake in eastern Austria at the Austrian-Hungarian border. They found the topic fascinating and decided to extend their range of study. Albert went to the United States to the Great Salt Lake to study how the plants there had adapted to the saline water, and Marianne went to Australia, where she studied different mango species, trees growing in salt water. She found that all the mangoes were rich in cyclitols or similar molecules.

Mangoes are interesting objects for study because they adapt to harsh environments. Besides the high salt concentration, as they grow at the edge of the water they are under constant mechanical stress. They are flooded every six hours, and they do not have access to oxygen at the root area. How they survive is an intriguing question for a botanist.

In time, Marianne extended her inquiries to see whether the cyclitols have a role in plants that live under desert conditions. She and Albert determined that the mechanisms of adaptation of the plants to extreme conditions are very much the same under the most varied circumstances. This mechanism is based on the accumulation of the cyclitols and similar molecular systems, which help to regulate the osmosis—the movement of the solution components through the cell membranes. If, for example, a cell swells because there is too much water, molecules like cyclitols open membrane channels through which they get out of the cell, taking water with them, and thus relieve the pressure within the cell. On the other hand, if conditions are too dry, these molecules let the cells dry out, and they themselves solidify and protect the cell membranes and proteins from desiccation.

Marianne has studied plant species all over the world. "Another of our interesting topics was the mistletoe; in fact, it has one of the highest concentrations of cyclitols among those plants we studied; they constitute about 25 percent of the dry material of their leaves. Mistletoes are also interesting from the point of view of being a parasite. We did a study in South Africa on how much carbon they extract from their hosts."[2]

Marianne is single. It was not by design, but she wanted to be a scientist and wanted to be good at it. She believes that it was probably easier for her to stay abroad for longer periods because she was free of family obligations. She has spent a lot of time participating in sports.

She does not think that she ever experienced outright discrimination. She was elected an associate member of the Austrian Academy of Sciences in 1997; she was the second woman ever elected. In 2001, she told me: "I was not yet promoted to full member. Somebody told me: 'You have to change your sex before you can become a full member.' At the moment they like to have me in the Academy, because they can show: 'Look, we have a woman in this body.'" However, she did get elected to full membership in the Academy in 2006.

About her election to deanship, she told me this:

> I am still amazed that this happened. When the time came to elect a new dean, a colleague told everyone that he wants to be it; so I just jokingly added, if that is so, I want to be the dean, too. Somehow my name got into the list and when the voting came, already in the first run I got the largest number of votes; I am pretty sure that all the students and technical staff who are represented in the committee voted for me; and eventually I got elected. It is a difficult job, but I believe that I already achieved several things. It is also true though that some of the older male professors did not accept me; I can see this by different small signs. For example, there are always farewell parties for professors who retire—sometimes I am invited but sometimes I am not, which I don't mind.

In spite of the story of how her election happened, this was not a spur-of-the-moment decision on her part. She wanted to have a say and an influence on the direction of her school. For example, she suspected that due to the importance of molecular biology, classical biology might be neglected.

> I am very keen in developing molecular biology but I did not want the traditional, organismic biology, plant biology, zoology, being completely turned down. And this was somewhat in the air. We have to give a solid base to young students in biology. Sometimes I have the feeling that we might at some stage have biologists who can only understand cell cultures and won't be able to distinguish between a corn and a reed sweetgrass. Just as an example, we will have to have good soil biologists. This will be very important for the future of mankind; we'll have to give agriculture more ecological guidelines.
>
> I have to try convincing people. Of course, this takes a lot of time but it is worth it. An important advantage of being a dean is that I have more possibilities to have contacts with industry and discuss my ideas with people whom I would not have met just being a professor. Discussing topics with the ministries is much more difficult and exhausting. I will do this for one term and although there is a possibility to extend it to one more, I won't do it. Then I will have the possibility for a one-year sabbatical and I will use that time to get back to my science in full gear.

MAXINE F. SINGER

Molecular biologist

Maxine Singer in 2000 in her office in Washington, DC. (photo by M. Hargittai)

"For her outstanding scientific accomplishments and her deep concern for the societal responsibility of the scientist"—thus ran the award citation when, in 1992, she received the National Medal of Science from President George Bush.[1] This quotation summarizes her achievements; both her science and her multifaceted accomplishments as a policy maker and science organizer are exceptional.

Maxine was born in 1931 in New York as Maxine Frank; her father was a lawyer and her mother was a homemaker until the war, when she went to work and liked it, so she continued afterwards. Maxine's interest in science started in high school, due to the influence of her science teachers. By the time she went to college, she knew what she wanted to do.

She graduated in 1952 from Swarthmore College, a small coeducational college; the years she spent there had the greatest impact on her development as an independent scientist[2]:

> Just by luck in my year the outstanding science students were women. There were six of us. We were close friends, we lived in the same dormitory and to a very large extent we educated one another. I tend to think that if as an undergraduate, I hadn't been in that group, I might not have had the will and the ambition to keep on going in science. I could easily have gone to medical school, maybe. It was the fact that I had that very strong group of friends for four years that made a tremendous difference. Thus, I think that my attitudes about science probably come much more from my undergraduate education than from my graduate education.

She got her PhD in biochemistry from Yale University. On the suggestion of her doctoral advisor, Joseph Fruton, she went into DNA research and joined the National Institutes of Health (NIH). During her time there, she was involved with molecular biology and biochemistry and studied different aspects of RNA and DNA. For some time, she worked with the 1968 Nobel laureate Marshall Nirenberg on deciphering the genetic code. When I asked her about what she considered as her most important achievement in science, this is what she told me: "Obviously, the work I did during the deciphering of the genetic code was very important, although I don't have any publications for that, because I was in the position to be able to make the polynucleotide that helped Marshall Nirenberg [to determine how the genetic information is written]. I spent probably a year and a half doing that. There weren't many people at that time who could have done that. So that was a big thing for me."

I wondered whether she knew already then how important that work was.

> Oh, sure, we all knew it was. And it was interesting, because Marshall asked me to become a formal collaborator, and I told him I didn't want to do that. When the person I had done my postdoc with a couple of years earlier heard about this, he came to me, looked at me and said, you just threw away your Nobel Prize. I said no, I didn't think so. If he got the Nobel Prize and I would have been his formal collaborator, I would be seen as somebody who worked for Marshall, whereas it was more important for me to keep my independence.

Maxine also mentioned two other important achievements:

> Another result concerns the recognition that there are some enzymes that work either to construct long polynucleotides or degrade them. They work by a mechanism that when you start building a chain, like if you make a string of beads, you can add a bead and then stop. Or you can keep on adding beads in a continuum. They begin to build a chain and they don't let loose on the chain they are building, the enzyme holds on to it and adds the next unit. That also turned out to be an important thing. In more recent years, the discovery of a human transposable element which makes up about 15% of the human genome is an important thing too.

After spending seventeen years at NIH, Maxine moved to the National Cancer Institute, where she had leading positions. In the meanwhile, she became more and more involved with ethical issues related to genetic research. From 1980, for eight years she was the director of the Laboratory of Biochemistry in the National Cancer Institute. In 1988, she became president of the Carnegie Institution in Washington, DC, a large private research organization in three fields: astronomy, biology, and the earth sciences. As president, she founded "First Light," a Saturday science school for children. This was the start of a new program called Carnegie Academy for Science Education. Its aim was to reach out to students and teachers of Washington-area schools to encourage their interest in science. In 2002, she founded a new department for research in global ecology. She retired from her position in 2002. She has been well recognized for her accomplishments; her distinctions include memberships in the National Academy of Sciences (1979) and the Pontifical Academy of Sciences (1986), the National Medal of Science (1992), and the Public Welfare Medal from the National Academy of Sciences (2007).

Going back a few decades, her role in the start of discussions about the possible consequences of genetic engineering deserves special mention. It was at a meeting on nucleic acids in 1973, where the potentially harmful effects of the recently developed recombinant DNA technologies came up. Such technologies cut and then join together the DNAs of different species. This raised serious concerns about the danger of creating new dangerous organisms. In 1975, together with Paul Berg and other leading scientists in the field, Singer organized the famous Asilomar Conference to discuss the potential hazards of this new technology. "The Asilomar Conference on recombinant DNA was the Woodstock of molecular biology: a defining moment for a generation, an unforgettable experience, a milestone in the history of science and society."[3] Among the participants were leading researchers in the field, also physicians and lawyers, to discuss whether the danger was real and, if so, what should be done to prevent any wrongdoing. Although most of the scientists thought that there was no real danger, they also understood that the stakes were too high to ignore the possibility. They prepared guidelines for future research—imposing self-restrictions on recombinant DNA research—and they agreed that the restrictions would be lifted gradually as new results proved the research safe. The Asilomar Conference was exemplary because the scientists understood the hazards and their responsibilities and acted upon them rather than waiting for legislative action.

During the following years, Singer remained active in keeping the public informed about genetic engineering, the human genome project, and other scientific issues. She testified before Congress and served as advisor on committees dealing with these topics.

Could it be that the caution in dealing with recombinant DNA was an overreaction? She did not think so: "I think that it is obvious that the concerns we had turned out not to be true. But we didn't know that, no one knew that, not even Jim [Watson] knew that. If they had been correct or if even a part of them had been correct, it would have been extremely difficult for biology to progress. By being cautious we gained some credibility and respect for the positions we've taken, and therefore I don't think it was an overreaction at all."

On whether the scientists have done an adequate job in informing the public about genetic engineering, this is what she had to say:

> Scientists have worked very hard at that; harder than they have worked at informing the public about almost any kind of science thing. Certainly, around the time of Asilomar, starting in 1973 and for a decade after that, there was a tremendous effort to educate the public, the congress, the legislatures. In more recent years, it has become more difficult for scientists to speak directly to the public. Much more of it is filtered though the media—of course, a lot was filtered through the media then but not quite everything. Today we have much more difficulty getting our views across without the complication of the media, which likes to make a big argument out of everything rather than just being informative. I think it's a problem now.

Singer has been a prolific writer. She has about a hundred scientific publications, and she has coauthored several books with the Nobel laureate Paul Berg.[4]

Singer was lucky in never having to experience sex discrimination, either as a student or later as a researcher. Her only noteworthy experience in this respect was this: "The first time that I ever thought I had a problem because I was a woman was when I had trouble recruiting postdocs for my lab at NIH. I went to see the head of our department and he told me: 'I've been trying to point postdocs in your direction but they don't want to work for a woman.' So this was a problem—but it very quickly went away."

Maxine Singer has had a spectacular career, as a scientist as well as a science administrator and policy maker—all that while having a happy family life with four children. To the question about the greatest challenge in her life, this was her answer: "I guess the greatest challenge is raising four kids and having them come out really wonderfully."

NATALIA TARASOVA

Chemist

Natalia Tarasova in 2013 in her laboratory in Moscow. (courtesy of N. Tarasova)

Natalia Tarasova has always been lucky, from having the right family background to her superb mentors and other supporters. She also has had the determination and stamina to make use of her luck and talent both in science and in her administrative positions. She has established herself as a leading scientist among Russian chemists in researching sustainable development, and was elected a corresponding member of the Russian Academy of Sciences.

According to the United Nations' definition, "Sustainable development is development that meets the needs of the present without compromising the ability of future generations to meet their own needs."[1] Natalia is now in charge of the Institute of Chemistry and Sustainable Development at the D. Mendeleev University of Chemical Technology of Russia in Moscow. It is one of the first institutes of its kind in the world and the leading institution in Russia. She has been involved with national as well as international organizations, and she is now vice president of the International Union of Pure and Applied Chemistry (IUPAC) for 2014 and president-elect for 2016.

Natalia Pavlovna Tarasova was born in 1948 in Moscow. Her parents, Raisa Tarasova (née Krivoruchko) and Pavel Tarasov, got married during World War II; her mother served as a surgeon at a front hospital, and her father was an artilleryman.

They were both war heroes. "After the war, my father worked at the Central Committee of the Communist Party of the Soviet Union and later in the Ministry of Culture. He passed away in 1969 at the age of 57. My mother was a very famous otorhinolaryngologist [ear, nose, and throat specialist]. She helped many singers and musicians. Her patients adored her. She restored hearing abilities. One of her last patients still helps me with housekeeping. Mother passed away in 2007 at the age of 87."[2]

The young Natalia had broad interests. That was the time of Gagarin's space flight and the discussions about the peaceful use of atomic energy, and she followed these developments with great enthusiasm. Beyond the school curriculum, she studied music and English. Her favorite subjects included mathematics, and she had an excellent math teacher. She was a theater buff, but her father warned her not to get involved with "bohemian" life, so she followed her other interests: "One of my mother's patients mentioned to her the Department of Radiation Chemistry at the Mendeleev Institute of Chemical Technology (as it was then). It seemed to be a nice combination of physics, math, and chemistry. As I graduated from school with a gold medal, I only had to pass one exam to be admitted to the institute, and I became a Mendeleev student."

In 1972, Natalia graduated, but stayed on as a junior researcher. She worked on radiation-induced synthesis of chemical compounds, and she defended her PhD-equivalent thesis in 1976. Still, she stayed on as a researcher—with a break in 1979, when she spent nine months in France, in the laboratory of Professor Claude Filliatre at the University of Bordeaux as a postdoc. At that time, this was an extraordinary opportunity for a student in the Soviet Union, as international travel was quite restricted. Even leading scientists of the country found it difficult to receive permission for foreign visits. Upon Natalia's return to Moscow, she enrolled in an evening program of applied mathematics and earned an additional degree. Her career was not at all a usual one.

> I have always loved math and I wanted to get a proper education in it. This was the time when the first computers appeared at the university. It took me four years, six days a week in the evenings, to get this diploma. It was worth it. During this time, I became associate professor at the newly organized Department of Industrial Ecology. The Rector of Mendeleev University (as it had become in the meantime), Professor Gennadi Yagodin, organized it. He invited professors from the different departments of the university because he understood the necessity of the interdisciplinary approach in this arising area. I was one of them.
>
> In 1988, Professor Dennis L. Meadows, one of the authors of *The Limits to Growth*[3] report to the Club of Rome,[a] visited Mendeleev University to get his honorary doctorate. He spoke about global models and the future predicament

[a] The Club of Rome is a global policy institute (think tank). Their report, *The Limits to Growth*, made a stir worldwide by forecasting that economic growth could not continue indefinitely on our finite planet. In their models, they considered as variables world population, industrialization, pollution, food production, and resource depletion.

> of human civilization. A year later, I visited him at Dartmouth College in the US. I became interested in the global problematic. Our then rector, Pavel D. Sarkisov, suggested that I organize a new department to teach sustainability for all the students of Mendeleev University. This was in 1992, and the Rio Summit [the UN's Conference on Environment and Development] had just happened. I accepted this challenge and this became the Department for the Problems of Sustainable Development.
>
> A few years later, I organized the Institute of Chemistry and Problems of Sustainable Development, which included beside my department six other departments: sociology, life safety, environmental sciences, and others. Last year my department was recognized as UNESCO Chair in Green Chemistry for Sustainable Development. We also work with the Russian Chemistry Society to introduce the ideas of Green Chemistry into industrial practice. Utilization of spent batteries is just one example.

Chemistry is often criticized for polluting the environment, and in the public eye, it often has a negative image. It is easy to overlook the myriads of substances from chemistry that have a positive impact, whether pharmaceuticals or other useful substances. Nonetheless, the issue of pollution is essential and needs to be taken seriously. Organizing a university institute to instruct future chemists to focus on environmental issues was a major step in creating the necessary workforce for attaining sustainability for our planet.

Natalia has been interested in radiation-induced synthesis in chemistry. Chemical reactions (that is, breaking and forming chemical bonds) may happen, among other processes, under the impact of nuclear radiation. For example, using isotopes such as cobalt-60, or shining an electron beam from an accelerator onto the reactant substances, may lead to accomplishing unique reactions.

> My basic research is dedicated to the investigation of synthesis of the polymeric forms of phosphorus and sulfur in different media under different types of radiation. It is possible to obtain samples of modified red (polymeric) phosphorus and sulfur and to "tune" their properties to the needs of a specific consumer. Another part of my research is in risk assessment and management as well as environmental quality control. We work on indices and indicators that might help to assess the human impact on the environment.

Natalia is married and has one son. Her husband is a soccer analyst and works in the Russian Soccer Union. He has a degree in electrotechnics and another in applied mathematics, as well as a PhD in applied modeling. Their son also has two degrees, in applied mathematics and in law, both from Moscow State University, where he works.

Natalia has led a privileged life, which included the support of her whole family. She stresses, "My mother was helping me until the very last day of her life. Her former patients help me even years after her death."

SHIRLEY M. TILGHMAN

Molecular biologist

Shirley Tilghman in 2001 in Princeton. (photo by M. Hargittai)

In the fall of 2012, Shirley Tilghman announced her resignation from the presidency of the prestigious Princeton University. She had served as its first female president for eleven years. During her tenure, she accomplished a lot—she helped create a new Neuroscience Institute and a new Center for the Arts, build a new residential college, change the disparate treatment of men and women faculty members, increase the number of students on financial aid, and we could continue the list. She did all this after having spent decades as one of the leaders in her research field, molecular biology. I visited Shirley in her office at Princeton back in 2001, just a few months after she started her presidency. I was curious about what was behind her drive to take this job in the middle of her successful career in science[1]:

> Probably what allowed me to take this job is self-confidence. There is nothing on paper that says that I should be able to do this job. I had no administrative

> experience of any substance; I was director of a research institute but that research institute had just been being built; it was not large. I had no experience in finance, I had no experience in management, I had no experience in hiring at the vice presidential level, so if you look at just the paper, there is nothing to say that I would be able to do this job. I think that the only reason for being here today is that I do have a lot of confidence that I will be able to learn this job.

I wondered where this self-confidence came from. “From my father. I have an older sister who is severely retarded and who lived in a home since she was quite young. So I lived as if I was the oldest and I got all the benefits that a firstborn gets from parents, which is a lot of attention. My father just believed in women; he believed that women could succeed, and women could be anything they wanted to be. He believed sentimentally that there was nothing I could not do. That made all the difference.”

Shirley was born in 1946 in Toronto, Canada, as Shirley Marie Caldwell. Her father worked for a bank, and the family moved frequently between the east and west coasts. Her interest in science started very early.

> I was always in love with numbers. Since I was a small child, instead of reading me bedtime stories, my father played with me mental math games. So I think that it was through liking mathematics and liking numbers that I was attracted to chemistry. This is what I studied at university. I look at chemistry as puzzle solving; I like being able to figure out a pathway from a substrate to a product, to find the minimum number of steps to get there. I think that it was the puzzle part of chemistry that appealed to me, which relates back to my liking mathematics.
>
> By the third year at university, I was bored with chemistry. It is not a reflection on anything in chemistry; I think that it was simply individual preferences, probably because I had the sense that I was not going to be a very good chemist. Therefore, I began looking for another area of science and it was at this time that I came across one of the greatest experiments of the twentieth century in molecular biology, which was by Matthew Meselson and Franklin Stahl, showing that DNA replicates semiconservatively.

Semiconservative replication means that first the double helix of the original DNA unwinds and the two strands separate; then molecules that build up the new DNA attach step by step to each of the two separated strands and build two new daughter DNA molecules, each of which consists of half of the old DNA and half new.[2] “I thought that that was the most breathtaking and beautiful experiment that I have ever seen and I decided then and there that I will be a molecular biologist.”

By the time Shirley finished Queen’s University in Kingston, Ontario, with a BSc degree, she understood the culture of the laboratory. She had the impression that once she started graduate school there would be virtually no chance to do anything else with her life other than being a molecular biologist. She did not welcome such a limitation, because she wanted to see some very different parts of the world and learn a culture that is very different from her own. This is how she decided to go to West Africa, and for three years she worked as a secondary-school teacher in Sierra Leone.

I asked her what her parents thought about this: "My parents were furious! My mother, who while I was at college, the whole time, every week, sent me what she used to call 'a care package,' with toilet paper, toothpaste, and other things that she was afraid I would not buy if I had to do that with my own money. When she heard about my going to Africa, she stopped sending them."

Shirley loved the time she spent in Africa. Sierra Leone had just become independent, it was a young country, and there was a sense of optimism in spite of the enormity of problems the new nation had to face. She indeed had the opportunity to learn, and I wondered whether she would suggest to young people that they get into similar adventures as she had. "Not only would, but I do. Whenever students ask me about my opinion on taking time out and do something different, I always encourage them to do so. I came back to graduate school with a completely different mindset. I was energized, I was ready to really get on with my studies, and I think if I had gone immediately after college, I would not have that. It was a terrific thing to do."

She did her doctoral studies in biochemistry at Temple University in Philadelphia, followed by postdoctoral studies at the National Institutes of Health (NIH) in Bethesda, where she worked with Philip Leder. A decade earlier, Leder had participated in important experiments with Marshall Nirenberg toward deciphering the genetic code. She greatly enjoyed working with Leder and considers her most important scientific achievement devising a method for cloning genes; that is what she did in Leder's lab. That time was the dawn of recombinant DNA.

> Our major problem was to figure out how we could clone a single gene out of about a hundred thousand genes, which is how many we thought there were there at the time. We worked with mouse genes. What came out of that cloning was the realization that genes were not contiguous with the mRNAs that were transcribed from them. The genes are interrupted; there are segments of the gene that have to be spliced together at the RNA level—that was a major discovery for which the Nobel Prize was awarded sometime in the 1990s.[a]
>
> . . . The questions we decided would be the most interesting ones were those surrounding the idea of integration and complexity. Molecular biology has been always reductionist. Now there was a chance to look at questions in an integrated way. To do that we needed tools that had never been developed for molecular biology . . . We needed computer science, physics, chemistry, for ways approaching problems that we have never used before. That was the challenge to put together people with different backgrounds, different training, who could think about how to study the whole that is in the parts.

[a] In 1993, Richard J. Roberts and Phillip A. Sharp shared the Nobel Prize in Physiology or Medicine for discovering the split genes.

She then worked as assistant professor at Temple University, continued as associate professor at the University of Pennsylvania, and in 1986 moved to Princeton University as professor of life sciences. She built up a laboratory, and in 1998 she founded an interdisciplinary institute called the Institute of Integrative Genomics. In 2001, I asked Shirley what would happen with her research program now that she was president. "The laboratory is still open; I still have graduate students, undergrads, and postdocs who are working in the lab. I spend one day a week in the lab—but I am not accepting any new people in the lab. The presidency is a full-time job and science is also a full-time job, and I am only one person, so I had to make a choice. I would get the lab closed within two years."

As she was the first female president of Princeton, I had to ask if the fact that she is a woman might have been a consideration.

> I hope that it was decided on merit alone and that such a superficial quality as my gender was not considered. In fact, I was originally on the search committee for the first four months and during that period of time the question of gender was never an issue. So I am fairly confident that gender did not play a role in the decision. It would bother me if it did. But I am very much in favor of affirmative action and what it does at its best is to give an equally qualified woman an equal chance to be considered rather than giving the less qualified woman a chance to be considered.

I wondered why, if affirmative action works at universities, so few women are in high positions in academia and so few are members of national academies (of course, we have seen quite some development since 2001).

> That is a long and deep question. Some of it lies with society; a lot of it with the culture of science; the culture is such that it is not welcoming to women. It is very difficult for a woman mathematician or a woman physicist—these are fields where women are strongly underrepresented—to get comfortable at a department where all the professors are male, and most of the students are male. This is so not only at a professorial level but also at the student level. It takes an extremely committed and extremely strong woman to strive in such an environment. Due to the fundamental culture, women get discouraged; they feel that this is a field where they are not welcome. How can we change the situation? One way would be to start it at the senior level and get more women professors for these departments, so that the young women interested in studying there would not be discouraged. But this is, of course, difficult because there are not that many senior-level scientists to choose from . . .

Shirley's marriage ended in divorce. She has two children, whom she brought up as a single mother. I asked her how difficult it was. "Terribly!" she responded. How did she manage? "I think by denying that there was a problem. If I had allowed myself to really see how difficult what I was doing was, I would have just collapsed." She did not have hired help, "because I never had enough money for that. I was living on one salary, I was not getting any child support; he just did not pay and I just decided that it was not worth going to court and becoming obsessed with this."

It was inevitable that we discussed many questions about the situation of women in science. There are very few high-positioned women in academia, and they are often overwhelmed with positions on boards, memberships on committees, and other commitments. She agreed with me that often it becomes difficult to keep your eyes on the ball; "it is one of the unfair burdens to any woman who is in an underrepresented profession," she said.

Another frequent topic in my discussions with successful women was whether they have to be aggressive in order to succeed. Shirley thinks so: "You have to be forceful. One of my great complaints has always been that women are not allowed to have as wide a personality range in science as men. Men are allowed to be anything; from quiet to over-the-top aggressive; that is never noted. But a woman who goes outside a pretty narrow range, either too quiet or too aggressive, is always commented negatively, in both directions."

As president, she was in a good position to improve the situation of women students as well as female faculty members. She achieved a great deal in this regard. One of the first things she did was to establish a committee to look into whether men and women faculty members were treated differently. They identified all the areas where women were in disadvantageous positions, such as salaries, space allocations, and nominations for prizes, and they made the necessary adjustments.

She received numerous awards and recognitions, among them the L'Oréal-UNESCO Award for Women in Science. She is a foreign associate (as a Canadian citizen) of the US National Academy of Sciences. She is also a fellow of the Royal Society (London) and of other learned societies. She has twenty-five honorary degrees from universities around the world.

When Shirley Tilghman was elected as the first woman president of Princeton University, a group of reporters from the college student paper went to her office and said to her in an accusatory tone: "According to the *New York Times*, you are a feminist and a liberal!" She was quite taken aback: "It was a shock to me to realize these students were seeing being a feminist as something that I should feel awkward about as opposed to feel proud about."[3] She told me: "I have been a feminist all my life, even when I perhaps did not know the meaning of this word. I absolutely believe in equality for women and equal opportunity for women. I share your view of frustration at how little progress we have made since the feminist revolution began in the late 1960s, early '70s. When I look at the quality of the women whom we educate at this university, it just seems inconceivable that this can continue."

FINAL THOUGHTS

Why have women been interested in science? The answer should be the same for men and women. But the complete question is, Why have women been interested in science despite often considerable difficulties and barriers? Because they could not live their lives confined to the restricted role unjustly assigned to them, which implied exclusion from these most exciting activities. Women's interests, just like men's, have been directed to every aspect of our world—from the vast universe to the tiny cells in our body, the smallest building units of matter, or how to make chemicals to cure a deadly disease. Women, just like men, have been curious and ambitious; they would like to probe into the unknown and make sense of something that nobody has yet understood. They have felt the thrill of starting this adventure every day with new plans and ideas that would bring them closer to their goal. The question is not so much why women, too, have been interested in these endeavors; rather, why should this be more difficult for them than for their male counterparts with the same interest and ambition? This book is more about success than failure, but the fact that its exploration as a book is necessary shows that it often requires extraordinary efforts on women's part to participate in this noble human endeavor that ideally should be gender-blind—but has not been.

In this book, I have described my conversations with prominent women, and introduced a few with whom I could not interact in person. Can we draw any conclusions from their stories? Being a chemist by profession, I am aware that I have to be careful not to try to do the work of a sociologist. I have learned from my sociologist daughter that such analyses require considerable special knowledge and expertise that I do not possess. Furthermore, sound conclusions should be drawn from statistically meaningful amounts of data. Rather than attempting a sociological evaluation, I restrict these final thoughts to single out and comment on some aspects of the lives of my heroes and share my impressions of what I gained from my encounters with them about their lives in science.

The expression "breaking boundaries" in the title of this book refers, on the one hand, to the multitude of scientific disciplines covered and, on the other, to the great number of countries these women scientists represent. The examples show that irrespective of their countries, their fields of research, and their particular circumstances, there are common lessons to learn from them. This is true even in historical perspective if we include my early examples from the middle of the twentieth century.

The distribution of women scientists appears uneven among different scientific fields. Biology has the most women, and physics and engineering still attract the fewest. Among the science Nobel Prizes, women have received most in biology (the formal designation of the category is "Physiology or Medicine") and fewest in physics. Nonetheless, we can find successful women in such areas as well. Even in our limited sample, there are excellent women representatives of theoretical physics, such as Mary Gaillard and Rohini Godbole, and experimental physics, such as Miriam Sarachick, a condensed matter physicist, and Catherine Bréchignac, who studies atomic clusters by experimental methods. For engineering, there are the aerospace engineer Yvonne Brill and the mechanical engineer Irina Goryacheva.

Women are still underrepresented in leadership positions in academia. Change comes slowly, but it is encouraging that there have already been successful women in the highest administrative positions. They have served as university deans, chancellors, or presidents (called rectors or vice chancellors in some countries) and been in charge of science academies and large research institutions. This shows that the "glass ceiling" hindering women from high-level positions is shaking.

Women started becoming professional scientists from the end of the nineteenth century, apart from rare earlier examples. Most of the early woman scientists had a scientist husband, which made it possible for the wife to work at her husband's institution, mostly, but not necessarily, directly with him. This usually meant no salary and no official position—but even this did not hold them back. Gerty Cori, one of the early Nobel laureates, is an example. Another Nobel winner, Maria Goeppert Mayer, who received the award for work independent of her husband, could only have research opportunities at universities where her husband was a professor.

The book features a section about husband-and-wife teams recognizing the importance of such collaborations. As often discussed in the literature, in the case of the Curies Pierre was the "thinker" and Marie the good experimenter; we could call her the "doer." In most husband-and-wife teams discussed, the situation was similar. John and Rita Cornforth and Jerome and Isabella Karle are examples. It added greatly to the success of this kind of cooperation that the spouses in these couples were often complementary to each other, helping to make their joint product exceptional. In other good working relationships, such as Szent-Györgyi and Banga or Berson and Yalow, again the woman was the good experimenter.

Several of my interviewees suffered from the consequences of the so-called antinepotism rules in the United States, while such rules existed. The restrictions affected not only those couples who wanted to work together, but often also those who wanted positions at the same university, even if at different departments. For example, the Yalows and the Dressselhauses had to move in order to be able to get jobs for both spouses. In these examples, the husbands fully supported their wives in their desire to be a scientist. Government laboratories were easier in this respect than universities; this is why some of the scientific couples ended up working at such institutions, for example, the Goldhabers at the Brookhaven National Laboratory and the Karles at the Naval Research Laboratory.

Women often had to follow their husbands when they received a better job offer. Jocelyn Bell Burnell's husband moved frequently because of his career, making it difficult for her to find a permanent job for herself. While their son was growing up, she worked only half time. Yvonne Brill had a very good job on the West Coast, but when her husband moved to the East Coast, she had to follow him and look for a new job for herself. Chien-Shiung Wu had a similar experience. Frances Kelsey moved several times at the beginning of her career before her family eventually settled in Maryland, and she could work at the Federal Drug Administration for the rest of her career.

Working together with one's husband often carries the risk that the woman's work will be attributed to the husband. Isabella Karle has been a much recognized scientist in her own right, yet when I asked her if this has ever happened to her, "I suppose so"[a] was her laconic answer. When I asked if this has ever happened the other way around, she answered: "Not often." Many couples realized this risk and tried to minimize it by, for example, alternating first place in the order of authorship. Eventually, they often developed independent research projects.

Pamela Matson, ecologist and dean of earth sciences at Stanford University, has often worked together with her husband. She remembered that earlier in her career, she overheard someone saying "Oh, all of her work is his, anyway." She was very angry. But they work on separate projects as well, and she is also very busy in the international community, so for her peers there has been no doubt about her expertise.

Rosalyn Yalow and Solomon Berson, not a married pair, worked together for decades very successfully. Outside of their lab, many people assumed that she was simply his assistant—which was not the case at all. When Berson died, it might have meant the end of her career. But she was strong and stubborn enough to prove that she was an equal partner; she continued as successfully as they did together. Eventually, she received the Nobel Prize for their joint work.

The condescending attitude toward women scientists does not change easily. As recently as 2010, a piece of joint work by Irène Curie and Frédéric Joliot was quoted as carried out by "Joliot and Curie," although in all the cited references the order of authors was "Curie I. and Joliot F."[1] Not even the alphabetical order required placing Joliot before Curie.

Several of the women I talked with were born in the 1920s and 1930s. Their start was often difficult. Especially for those living in the United States, getting into graduate school was a challenge. Miriam Sarachik's husband had to be persuaded to go to graduate school, when she was "just dying to do the same thing! But it was not something that women did." Eventually, she succeeded.

World War II made a difference for women in academia in the United States. Many male scientists, especially physicists and chemists, were away on war-related projects, and the universities needed women to teach. It was during this time that Goeppert Mayer, for the first time in her life, was offered a university job. Chien-Shiung Wu became a professor at Princeton University at this time—when women were not even allowed to study there. After the war, as the men returned, things reverted to the previous situation, but the precedents lingered. Lasting improvement in the conditions of women scientists began at the time of the women's movement in the 1960s.

In the last decades of the twentieth century, women in science still experienced various kinds of discrimination, such as lower salaries, smaller lab spaces, and

[a] Here and below for references to quotations, see the respective chapters in this book.

delayed promotions compared to their male colleagues, not to mention their very small numbers in certain fields. It was a major step when, in 1995, a committee was formed at MIT, chaired by Professor Nancy Hopkins, to investigate the status of senior women faculty at that institution. The results showed serious inequalities between the male and female senior faculty in almost every aspect. The school made changes and published a progress report in 1999.[2] This made national news, and it motivated other institutions to look into the problems in their own schools.[3] MIT published a follow-up report in 2011 showing remarkable progress at this institution.[4] Whereas the number of women faculty remained unchanged from 1985 to 1994, by 1999 it had increased from 8 percent to 13 percent, and to 19 percent by 2011.

Discrimination concerning women in science still exists, even though its level has diminished and it is often manifested in more subtle ways than before. Considering that women make up half the population, there are still conspicuously few women scientists, especially in the higher strata of academia.

The inevitable conflict between work and family responsibilities remains one of the most difficult hurdles, if not the most difficult, to overcome. In 2002, I talked about this with Jim Watson of double helix fame, who is known for having made controversial statements, but also for having practiced fairness while in leadership positions. He told me[5]:

> [My] old friend, Arthur Kornberg, who is certainly not a feminist . . . says basically that women can never be equal in universities because they want part-time positions. He was at a medical school with twenty-four women and the families were opting for not working for eighty hours a week whereas men in a similar position would work eighty hours a week. I think you have to give the jobs to people who do the best science. If doing the best science is the result partly of working eighty hours a week, someone who wants to work forty cannot compete with the one who works eighty hours. There are all sorts of schemes of how to let women back in after the children are no longer infants and things like that. I would certainly favor those sorts of things.
>
> *Are you saying that the situation is hopeless for women who would like to have both a career in science and a family?*
>
> Some do it; Dorothy Hodgkin did, but she had servants. There's also daycare. You begin to do things. But the quality of really good science is pretty extraordinary obsession. It's hard to be obsessed about two things at the same time; about your family and your work. I guess I'm just saying that in a science department you could be fair and have the number of women to be half of that of men. Some will be able to arrange their lives and have the right personality for it; others will say, it's just asking too much.
>
> . . . Women want to have children in many cases . . . You can't change the essential difference between men and women. You can't make some things equal when you don't have the basis for it. Given everything in nature, it's harder for women to be in the top 1 percent of the field.

This was not very encouraging, especially from a man whom his former graduate student, now MIT professor of molecular biology Nancy Hopkins, called the first feminist she ever met.

Similar problems have marred the situation of women scientists in European countries. Christiane Nüsslein-Volhard described in 2001 how many different ways and how often she was discriminated against because she was a woman. Today this problem is taken seriously and is investigated at many levels. The European Commission has been publishing a book biannually since 2003 that discusses all aspects of women in science, both in academia and in industry. The latest volume appeared in 2013.[6] In its introduction, the then commissioner for research, innovation, and science, Máire Geoghegan-Quinn, writes that in spite of all efforts, the gender imbalance is still large and only very slow progress can be seen. Although women outnumber men in the higher-education sector, still only 10 percent of university rectors are women, and a similarly large imbalance is characteristic of the decision-making bodies.

Looking further into the question of family versus work, I note that from among the women scientists figuring in this book, only a few were not married. An example of the few was Rita Levi-Montalcini, whose father did not want her to study because that would interfere with her life as a wife and mother. When she turned twenty-one, she told him: "I do not care about being a wife or a mother, I want to study." Others felt that pursuing a science career was not in conflict with having a family and children—and managed both. In fact, quite a few had three or four children. It does not come as a great surprise that when asked about the greatest challenge in their life, they inevitably mentioned the problem of adjusting family obligations and work.

Here are just a few quotes to illustrate this: Yvonne Brill: "just managing everything . . . getting everything done at home in a twenty-four-hour day is a challenge in itself." Rita Cornforth: "It was difficult, of course, and by the time I had three children I sometimes thought: 'If I hadn't embarked on this I wouldn't, but I can't give up now.' . . . I found it easier to put chemistry out of my mind when I was at home than to put the children out of my mind when I was in the lab." Mildred Dresselhaus: "There was always a problem in organizing one's family in the morning. . . . They didn't like that I came to work at 8:30 instead of 8 o'clock. My oldest child was less than five years [old]. I had a baby essentially every year and it was very hard to make everything work out for an 8 am arrival. The people who were judging me were all bachelors." Anne McLaren: "Time. Time. Organization of time."

The most important factor concerning science *and* family was having a supportive husband. A curious case of the supportive husband is the one who wants his wife to be a successful scientist but does not help with the household duties at all. Mildred Cohn's husband, the physicist Henry Primakoff, never took a job before making sure that his wife would also find a research position at the institution. But when it came to the home, whenever she asked him to do something, he always answered: "hire somebody." For George Klein, it was imperative that his wife be a scientist. But he did not help at all with household duties and bringing up their three children. Eva had a hard time with balancing science and family, and still has a bad conscience about perhaps not being there for their children when they needed a parent.

Fortunately, in most cases the husbands were fully supportive and participated in household duties. Most families hired outside help when their children were small—with two salaries, this was affordable in most cases. Two women from Scandinavia, Marit Traetteberg of Norway and Kerstin Fredga of Sweden, told me that when their children were young, together with other young scientists at the university, they actually built a kindergarten so that they could work. Indian and Russian families often seem to count on grandparents for child-care assistance.

Even with a supportive husband, the problem of having young children and keeping up with research persists. Some women decided that they would have children only after they established themselves. Rosalyn Yalow waited until she felt that she was irreplaceable. Similarly, Margarita Salas and her husband waited to have a child until her laboratory was well established in Madrid.

Children react in different ways to having a working mother. For some families, this meant a persistent problem. Maria Goeppert Mayer's children suffered from a rude babysitter, and their mother did not notice. When Maria's daughter herself had a child, she made sure that she was first a mother. Catherine Bréchignac knows that her children felt ignored by her and she regrets it; fortunately, they eventually developed a good relationship. Quite a few of my interviewees mentioned that science talk during dinner was not well tolerated by their children. There were opposite reactions as well. Marye Ann Fox's children thought that the conversation at their dinner table was much more interesting than at their friends' houses.

Mildred Cohn's daughter complained that she was the only one in school whose mother worked. But when she grew up, she became a psychologist and wrote a paper about the children of working mothers, with the conclusion that it did not make any difference for them. All three Cohn children became scientists. Quite often, the children of scientists become scientists themselves. In this regard, it is worthwhile to read the relevant part of the life story of Vera Rubin, whose children are all scientists. One of her sons wrote, "I think it's no coincidence that the four children all ended up doing science. A pervasive early memory of mine is of my mother and father with their work spread out along the very long dining room table . . . At some point I grew old enough to realize that if what they really wanted to do after dinner was the same thing they did all day at work, then they must have pretty good jobs." Another quote from another of her sons: "There was never ever pressure to become a scientist, but it did seem like the natural thing to do. . . . I've learned that . . . having parents, who understand and encourage such a life is an advantage most of my colleagues didn't have."

My book is focused on the struggles and triumphs of women scientists through the examples of some of their most successful and prominent representatives. The goal was to pay tribute to these women and present them as role models for budding scientists—male and female alike. The book aims to encourage dedicated and intelligent young people to embark on a career in science. I did not try to idealize the women scientists I introduced; rather, I tried to show them as they are—human beings who have merits and shortcomings. Their common traits, their love of and dedication to science, shine through all their activities. In a few decades, there may no longer be need to write special books about women scientists, but until then I am offering this volume as my contribution toward this goal.

NOTES

Introduction

1. Caroline L. Herzenberg, Susan V. Meschel, and James A. Altena, "Women Scientists and Physicians of Antiquity and the Middle Ages," *J. Chem. Ed. 61* (1991) 101–105.
2. Sue Vilhauer Rosser, *Women, Science, and Myth: Gender Beliefs from Antiquity to the Present* (Santa Barbara, CA: ABC-CLIO: 2008), 22.
3. http://web.mit.edu/fnl/women/women.html (accessed on February 1, 2014)

Husband and Wife Teams

1. C. Kimberling, "Emmy Noether and Her Influence," in *Emmy Noether: A Tribute to Her Life and Work*, ed. J. W. Brewer and M. K. Smith, New York: Marcel Dekker, 1981, 14.

The Curie "Dynasty"

1. Marie Curie, *Pierre Curie*, New York: Dover Publications, 1963.
2. Helena M. Pycior, "Pierre Curie and 'His Eminent Collaborator Mme Curie': Complementary Partners," in *Creative Couples in the Sciences*, ed. H. M. Pycior, N. G. Slack, and P. G. Abir-Am, New Brunswick, NJ: Rutgers University Press, 1996, 39–56, 40.
3. R. K. Merton, "The Matthew Effect in Science," *Science* 159 (1968): 56–63.
4. Margaret W. Rossiter, "The ~~Matthew~~ Matilda Effect of Science," *Social Studies of Science* 23 (1993): 325–341.
5. Magdolna Hargittai, "Valentine Telegdi," in Magdolna Hargittai and Istvan Hargittai, *Candid Science IV: Conversations with Famous Physicists*, London: Imperial College Press, 2004, 189.
6. Pycior et al., *Creative Couples*, 46. Translation by Helena Pycior from Irène Joliot-Curie, "Marie Curie, ma mère," *Europe* 108 (December 1954): 90.
7. Ibid., 300.
8. Ibid., 46.
9. "Nobel Lecture: Radioactive Substances, Especially Radium," *Nobelprize.org*, http://www.nobelprize.org/nobel_prizes/physics/laureates/1903/pierre-curie-lecture.html.
10. Private communication from Anders Bárány, University of Stockholm, May 2011.
11. Ibid.
12. Interview with Hélène Langevin-Joliot, accessed August 15, 2013, http://www.eurekalert.org/features/doe/2003-07/djna-mp071103.php.
13. James Chadwick, *Nature* 177 (1956): 964.

Gerty and Carl Cori

1. Carl Cori, Banquet Speech at the Nobel Banquet in Stockholm, December 10, 1947, in *Les Prix Nobel en 1947*, ed. Arne Holmberg, Stockholm: [Nobel Foundation], 1948, available online at http://www.nobelprize.org/nobel_prizes/medicine/laureates/1947/cori-cf-speech.html.
2. Istvan Hargittai, "Arthur Kornberg," in Istvan Hargittai, *Candid Science II: Conversation with Famous Biomedical Scientists*, ed. Magdolna Hargittai, London: Imperial College Press, 2002, 51–71, 58.
3. Magdolna Hargittai, "Mildred Cohn," in Istvan Hargittai, *Candid Science III: More Conversations with Famous Chemists*, ed. Magdolna Hargittai, London: Imperial College Press, 2003, 251–267, 258.
4. Sharon Bertsch McGrayne, *Nobel Prize Women in Science: Their Lives, Struggles and Momentous Discoveries*, 2nd ed., Secaucus, NJ: Carol Publishing Group, 1998, 93.
5. H. Theorell, Presentation Speech, *Nobel Lectures, Physiology or Medicine 1942–1962*, Amsterdam: Elsevier, 1964, available online at http://www.nobelprize.org/nobel_prizes/medicine/laureates/1947/press.html.
6. Hargittai, "Arthur Kornberg," 58.
7. McGrayne, *Nobel Prize Women in Science*, 112.
8. Istvan Hargittai, "Osamu Hayaishi," in Istvan Hargittai and Magdolna Hargittai, *Candid Science VI: More Conversations with Famous Scientists*, London: Imperial College Press, 2006, 361–387, 375.

Ilona Banga and József Baló

1. Conversation with Ilona Banga's son, Dr. Mátyás Baló, in Budapest, 2013.
2. I. Banga and A. Szent-Györgyi, "CCXIV. The Large Scale Preparation of Ascorbic Acid from Hungarian Pepper (*Capsicum annuum*)," *Biochemical Journal* 28 (1934): 1625–1628.
3. V. A. Engelhardt and M. N. Ljubimova, "Myosine and Adenosinetriphosphatase," *Nature* 144 (1939): 668–669.
4. Ralph W. Moss, *Free Radical: Albert Szent Gyorgyi and the Battle over Vitamin C*, New York: Paragon House, 1988, 121.
5. A. Szent-Györgyi, "The Contraction of Myosin Threads," *Stud. Inst. Med. Chem. Univ. Szeged* I (1941–1942): 6–15; A. Szent-Györgyi, "Discussion," *Stud. Inst. Med. Chem. Univ. Szeged* I (1941–1942): 67–71; B. F. Straub, "Actin," *Stud. Inst. Med. Chem. Univ. Szeged* II (1942): 1–15.
6. S. V. Perry, "When Was Actin First Extracted from Muscle?" *Journal of Muscle Research and Cell Motility* 24 (2003): 597–599.
7. I. Banga and J. Balo, "Elastin and Elastase," *Nature* 171 (1952): 44.
8. I. Banga, *Structure and Function of Elastin and Collagen*, Budapest: Akadémiai Kiadó, 1966.
9. F. Guba, "Megkésett megemlékezés dr. Balóné, dr. Banga Ilonáról" [Belated reminiscences about Dr. Mrs. Balo, Dr. Ilona Banga], *Vitalitas* 141 (2000), http://www.vitalitas.hu/olvasosarok/online/oh/2000/40/51.htm.

Rita and John W. Cornforth

1. Magdolna Hargittai, interview with Rita Cornforth in correspondence, September 7, 2000. All unreferenced quotations are from this correspondence.
2. John Warcup Cornforth, "Asymmetry and Enzyme Action," Nobel Lecture, December 12, 1975, in *Nobel Lectures, Chemistry 1971–1980*, ed. Sture Forsén, Singapore: World Scientific, 1993.
3. Istvan Hargittai, "John W. Cornforth," in Istvan Hargittai, *Candid Science: Conversations with Famous Chemists*, ed. Magdolna Hargittai, London: Imperial College Press, 2000.

Jane M. and Donald J. Cram

1. Istvan Hargittai, "Donald J. Cram," in *Candid Science III*, 178–197, 196.
2. D. J. Cram and J. M. Cram, *Container Molecules and Their Guests*, Boca Raton, FL: CRC Press, 1994.
3. Hargittai, "Donald J. Cram," 195.
4. Ibid., 196.

Mildred and Gene Dresselhaus

1. Magdolna Hargittai, "Mildred S. Dresselhaus," in *Candid Science IV*, 546–569, 548.
2. Ibid., 549.
3. Ibid., 550.
4. Ibid., 552.
5. Ibid., 550–551.
6. H. W. Kroto, J. R. Heath, S. C. O'Brien, R. F. Curl, and R. E. Smalley, "C_{60}: Buckminsterfullerene," *Nature (London)* 318 (1985): 162–163.
7. Hargittai, "Mildred S. Dresselhaus," 562.
8. M. Cimonds, "Queen of Carbon Science: Kavli Prize Winner Is a Nanoscience Pioneer," *US News & World Report*, July 27, 2012.
9. G. Dresselhaus, "Mildred Spiewak Dresselhaus," http://mgm.mit.edu/group/millie.html.
10. "A Study on the Status of Women Faculty in Science at MIT," MIT 1999, http://web.mit.edu/fnl/women/women.html#The Study.
11. "MIT Profiles: Mildred Dresselhaus," *MIT Faculty Newsletter* 18.3 (January/February 2006), available online at http://web.mit.edu/fnl/volume/183/dresselhaus.html.
12. Hargittai, "Mildred S.Dresselhaus," 557.
13. Ibid., 558.
14. Ibid., 552.

Gertrude Scharff and Maurice Goldhaber

1. P. D. Bond and E. Henley, "Gertrude Scharff Goldhaber (1911–1998)," *Biographical Memoirs* 77 (1999): 1–14.
2. W. Saxon, "Gertrude Scharff Goldhaber, 86, Crucial Scientist in Nuclear Fission," *New York Times*, February 6, 1998.

3. Conversation with Maurice Goldhaber on November 8, 2001, at Brookhaven National Laboratory; see Istvan Hargittai, "Maurice Goldhaber," in *Candid Science IV*, 214–231. Additional unreferenced quotations are from this source.
4. Bond and Henley, "Gertrude Scharff Goldhaber," 8.

Isabella and Jerome Karle

1. Magdolna Hargittai, "Jerome Karle," in *Candid Science VI*, 422–437, 427–428.
2. Magdolna Hargittai, "Isabella Karle," in *Candid Science VI*, 402–421, 404.
3. Istvan Hargittai, "Herbert A. Hauptman," in *Candid Science III*, 292–317, 301–302.
4. Istvan Hargittai, "Alan L. Mackay," in Balazs Hargittai and Istvan Hargittai, *Candid Science V: Conversations with Famous Scientists*, London: Imperial College Press, 2005, 56–75, 72.
5. Istvan Hargittai, *The DNA Doctor: Candid Conversations with James D. Watson*, Singapore: World Scientific, 2007, 54.
6. Hargittai, *Candid Science VI*, 408–409.
7. Ibid., 409.
8. Ibid., 407–408.
9. Donna McKinney, "Jerome and Isabella Karle Retire from NRL Following Six Decades of Scientific Exploration," USNRL press release, July 21, 2009, accessed February 21, 2013, http://www.nrl.navy.mil/media/news-releases/2009/jerome-and-isabella-karle-retire-from-nrl-following-six-decades-of-scientific-exploration.

Eva and George Klein

1. George Klein about Eva Klein, *MTC News* 55 (2005): 6.
2. Istvan Hargittai, "George Klein," in *Candid Science II*, 416–441, 429.
3. Magdolna Hargittai, conversations with Eva Klein in Budapest between 2000 and 2003. All unreferenced quotations are from these conversations.

Sylvy and Arthur Kornberg

1. Arthur Kornberg, *For the Love of Enzymes: The Odyssey of a Biochemist*, Cambridge, MA: Harvard University Press, 1991, 172.
2. Ibid.
3. Istvan Hargittai, "Arthur Kornberg," in *Candid Science II*, 50–71, 54–55.
4. E-mail correspondence with Roger Kornberg, April 2013.
5. A. Kornberg, S. R. Kornberg, and E. S. Simms, "Metaphosphate Synthesis by an Enzyme from Escherichia coli," *Biochim. Biophys. Acta* 20 (1956): 215–227.
6. E-mail correspondence with Roger Kornberg.
7. Hargittai, *Candid Science II*, 63–64.
8. Hargittai, *Candid Science II*, 64.
9. "They Helped Husbands to Nobel Prize," *Miami News*, November 8, 1959.
10. E-mail correspondence with Roger Kornberg.
11. Kornberg, *For the Love of Enzymes*, 172.

Militza N. Lyubimova and Vladimir A. Engelhardt

1. A. V. Tichonova and A. V. Engelhardt, "Rol' roda Engelhardtov v istorii Rossii" [Role of the Engelhardt family in the history of Russia].
2. *Stanovlenie i dostizheniya biokhimicheskoi skoli Kazanskovo universiteta* [The formation and achievements of the biochemical school of Kazan University], Kazan: Otechestvo, 2009, 1–267, 27.
3. W. A. Engelhardt, "Life and Science," *Ann. Rev. Biochem.* 51 (1982): 1–19, 18.
4. Engelhardt, "Life and Science," 11.
5. Albert Szent-Gyorgyi, *Chemistry of Muscular Contraction*, New York: Academic Press, 1947.
6. Albert Szent-Gyorgyi, O myshechnoi deyatelnosti [About muscle operations], trans. Militza N. Lyubimova, Moscow: Medgiz, 1947.
7. The Nomination Database for the Nobel Prize in Physiology and Medicine, 1901–1953, http://www.nobelprize.org/nobel_prizes/medicine/nomination/country.html.
8. A. D. Mirzabekov, *Akademik Aleksandr Aleksandrovich Baev: Ocherki, perepiska, vospominaniia* [Notes, correspondence, remembrances], Moscow: Nauka, 1997.
9. Engelhardt, "Life and Science," 2.

Ida and Walter Noddack

1. Ida Noddack, "Uber das Element 93," *Zeitschrift fur Angewandte Chemie* 47 (1934): 653.
2. There is rich literature evidencing Ida Noddack's story, of which here only a few references are mentioned.
3. J.-P. Adloff, private communication to the author by e-mail, September 15, 2012.
4. B. Van Tiggelen and A. Lykknes, "Ida and Walter Noddack Through Better and Worse: An *Arbeitsgemeinschaft* in Chemistry," in *For Better or Worse: Collaborative Couples in the Sciences*, ed. A. Lykknes et al., Basel: Springer, 2012, 103–147, 113.
5. E. B. Hook, "Interdisciplinary Dissonance and Prematurity: Ida Noddack's Suggestion of Nuclear Fission," in *Prematurity in Scientific Discovery: On Resistance and Neglect*, ed. E. B. Hook, Oakland, CA: University of California Press, 2002, 124–148.
6. Laura Fermi, *Atoms in the Family: My Life with Enrico Fermi*, Chicago: University of Chicago Press, 1954, 157.
7. Ida Noddack, "Remarks on the Work of O. Hahn, L. Meitner, and F. Strassmann on Products Formed in Irradiation of Uranium," *Naturwissenschaften* 27 (1939): 212–213.

Further Comments

1. Lily Yan's biography on the Grube Foundation website, accessed on August 18, 2013, http://gruber.yale.edu/neuroscience/lily-jan.
2. Y. Bhattacharjee, "The Cost of a Genuine Collaboration," *Science* 320 (2008): 859.
3. E-mail letter from Nancy Jenkins, July 21, 2011. All unreferenced quotations are from this correspondence.
4. J. Kaiser "Texas's $3 Billion Fund Lures Scientific Heavyweights," *Science* 332 (2011): 1019–1020.

At the Top

1. James D. Watson, *The Double Helix*, New York: Atheneum, 1968.
2. Brenda Maddox, *Rosalind Franklin: The Dark Lady of DNA*, New York: Harper Perennial, 2003; Anne Sayre, *Rosalind Franklin and DNA*, New York: W.W. Norton, 2000.
3. Ruth Lewin Sime, *Lise Meitner: A Life in Physics*, Berkeley: University of California Press, 1997; Patricia Rife and John A. Wheeler, *Lise Meitner and the Dawn of the Nuclear Age*, Boston: Birkhauser, 1999.
4. Ingmar Bergström, "Lise Meitner och atomkärnans klyvning," in *Kungl. Vetenskapsakademiens årsberättelse 1999* (Stockholm: Kungl. Vetenskapsakademien, 2000), pp. 17-25. See also Istvan Hargittai, *The Road to Stockholm*, Oxford and New York: Oxford University Press, 2007, 232–236, 299–300, notes 51 and 52. Professor Ingmar Bergström has graciously donated to my husband the English translation of the published version of his lecture and of his original lecture.
5. Georgina Ferry, *Dorothy Hodgkin: A Life*, London: Granta Books, 1998, 402–403.
6. Patricia Parratt Craig, *Jumping Genes: Barbara McClintock's Scientific Legacy*, Perspectives in Science 6, Washington, DC: Carnegie Institution, 1994, 6.
7. Fred Hutchinson Cancer Research Center, "Dr. Linda Buck, 2004 Nobel Laureate," https://www.fhcrc.org/en/about/honors-awards/nobel-laureates/linda-buck.html, accessed January 17, 2014.
8. "The 2008 Nobel Prize in Physiology or Medicine—Press Release," Nobelprize.org, January 17, 2014, http://www.nobelprize.org/nobel_prizes/medicine/laureates/2008/press.html.
9. "The 2009 Nobel Prize in Physiology or Medicine—Press Release," *Nobelprize.org*, January 22, 2014, http://www.nobelprize.org/nobel_prizes/medicine/laureates/2009/press.html.

Jocelyn Bell Burnell

1. Magdolna Hargittai, "Freeman Dyson," in Magdolna Hargittai and Istvan Hargittai, *Candid Science IV: Conversations with Famous Physicists*, London: Imperial College Press, 2004, 440–477, 477.
2. Magdolna Hargittai, "Jocelyn Bell Burnell," in *Candid Science IV*, 638–655, 640.
3. Istvan Hargittai, "Anthony Hewish," in *Candid Science IV*, 626–637, 632.
4. Hargittai, "Jocelyn Bell Burnell," 641.
5. Ibid., 652–653.
6. Magdolna Hargittai, "Joseph H. Taylor," in *Candid Science IV*, 656–669, 661.
7. Hargittai, "Jocelyn Bell Burnell," 654.
8. Hargittai, *Road to Stockholm*, 7.
9. Hargittai, "Anthony Hewish," 633.
10. "Jocelyn Bell: The True Star," *Belfast Telegraph*, June 13, 2007, http://www.belfasttelegraph.co.uk/lifestyle/jocelyn-bell-the-true-star-13450159.html#ixzz2Hn2jvFqK.
11. Hargittai, *Road to Stockholm*, 240.
12. Hargittai, "Jocelyn Bell Burnell," 648.
13. Ibid., 646.

Yvonne Brill

1. United States Patent and Trademark Office, "President Obama Honors Nation's Top Scientists and Innovators," http://www.uspto.gov/about/nmti/NMTI_Announcement.jsp, accessed on April 22, 2012.
2. Magdolna Hargittai, conversation with Yvonne Brill, on April 28, 2000 at her home in Skillman, New Jersey. All unreferenced quotations are from this conversation.
3. Monica Hesse, "The National Inventors Hall of Fame Inducts 16 for Its 2010 Class," *The Washington Post*, April 1, 2010, http://www.washingtonpost.com/wp-dyn/content/article/2010/03/31/AR2010033102355.html, accessedAugust 10, 2013.
4. Douglas Martin, "Yvonne Brill, a Pioneering Rocket Scientist, Dies at 88," *New York Times*, March 30, 2013, http://www.nytimes.com/2013/03/31/science/space/yvonne-brill-rocket-scientist-dies-at-88.html?pagewanted=all&_r=0, accessed August 10, 2013.

Mildred Cohn

1. Istvan Hargittai, "Paul Boyer," in Istvan Hargittai, *Candid Science III: More Conversations with Famous Chemists,* ed. Magdolna Hargittai, London: Imperial College Press, 268–279, 274.
2. Magdolna Hargittai, "Mildred Cohn," in *Candid Science III,* 250–267, 257.
3. Ibid., 262.
4. Ibid., 263.
5. Ibid., 263.
6. Ibid., 266.

Gertrude Elion

1. Istvan Hargittai, "Gertrude B. Elion," in Istvan Hargittai, *Candid Science: Conversations with Famous Chemists*, ed. Magdolna Hargittai, London: Imperial College Press, 2000, 54–71, 71.
2. Ibid., 68.
3. Ibid., 59–60.
4. George Hitchings, "Banquet Speech," in *Les Prix Nobel 1988: Nobel Prizes, Presentations, Biographies and Lectures*, ed. Tore Frängsmyr, Stockholm: Almqvist & Wiksell International, 1989, available at http://www.nobelprize.org/nobel_prizes/medicine/laureates/1988/hitchings-speech.html.
5. Hargittai, "Gertrude B. Elion," 69.
6. Ibid., 70.

Mary Gaillard

1. Roger Bingham, "The Science Studio: Interview with Leon Lederman," http://thescie ncenetwork.org/media/videos/2/Transcript.pdf, accessed October 20, 2013.
2. Magdolna Hargittai, conversation with Mary Gaillard, February 19, 2004, Berkeley, California. All unreferenced quotations are from this conversation.
3. Mary, K. Gaillard, "Report on Women Scientific Careers at CERN," 1980, available at http://ccdb5fs.kek.jp/cgi-bin/img/allpdf?198006143.

4. Gaillard, "Report on Women," 4.
5. M. K. Gaillard and B. W. Lee, "Rare Decay Modes of the K Mesons in Gauge Theories," *Phys. Rev.* D10 (1974): 897.

Maria Goeppert Mayer

1. Joan Dash, *A Life of One's Own: Three Gifted Women and the Men They Married*, New York: Harper and Row, 1973, 238.
2. Lisa Yount, "Maria Goeppert Mayer," in *Contemporary Women Scientists*, New York: Facts on File, 1994, 13–25, 15.
3. Robert G. Sachs, "Maria Goeppert Mayer: 1906–1972," in *Remembering the University of Chicago: Teachers, Scientists, and Scholars*, ed. Edward Shils, Chicago: University of Chicago Press, 1991, 317–337, 320.
4. Peter Mayer, *Son of (Entropy)2*, Bloomington, IN: AuthorHause, 2011, 7.
5. Joseph Mayer and Maria Goeppert Mayer, *Statistical Mechanics*, New York: John Wiley and Sons, 1940.
6. Sachs, "Maria Goppert Mayer," 322.
7. Edward Teller with Judith Shoolery, *Memoirs: A Twentieth-Century Journey in Science and Politics*, Cambridge, MA: Perseus, 2001, 126.
8. Vivian Gornick, *Women in Science: Then and Now*, New York: The Feminist Press, 2009, 26.
9. Yount, *Contemporary Women Scientists*, 18.
10. Teller, *Memoirs*, 188.
11. Yount, *Contemporary Women Scientists*, 19.
12. Sharon Bertsch McGrayne, *Nobel Prize Women in Science: Their Lives, Struggles, and Momentous Discoveries*, Secaucus, NJ: Carol, 1993, 196.
13. Robert G. Sachs, "Maria Goeppert Mayer: 1906–1972," in *Biographical Memoirs*, Vol. 50, Washington, DC: National Academy of Sciences, 1979, 309–328, 322, available at http://www.nasonline.org/publications/biographical-memoirs/memoir-pdfs/mayer-maria.pdf.
14. Sachs, "Maria Goeppert Mayer," 333.
15. Yount, *Contemporary Women Scientists*, 23.
16. Andrea Gabor, *Einstein's Wife*, New York: Penguin, 1995, 141.
17. Mayer, *Son of (Entropy)2*, 4.
18. Nancy Thorndike Greenspan, *The End of the Certain World: The Life and Science of Max Born*, New York: Basic Books, 2005, 158.
19. Istvan Hargittai, *Judging Edward Teller: A Closer Look at One of the Most Influential Scientists of the Twentieth Century*, Amherst, NY: Prometheus, 2010.
20. Maria Goeppert Mayer and Hand D. Jensen, *Elementary Theory of Nuclear Shell Structure*, New York: John Wiley and Sons, 1955.
21. Dash, *A Life of One's Own*, 276.
22. Quoted in Gabor, *Einstein's Wife*, 138.
23. Karen Johnson, "Science at the Breakfast Table," *Phys. Perspect.* 1 (1999): 22–34, 32.
24. Gabor, *Einstein's Wife*, 103.

Darleane C. Hoffman

1. Magdolna Hargittai, "Darleane C. Hoffman," in Istvan Hargittai and Magdolna Hargittai, *Candid Science VI: More Conversations with Famous Scientists,* Imperial College Press, 2006, 458–479, 466.
2. Ibid., 472–473.
3. Ibid., 467.
4. Ibid., 478–479.

Vilma Hugonnai

1. Sources of information about Hugonnai:
   1. Margit Balogh and Maria Palasik, eds., *Nők a magyar tudományban* [Women in Hungarian Science], Budapest: Napvilág Kiadó, 2010, 316.
   2. "Hugonnai Vilma," *Wikipédia* [Hungarian Wikipedia], last modified April 8, 2013, accessed August 2, 2013, http://hu.wikipedia.org/wiki/Hugonnai_Vilma.
   3. Erzsébet Kertész, *Vilma doktorasszony: Az első magyar orvosnő életregénye* [Doctor Vilma: Biography of the first Hungarian medical doctor], Budapest: Jelenkor Kiadó, 1998.

Frances Oldham Kelsey

1. Morton Mintz, "'Heroin' of FDA Keeps Bad Drug Off of Market," *Washington Post,* July 15, 1962.
2. Magdolna Hargittai, conversation with Frances O. Kelsey on April 16, 2000.
3. Frances Kelsey, assembly talk at the National Cathedral School, January 31, 1967, private communication from Frances O. Kelsey, April 16, 2000.
4. Frances O. Kelsey, "Denial of Approval for Thalidomide in the U.S.," in *Medicine and Health Since World War II: Four Federal Achievements*, Bethesda, MD: National Library of Medicine, 1993, 8.
5. Daniel Carpenter, *Reputation and Power: Organizational Image and Pharmaceutical Regulation at the FDA*, Princeton, NJ: Princeton University Press, 2010, 230.
6. Ibid., 248.
7. Angus Crawford, "Brazil's New Generation of Thalidomide Babies," *BBC News Magazine*, July 24, 2013, http://www.bbc.co.uk/news/magazine-23418102; Winerip, Michael, "The Death and Afterlife of Thalidomide," *New York Times*, September 23, 2013, http://www.nytimes.com/2013/09/23/booming/the-death-and-afterlife-of-thalidomide.html; both accessed November 8, 2013.
8. Gardiner Harris, "The Public's Quiet Savior From Harmful Medicines," *New York Times*, September 23, 2013. http://www.nytimes.com/2010/09/14/health/14kelsey.html, accessed November 8, 2013.

Olga Kennard

1. Magdolna Hargittai, conversation with Olga Kennard, March 2, 2000. All unreferenced quotations are from this conversation.

2. http://www.ccdc.cam.ac.uk/Solutions/CSDSystem/Pages/CSD.aspx, accessed April 9, 2013.

Reiko Kuroda

1. Istvan Hargittai, "Reiko Kuroda," in *Candid Science III*, 466–471, 467.
2. Magdolna Hargittai, conversation with Reiko Kuroda, September 16, 2000. All unreferenced quotations are from this conversation.
3. Correspondence with Reiko Kuroda, 2013.
4. Hargittai, "Reiko Kuroda," 471.

Nicole M. Le Douarin

1. Inamori foundation, "Nicole Marthe Le Douarin," http://www.inamori-f.or.jp/laureates/k02_a_nicole/ctn_e.html, accessed March 27, 2013.
2. Magdolna Hargittai, conversation with Nicole Le Douarin, October 24, 2000. All unreferenced quotations are from this conversation.
3. "Nicole Le Douarin (Collège de France) Part 1: The Neural Crest in Vertebrate Development," video, http://www.youtube.com/watch?v=Our-x4WS4JI, accessed March 27, 2012.
4. Nicole M. Le Douarin, *The Neural Crest*, Cambridge, UK: Cambridge University Press, 1982, 2000.

Rita Levi-Montalcini

1. Magdolna Hargittai, "Rita Levi-Montalcini," in Istvan Hargittai, *Candid Science II: Conversations with Famous Biomedical Scientists*, ed. Magdolna Hargittai, London: Imperial College Press, 2003, 364–373, 367.
2. "Rita Levi-Montalcini—Biographical," *Nobelprize.org*, December 29, 2013, http://www.nobelprize.org/nobel_prizes/medicine/laureates/1986/levi-montalcini-bio.html.
3. Hargittai, "Rita Levi-Montalcini," 371.
4. Rita Levi-Montalcini, *In Praise of Imperfection*, trans. Luigi Attardi, New York: Basic Books, 1989.
5. Rita Levi-Montalcini, "From Turin to Stockholm via St. Louis and Rio de Janeiro," *Science* 287 (2000): 809.
6. Joahim Peitzsch, "Neighbourhood Growth Scheme," accessed April 1, 2013, http://www.nobelprize.org/nobel_prizes/medicine/laureates/1986/speedread.html.
7. Tore Frängsmyr and Jan Lindsten, ed., *Nobel Lectures, Physiology or Medicine 1981-1990*, Singapore: World Scientific, 1993.
8. John Harris, "The Question: Is Love Just a Chemical?" *The Guardian*, November 29, 2005, http://www.guardian.co.uk/education/2005/nov/29/research.highereducation1, accessed April 1, 2013.
9. Rita Levi-Montalcini, *The Saga of the Nerve-Growth Factor*, Singapore: World Scientific, 1997.
10. David Ottoson, "The Unravelling of the Code of Nerve Growth: A Modern Saga of the Dedication to Science," *Brain Research Bulletin* 50 (1999): 473–474, 473.
11. Hargittai, "Rita Levi-Montalcini," 371–372.
12. Alison Abbott, "One Hundred Years of Rita," *Nature* 458 (April 2, 2009): 564–567.

13. Hargittai, "Rita Levi-Montalcini," 374–375.

Jennifer L. McKimm-Breschkin

1. J. L. McKimm-Breschkin, "Influenza: A Cure from Structural Chemistry," *Chemical Intelligencer* 6 (2000): 43–46, 45.
2. McKimm-Breschkin, "Influenza," 46.

Anne McLaren

1. Paul Burgoyne, "Anne McLaren 1927–2007," *Nature Genetics*, 39 (2007): 1041.
2. H. M. Blau, "Anne McLaren (1927–2007)," *Differentiation* 75 (2007): 899–901, 900.
3. Magdolna Hargittai, conversation with Anne McLaren, Cambridge, June 30, 2004. All unreferenced quotations are from this conversation.
4. J. Biggers, "Dame Anne McLaren," *The Guardian*, July 9, 2007.
5. A. McLaren and D. Michie, "Current Trend of Genetical Research in Hungary," *Nature* 174 (1954): 390–391.
6. A. Murray, "Letter: Donald Michie and Anne McLaren," *The Guardian*, July 10, 2007.
7. S.G. Vasetzky, A. P. Dyban, and A. V. Zelenin, "Anne McLaren (1927–2007)" *Russ. J. Developmental Biology* 39 (2008): 125–126.
8. Gurdon Institute, "Anne McLaren DBE, DPhil, FRS, FRCOG: April 26th 1927 - July 7th 2007," http://www2.gurdon.cam.ac.uk/anne-mclaren.html, accessed August 3, 2013.
9. Biggers, "Dame Anne McLaren."

Christiane Nüsslein-Volhard

1. Editorial, "Pattern Recognition and Gestalt Psychology: The Day Nüsslein-Volhard Shouted 'Toll!'" *FASEB Journal* 24 (2010): 2137–2141, available at www.fasebj.org/content/24/7/2137.full.pdf.
2. Magdolna Hargittai, "Christiane Nüsslein-Volhard," in *Candid Science VI*, 134–151, 137.
3. Ibid., 144.
4. Ibid., 145–146.
5. Ibid., 147.
6. C. Nüsslein-Volhard, "Women in Science—Passion and Prejudice," *Cell* 18 (2008): R185–187.
7. Nüsslein-Volhard, "Women in Science," R185.

Sigrid Peyerimhoff

1. P. A. M. Dirac, "Quantum Mechanics of Many-Electron Systems," *Proc. Roy. Soc. A*, 123 (1929): 714–733.
2. Magdolna Hargittai, conversation with Sigrid Peyerimhoff, Bonn, June 18, 1999. All unreferenced quotations are from this conversation.

Miriam Rothschild

1. Douglas Martin, "Miriam Rothschild, High-Spirited Naturalist, Dies at 96," *New York Times*, January 25, 2005.
2. Eugene Garfield, "A Tribute to Miriam Rothschild: Entomologist *Extraordinaire*," in *Essays of an Information Scientist*, Vol. 7, Philadelphia: ISI Press, 1984, 120–127, available at http://www.garfield.library.upenn.edu/essays/v7p120y1984.pdf.
3. Miriam Rothschild and Peter Marren, *Rothschild's Reserves: Time and Fragile Nature*, Colchester, Essex, UK: Harley Books, 1997, 6.
4. Christopher Sykes, *Seven Wonders of the World: Miriam Rothschild,* documentary film, part 1 of 3, available online at https://www.youtube.com/watch?v=K2VaTmrsFLg. All three parts accessed March 30, 2013.
5. Magdolna Hargittai, conversation with Miriam Rothschild, Ashton Wold, spring 2002. All unreferenced quotations are from this conversation.
6. Sykes, *Seven Wonders of the World*, part 2 of 3, available online at https://www.youtube.com/watch?v=fec8DCl0hgo.
7. Peter Marren, "Dame Miriam Rothschild: Expert on Fleas and Energetic Campaigner for Nature Conservation," *The Independent*, January 22, 2005, available at http://www.independent.co.uk/news/obituaries/dame-miriam-rothschild-6154388.html.
8. Michael Downes, "Dame Miriam Rothschild DBE FRS," http://downesmichael.blogspot.hu/2011/03/dame-miriam-rothschild-dbe-frs.html, accessed April 2, 2013.
9. Miriam Rothschild, "My First Book," *The Author*, Spring 1994, 11.
10. Ibid.
11. Miriam Rothschild and Theresa Clay, *Fleas, Flukes & Cuckoos: A Study of Bird Parasites*, London: Collins, 1952, 56. Cited in Garfield, "A Tribute to Miriam Rothschild," 122.
12. Quoted in Martin, "Miriam Rothschild."
13. George H. E. Hopkins and Miriam Rothschild, *An Illustrated Catalogue of the Rothschild Collection of Fleas (Siphonaptera) in the British Museum (Natural History)*, 5 vols., London: Trustees of the British Museum, Natural History, 1953–1987.
14. Sykes, *Seven Wonders of the World*, part 2.
15. Rothschild and Marren, *Rothschild's Reserves.*
16. Miriam Rothschild, *Dear Lord Rothschild: Birds, Butterflies and History*, London/Melbourne/Sydney: Hutchinson, 1983.
17. Garfield, "Tribute to Miriam Rothschild," 120

Vera Rubin

1. Magdolna Hargittai, "Vera C. Rubin," in Balazs Hargittai and Istvan Hargittai, *Candid Science V: Conversations with Famous Scientists*, 246–265, 248.
2. Ibid., 253.
3. Ibid., 253.
4. Ibid., 254.
5. Ibid., 256.
6. Ibid., 256.
7. Ibid., 257–258.
8. Ibid., 260.

9. Vera Rubin, "An Interesting Voyage," *Ann. Rev. Astro. Astrophys.* 49 (2011): 1–28, 26–27.
10. Hargittai, "Vera C. Rubin," 265.

Margarita Salas

1. Jesus Avila and Federico Mayor Jr., "Obituary: Eladio Viñuela (1937–1999)," *Nature* 400 (1999): 822.
2. Margarita Salas, "40 Years with Bacteriophage ø29" *Ann. Rev. Microbiol.* 61 (2007): 1–22, 2.
3. Magdolna Hargittai, correspondence with Margarita Salas, September 2013. All unreferenced quotations are from this correspondence.
4. Salas, "40 Years with Bacteriophage ø29," 6.
5. *She Figures 2012: Gender in Research and Innovation*, European Commission, 2013 (based on 2010 data), available at http://ec.europa.eu/research/science-society/document_library/pdf_06/she-figures-2012_en.pdf.

Myriam P. Sarachik

1. Magdolna Hargittai, conversations with Myriam Sarachik, October 2000 and September 2008. All quotations are from these conversations.

Marit Traetteberg

1. Magdolna Hargittai, conversation with Marit Traetteberg in Trondheim, Norway, September 15, 1996. All quotations are from this conversation.

Chien-Shiung Wu

1. Noemie Benczer-Koller, "Chien-Shiung Wu (1912–1997)," in *Biographical Memoirs* Washington, DC: National Academy of Sciences, 2009, 1–17, available at http://www.nasonline.org/publications/biographical-memoirs/memoir-pdfs/wu-chien-shiung.pdf.
2. For more detail about this experiment, see Magdolna Hargittai, "Credit Where Credit's Due?" *Physics World*, September 2012, 38–42.
3. C.N. Yang, "The Law of Parity Conservation and Other Symmetry Laws of Physics," in *Nobel Lectures Physics, 1942–1962*, Singapore: World Scientific, 1964, available at http://www.nobelprize.org/nobel_prizes/physics/laureates/1957/yang-lecture.pdf.
4. C.-S. Wu, "Discovery Story I: One Researcher's Personal Account," in *Adventures in Experimental Physics: Gamma Volume*, ed. Bogdan Maglich, Princeton, NJ: World Science Education, 1973, 101–123.
5. See also E. Ambler, M. A. Grace, H. Halban, N. Kurti, H. Durand, C. E. Johnson, and H. E. Lemmer, "Nuclear Polarization of Cobalt-60," *Philosophical Magazine* 44 (1953): 216–218.
6. C. S. Wu, E. Ambler, R. W. Hayward, D. D. Hoppes, and R. P. Hudson, "Experimental Test of Parity Violation in Beta Decay," *Phys. Rev.* 105 (1957): 1413–1415.

7. R. L. Garwin, L. M. Lederman, and M. Weinrich, “Observation of the Failure of Conservation of Parity and Charge Conjugation in Meson Decays,” *Phys. Rev.* 104(1957): 1415–1417.
8. J. I. Friedman and V. L. Telegdi, “Nuclear Emulsion Evidence for Parity Nonconservation in the Decay Chain $\pi+\rightarrow\mu+\rightarrow e+$,” *Phys. Rev.* 105 (1957): 1681–1682.
9. Anders Bárány, private communication, March 20, 2012.
10. Magdolna Hargittai and Istvan Hargittai, “Leon M. Lederman,” in *Candid Science IV*, 142–159.
11. Hargittai, “Credit Where Credit’s Due?”
12. Hargittai, “Jerome I. Friedman,” in *Candid Science IV*, 64–79, 74.
13. Hargittai, “Val L. Fitch,” in *Candid Science IV*, 192–213, 206–207.
14. Wu, “One Researcher’s Personal Account,” 102.
15. R. L. Garwin and T. D. Lee, “Chien-Shiung Wu,” obituary, *Physics Today* 50 (1997): 120–122, 121.

Rosalyn Yalow

1. Istvan Hargittai, “Rosalyn Yalow,” in *Candid Science II*, 518–523, 520.
2. Joan Dash, *The Triumph of Discovery: Women Scientists Who Won the Nobel Prize*, Englewood Cliffs, NJ: Julian Messner, 1991, 42.
3. Denis Gellene, “Rosalyn S. Yalow, Nobel Medical Physicist, Dies at 89,” *New York Times*, June 1, 2011.
4. Hargittai, “Rosalyn Yalow,” 521.
5. Ibid., 521.
6. Ibid., 523.
7. Dash, *Triumph of Discovery*, 50.
8. Ibid., 50.
9. McGrayne, *Nobel Prize Women in Science*, 343.
10. In fact, their discovery was so novel that the journal *Science* rejected the paper in which they described it, and the journal that eventually accepted it, *The Journal of Clinical Investigation*, did so only on the condition that the expression “insulin-binding antibody” was deleted from the title: S. A. Berson, R. S. Yalow, A. Bauman, M. A. Rothschild, and K. Newerly, “Insulin-I^{131} Metabolism in Human Subjects: Demonstration of Insulin Binding Globulin in the Circulation of Insulin-Treated Subjects,” *J. Clin. Invest.* 35 (1956): 170–190.
11. Eugene Straus, *Rosalyn Yalow Nobel Laureate: Her Life and Work in Medicine*, New York: Plenum Trade, 1998. 151.
12. Hargittai, “Rosalyn Yalow,” 523.
13. Magdolna Hargittai, “Mildred S. Dresselhaus,” in *Candid Science IV*, 546–569, 548.
14. Straus, *Rosalyn Yalow Nobel Laureate*, 244.
15. Ibid., 247.
16. R. S. Yalow, “Radioimmunoassay: A Probe for Fine Structure of Biologic Systems,” Nobel Lecture, December 8, 1977, 447–469. 449.
17. Straus, *Rosalyn Yalow Nobel Laureate*, chap. 13.
18. Hargittai, “Rosalyn Yalow,” 522.
19. Ibid., 522.
20. Dash, *Triumph of Discovery*, 53.

21. Hargittai, "Mildred S. Dresselhaus," 549.
22. Straus, *Rosalyn Yalow Nobel Laureate*, 224.

Ada Yonath

1. Royal Swedish Academy of Sciences, "The Key to Life at the Atomic Level," 2009, http://www.nobelprize.org/nobel_prizes/chemistry/laureates/2009/popular-chemistryprize2009.pdf.
2. Magdolna Hargittai, "Ada Yonath," in *Candid Science VI*, 389–401, 394.
3. Ibid., 391.
4. Ibid., 391–393.
5. Ibid., 393.
6. Ibid., 397.
7. Ibid.
8. Ibid., 400.
9. Lou Woodley, "An Interview with Ada Yonath," July 20, 2010, accessed August 17, 2013, http://lindau.nature.com/lindau/2010/07/an-interview-with-ada-yonath/.
10. Kathleen Raven, "Ada Yonath and the Female Question," July 11, 2013, accessed August 17, 2013, http://lindau.nature.com/lindau/2013/07/ada-yonath-and-the-female-question/.

Women Scientists in Russia

1. Michael D. Gordin, Karl Hall, and Alexei Kojevnikov, *Intelligentsia Science: The Russian Century, 1860–1960*, Chicago: University of Chicago Press, 2008.
2. Ann Hibner Koblitz, *Science, Women, and Revolution in Russia*, Amsterdam: Harwood Academic, 2000.
3. Olga Valkova, "The Conquest of Science: Women and Science in Russia: 1860–1940," in *Intelligentsia Science*, 136–165, 136.
4. Svetlana A. Sycheva, *Zhenshchiny v rossiiskoi nauke: Rol' i sotsial'nyi status* [Women in Russian science: Their role and social status], Moscow: NIA-Priroda, 2005.
5. Koblitz, "The Mythification of Sofia Kovalevskaia," in *Science, Women, and Revolution*, 105–135, 127.
6. Valkova, "Conquest of Science," 143–147, 163–164.
7. A. S. Kotelnikova and V. G. Tronev, "Issledovanie kompleksnykh soedinenii dvukhva-lentnogo reniya," *Zh. Neorg. Khim.* 3 (1958): 1008.
8. See I. Hargittai, "The Beginnings of Multiple Metal-Metal Bonds," in *Candid Science*, 246–249.

Irina P. Beletskaya

1. Magdolna Hargittai, correspondence with Irina Beletskaya. All quotations are from this correspondence.

Rakhil Kh. Freidlina

1. A. B. Terentiev, "Kratkii ocherk nauchnoi deyatelnosti chlena-korrespondenta AN SSSR R. Kh. Freidlinoi" (Brief Review of the Scientific Activities of Corresponding Member of the Soviet Academy of Sciences R. Kh. Freidlina.), in L. A. Kalashnikova,

and N. M. Anserova, eds., *Rakhil Khatskelevna Freidlina (1906–1986)*, Moscow: Nauka, 2004, 5–42, 37. Here Terentiev refers to the reminiscences of one of Freidlina's closest pupils and associates, Emma M. Brainina.
2. Terentiev, "Brief Review," 38.

Elena G. Galpern

1. Istvan Hargittai and Magdolna Hargittai, *In Our Own Image: Personal Symmetry in Discovery*, New York: Plenum/Kluwer, 2000, 52–80; see, in particular, 62.
2. D. A. Bochvar and E. G. Galpern, "O gipoteticheskikh sistemakh: Karbododekaedre, s-ikosaedre i karbo-s-ikosaedre," *Doklady Akademii nauk SSSR* 209 (1973): 610–612.

Irina G. Goryacheva

1. 2009 Tribology Gold Medal, http://www.imeche.org/knowledge/industries/tribology/prizes-and-awards/all-tribology-gold-medal-laureates/2009TGM, accessed September 27, 2013.
2. All unreferenced quotations are from our correspondence in 2013.
3. 2009 Tribology Gold Medal.

Antonia F. Prikhotko

1. B. S. Gorobets, *Sekretnye fiziki iz atomnogo proekta SSSR: Semya Leipunskikh* [Classified physicists from the Soviet atomic project: The Leipunskii family], Moscow: Librokom, 2009, 140.
2. Gorobets, *Sekretnye fiziki*, 87.

Women Scientists in India

1. "List of countries by literacy rate," *Wikipedia*, http://en.wikipedia.org/wiki/List_of_countries_by_literacy_rate, accessed August 3, 2013.
2. Rohini Godbole and Ramakrishna Ramaswamy, eds., *Lilavati's Daughters: The Women Scientists of India*, Bangalore: Indian Academy of Sciences, 2008.
3. P. Thakar, "Amandi Gopal," in *Lilavati's Daughters*, 13–16.
4. C. V. Subramanian, "Edavaleth Kakkat Janaki Ammal," in *Lilavati's Daughters*, 1–4.
5. C. D. Darlington and E. K. Janaki Ammal, *The Chromosome Atlas of Cultivated Plants*, London: Allen & Unwin, 1945.
6. Subramanian, in *Lilavati's Daughters*, 4.
7. S. C. Pakrashi, "Asima Chatterjee," in *Lilavati's Daughters*, 9–12.
8. Asima Chatterjee and Satyesh Chandra Pakrashi, *The Treatise on Indian Medicinal Plants*, 6 vols, New Delhi: National Institute of Science Communication and Information Resources, 1991–2001.

Charusita Chakravarty

1. Magdolna Hargittai, conversation with Charusita Chakravarty, September 27, 2011, Delhi.

Rohini Godbole

1. G. N. Prashanth, "IISc Prof Does India Proud at CERN," *The Times of India*, November 9, 2011.
2. Another example: "Chasing the One Trillion Trillionth of a second," *The Hindu*, January 2012.
3. Malini Nair, "The Higgs Hunt Is Over, but a New Journey Has Begun," *The Times of India*, July 7, 2012.
4. Magdolna Hargittai, conversation with Rohini Godbole, September 21, 2012, Bangalore.

Shobhana Narasimhan

1. Magdolna Hargittai, conversation with Shobhana Narasimhan, September 20, 2012, Bangalore.

Sulabha Pathak

1. Magdolna Hargittai, conversation with Sulabha Pathak, September 23, 2012, Mumbai.
2. S. Pathak and U. Palan, *Immunology: Essential and Fundamental*, Enfield, NH: Science Publishers, 2005; rev. ed., Tunbridge Wells, UK: Anshan, 2011.

Riddhi Shah

1. Riddhi Shah, "New Challenges Ahead" in *Lilavati's Daughters*, 276.
2. Magdolna Hargittai, conversation with Riddhi Shah on September 27, 2011, New Delhi.

Shobhona Sharma

1. Magdolna Hargittai, conversation with Shobhona Sharma, September 23, 2012, Mumbai.

Vidita Vaidya

1. In an e-mail from Vidita Vaidya, January 31, 2012.

Further Comments

1. I express my thanks to Shobhana Narasimhan and Rohini Godbole for helpful discussions of these issues.
2. Information from Rohini Godbole.
3. Hargittai, conversation with Charusita Chakravarty.
4. Ibid.
5. Ibid.

Women Scientists in Turkey

1. Çiğdem Kağıtçıbaşı, "Caution: Men at Work," *Nature* 456 (2008): 12–14, accessed September 14, 2013, http://www.nature.com/nature/journal/v456/n1s/full/twas08.12a.html. Referring to Blitz, R.C. 1970.
2. *She Figures 2012*, Table 3.1, p. 90, http://ec.europa.eu/research/science-society/document_library/pdf_06/she-figures-2012_en.pdf.
3. Based mostly on the following literature: A. Tatli, M. F. Özbilgin, and F. Küskü, Gendered Occupational Outcomes from Multilevel Perspectives: The Case of Professional Training and Work in Turkey, Chapter 10 in *Gender and Occupational Outcomes: Longitudinal Assessment of Individual, Social, and Cultural Influences*, ed. H. M. G. Watt and J. S. Eccles, Washington, DC: American Psychological Association, 2008, 405–449, available at http://www.academia.edu/345483/Gendered_Occupational_Outcomes_the_case_of_professional_training_and_work_in_Turkey, accessed Sept. 15, 2013; "Asia: Shaking up Tradition," in *Beating the Odds: Remarkable Women in Science*, Science/AAAS and L'Oreal Foundation, 2008, available at http://sciencecareers.sciencemag.org/tools_tips/outreach/loreal_wis/asia_shaking_up_tradition, accessed September 13, 2013; and Gülsün Sağlamer, "Women Academics in Science and Technology with Special Reference to Turkey," in *Women Status in the Mediterranean: Their Rights and Sustainable Development*, ed. L. Ambrosi, G. Trisorio-Liuzzi, R. Quagliariello, L. Santelli Beccegato, C. Di Benedetta, and F. Losurdo, Bari, Italy: Mediterranean Agronomic Institute of Bari, 2009, 45–61.
4. Kağıtçıbaşı, "Caution: Men at Work," 12
5. *She Figures 2012*, Table 3.2, p. 93.
6. "Employment rate, by sex," 2012, http://epp.eurostat.ec.europa.eu/tgm/table.do?tab=table&language=en&pcode=tsdec420&tableSelection=3&footnotes=yes&labeling=labels&plugin=1, accessed September 15, 2013.
7. Women in Statistics, 2012, http://www.turkstat.gov.tr/Kitap.do?metod=KitapDetay&KT_ID=11&KITAP_ID=238, accessed September 13, 2014. (from this site a pdf file can be downloaded)

Sezer Şener Komsuoğlu

1. Magdolna Hargittai, conversation with Sezer Komsuoğlu, November 12, 2008, Istanbul. All quotations are from this conversation.

Gülsün Sağlamer

1. Magdolna Hargittai, conversation with Gülsün Sağlamer, November 13, 2008, Istanbul. All quotations are from this conversation.

Ayhan Ulubelen

1. G. Topçu, N. Gören, and A. Öksüz, "Editorial: Special issue in Honour of Professor Aylan Ulubelen," *Phytochemistry Letters* 4 (2011): 389–390, 389.
2. Istvan Hargittai, "Ayhan Ulubelen," in *Candid Science*, 114–121, 115.
3. Ibid., 118–119.

4. Ibid., 119.
5. Ibid., 116.
6. Correspondence with A. Ulubelen, 2013.

In High Positions

1. Magdolna Hargittai, conversation with James D. Watson, Cold Spring Harbor Laboratory, March 14, 2002.
2. Lynn Harris, "Heads of the Class: The Female Presidents of the Ivy League," http://www.glamour.com/inspired/women-of-the-year/2007/11/female-presidents-of-ivy-league-schools, accessed April 14, 2014.
3. Sarah Gibbard Cook, "Women Presidents: Now 26.4% but Still Underrepresented," *Women in Higher Education*, 21(5) (2012) 1–3, accessed September 13, 2014, http://onlinelibrary.wiley.com/doi/10.1002/whe.10322/full.
4. *She Figures 2012: Gender in Research and Innovation*, http://ec.europa.eu/research/science-society/document_library/pdf_06/she-figures-2012_en.pdf, Figure 4-1 and Table 4-1, pp. 115–116, accessed May 17, 2013.
5. Paul Bateman, "Why Are There So Few Female Vice-Chancellors?" *The Times Higher Education*, August 22, 2013, http://www.timeshighereducation.co.uk/features/why-are-there-so-few-female-vice-chancellors/2006576.article, accessed March 4, 2014.
6. Ramakrishna Ramaswamy, private communication, March 4, 2014.
7. *She Figures* 2012, 116.

Catherine Bréchignac

1. Magdolna Hargittai, "Catherine Bréchignac," in Magdolna Hargittai and Istvan Hargittai, *Candid Science IV: Conversations with Famous Physicists*, London: Imperial College Press, 2004, 570–585, 579.
2. Ibid., 571–572.
3. Ibid., 572.
4. Ibid., 572.
5. "At the Centre of Revolution in Research," *The Times Higher Education*, April 14, 2000, http://www.timeshighereducation.co.uk/story.asp?storyCode=151148§ioncode=26, accessed January 21, 2013.
6. "Catherine Bréchignac," curriculum vitae, http://www.academie-sciences.fr/academie/membre/BrechignacC_bio0810.pdf, accessed January 7, 2014.
7. Hargittai, "Catherine Bréchignac," 578.
8. Ibid., 580.
9. Ibid., 581.
10. Ibid., 584.
11. Ibid., 583–584.

France A. Cordova

1. Magdolna Hargittai, conversation with France Cordova, Fort Lee, NJ, April 2, 2008. All unreferenced quotations are from this conversation.

2. "Outgoing President of Purdue University, Incoming Chair of the Smithsonian," http://articles.washingtonpost.com/2012-09-10/news/35495344_1_uc-riverside-nasa-physics-and-astronomy, accessed July 23, 2013.
3. "Series Overview," pbs.org, http://www.pbs.org/breakthrough/resource/prelease.htm, accessed July 23, 2013.
4. "The 2000 Kilby Laureates," http://www.kilby.org/kl_past_laureates.html, accessed July 23, 2013.

Marye Anne Fox

1. "Seventh Chancellor Guided UC San Diego to Historic Growth and Scholarly Achievement," http://mafox.ucsd.edu/, accessed October 19, 2013.
2. Magdolna Hargittai, conversation with Marye Anne Fox, May 12, 2000. All unreferenced quotations are from this conversation.

Kerstin Fredga

1. Magdolna Hargittai, conversation with Kerstin Fredga, Stockholm, September 15, 2000. All quotations are from this conversation.

Claudie Haigneré

1. Magdolna Hargittai, conversation with Claudie Haigneré, July 1, 2003. All unreferenced quotations are from this conversation.
2. "Expedition Three Crew, Soyuz 3 Taxi Flight Crew: Claudie Haigneré (formerly André-Deshays)," http://spaceflight.nasa.gov/station/crew/exp3/taxi3/haignere.html.
3. Eduard Launet, "Claudie Haigneré, sortie du trou noir" [Claude Haigneré, out of a black hole], *Libération*, May 29, 2009, trans. Janet Denlinger, http://www.liberation.fr/culture/0101570139-claudie-haignere-sortie-du-trou-noir, accessed July 28, 2013.

Helena Illnerová

1. Magdolna Hargittai, conversation with Helena Illnerová, Prague, September 14, 2001. All unreferenced quotations are from this conversation.
2. Christian Falvey, "Helena Illnerová, the Leading Lady of Czech Science," Radio Praha interview, September 2, 2010, http://www.radio.cz/en/section/special/helena-illnerova-the-leading-lady-in-czech-science, accessed January 19, 2014.

Chulabhorn Mahidol

1. Istvan Hargittai and Magdolna Hargittai, "Royal Chemistry: Princess Chulabhorn of Thailand," *The Chemical Intelligencer* 6 (2000): 25–28, 25.
2. Hargittai, "Royal Chemistry," 26.
3. C. Mahidol, H. Prawat, and S. Ruchirawat, "Bioactive Natural Products from Thai Medicinal Plants," in *Phytochemical Diversity: A Source of New Industrial Products*, edited by Stephen Wrigley, London: Royal Society of Chemistry, 1997, 96–105.

4. Hargittai, "Royal Chemistry," 27.

Pamela Matson

1. Magdolna Hargittai, conversation with Pamela Matson, Palo Alto, California, on April 16, 2009. All unreferenced quotations are from this conversation.
2. Pamela Matson and Walter Falcon, "Why the Yaqui Valley? An Introduction," in *Seeds of Sustainability: Lessons from the Birthplace of the Green Revolution in Agriculture,"* ed. Pamela Matson, Washington, DC: Island Press, 2011, 2.
3. Matson and Walter, "Why the Yaqui Valley?" 3–4.
4. Neeraja Sankaran, "Scientist Recipients of MacArthur Fellowships an Eclectic Collection, *The Scientist*, September 4, 1995, http://www.the-scientist.com/?articles.view/articleNo/17550/title/Scientist-Recipients-Of-MacArthur-Fellowships-An-Eclectic-Collection/, accessed October 31, 2013.

Kathleen Ollerenshaw

1. Magdolna Hargittai, conversation with Kathleen Ollerenshaw, April 29, 2003. All unreferenced quotations are from this conversation.
2. Kathleen Ollerenshaw, *To Talk of Many Things: An Autobiography*, Manchester, UK: Manchester University Press, 2004.
3. Catherine Felgate and Edmund Robertson, "Kathleen Timpson Ollerenshaw," http://www-history.mcs.st-andrews.ac.uk/Printonly/Ollerenshaw.html, accessed April 12, 2013.
4. Ollerenshaw, *To Talk of Many Things*, 72.
5. Kathleen Ollerenshaw, *Education for Girls*, London: The Conservative Political Centre, 1958.
6. Kathleen Ollerenshaw and Hermann Bondi, *Magic Squares of Order Four*, London: Royal Society, 1982.
7. Kathleen Ollerenshaw, David Bree, and Hermann Bondi, *Most-perfect Pandiagonal Magic Squares: Their Construction and Enumeration*, Southend-on-Sea, UK: Institute of Mathematics and its Applications, 1998, 186.
8. Ollerenshaw, *To Talk of Many Things*, 229.

Marianne Popp

1. Department of Terrestrial Ecosystem Research, Faculty of Life Sciences, University of Vienna, "Mission Statement," http://131.130.57.230/cms/index.php?id=90, accessed July 25, 2013.
2. Magdolna Hargittai, conversation with Marianne Popp, September 6, 2001. All unreferenced quotations are from this conversation.

Maxine F. Singer

1. National Science Foundation, "The President's National Medal of Science: Recipient Details; Maxine F Singer," http://www.nsf.gov/od/nms/recip_details.jsp?recip_id=327, accessed April 13, 2014.

2. Magdolna Hargittai, conversation with Maxine Singer on May 16, 2000, Washington, D.C. All unreferenced quotations are from this conversation.
3. M. Barinaga, "Asilomar Revisited: Lessons for Today?" *Science* 287(2000): 1584–1585.
4. Maxine Singer and Paul Berg, *Genes & Genomes*, Mill Valley, CA: University Science Books, 1991; M. Singer and P. Berg, *Exploring Genetic Mechanisms*, Mill Valley, CA: University Science Books, 1997; Paul Berg and Maxine Singer, *George Beadle: An Uncommon Farmer: The Emergence of Genetics in the 20st Century*, Cold Spring Harbor, NY: Cold Spring Harbor Laboratory Press, 2005; P. Berg and M. Singer, *Dealing with Genes: The Language of Heredity*, Mill Valley, CA: University Science Books, 2008.

Natalia Tarasova

1. NGO Committee on Education, *Our Common Future*, chap. 2: "Towards Sustainable Development," http://www.un-documents.net/ocf-02.htm, accessed October 27, 2013.
2. Magdolna Hargittai, correspondence with Natalia Tarasova, October 2013. All unreferenced quotations are from this correspondence.
3. Donella H. Meadows, Dennis L. Meadows, Jorgen Randers, and Williams W. Behrens III, *The Limits to Growth: A Report for the Club of Rome's Project on the Predicament of Mankind*, New York: Universe Books, 1972.

Shirley Tilghman

1. Magdolna Hargittai, conversation with Shirley Tilghman, October 25, 2001. All unreferenced quotations are from this conversation.
2. Frederick L. Holmes, *Meselson, Stahl, and the Replication of DNA: A History of "The Most Beautiful Experiment in Biology,"* New Haven: Yale University Press, 2001.
3. "MAKERS Profile: Molecular Biologist & Princeton President," video, http://www.makers.com/shirley-tilghman, accessed October 17, 2013.

Final Thoughts

1. Matteo Leone and Nadia Robotti, "Frédéric Joliot, Irène Curie and the Early History of the Positron (1932–33)," *Eur. J. Physics* 31 (2010): 975–987.
2. "A Study on the Status of Women Faculty in Science at MIT," special issue, *MIT Faculty Newsletter* 9.4 (March 1999), http://web.mit.edu/fnl/women/women.html, accessed April 20, 2014.
3. For example, Carey Goldberg, "M.I.T. Admits Discrimination against Female Professors," *New York Times*, March 23, 1999, http://www.nytimes.com/1999/03/23/us/mit-admits-discrimination-against-female-professors.html.
4. "A Report on the Status of Women Faculty in the Schools of Science and Engineering at MIT, 2011," http://web.mit.edu/faculty/reports/pdf/women_faculty.pdf, accessed April 20, 2014.
5. Magdolna Hargittai, conversation with James D. Watson, Cold Spring Harbor, March 14, 2002.
6. *She Figures 2012: Gender in Research and Innovation, Statistics and Indicators*, European Commission, 2013, available at http://ec.europa.eu/research/science-society/document_library/pdf_06/she-figures-2012_en.pdf.

INDEX